AAPG REPRINT SERIES NO. 7

Sandstone Reservoirs and Stratigraphic Concepts

Selected Papers Reprinted from

AAPG BULLETIN and MEMOIR 18

Published by The American Association of Petroleum Geologists
Tulsa, Oklahoma, U.S.A., 1973

Contents

Preface. *Robert J. Weimer* 3

Time-Stratigraphic Analysis and Petroleum Accumulations, Patrick Draw Field,
  Sweetwater County, Wyoming. *Robert J. Weimer* 4

Sand Bodies and Sedimentary Environments: Review. *Paul Edwin Potter* 30

Comparison of Marine-Bar with Valley-Fill Stratigraphic Traps, Western Nebraska.
  F. A. Exum and *J. C. Harms* 59

Bell Creek Field, Montana: A Rich Stratigraphic Trap. *Alexander A. McGregor* and
  Charles A. Biggs 77

Depositional Systems in Wilcox Group (Eocene) of Texas and Their Relation to
  Occurrence of Oil and Gas. *W. L. Fisher* and *J. H. McGowen* 96

Recognition of Barrier Environments. *David K. Davies, Frank G. Ethridge,* and
  Robert R. Berg 121

Genetic Units in Delta Prospecting. *Daniel A. Busch* 137

Geometry of Sandstone Reservoir Bodies. *Rufus J. LeBlanc* 155

Copyright 1973 by The American Association of Petroleum Geologists.
All Rights Reserved. Published November 1973. Fifth printing November 1979.
Library of Congress Card No. 73-89375. ISBN: 0-89181-532-5.

Sandstone Reservoirs and Stratigraphic Concepts I: Preface

The AAPG Reprint Series provides an invaluable service to students of sedimentary rocks by collecting for study and review important topical papers previously published in the *Bulletin* and other AAPG publications. The papers describing sandstones in Reprint Series Nos. 7 and 8 should be most welcome. They will be extremely useful to the explorationist searching for the elusive stratigraphic trap, or to the earth scientist working to interpret the geologic events recorded by accumulations of detrital sedimentary rocks. The data presented by these papers have both immense practical and theoretical value.

A diversity of subjects relating to sandstones is presented in this volume, including new stratigraphic concepts, a cross section of exploration practices, interpretations of environments of deposition, and typical conditions of petroleum entrapment. Core studies of rocks exhibiting reservoir and nonreservoir properties are an integral part of several papers. The geologic examples have a broad geographic and stratigraphic distribution and describe petroleum-producing sandstones deposited under fluvial, deltaic, and marine conditions of sedimentation. The summary paper by LeBlanc reviews sandstone distribution in modern depositional systems and gives criteria for identification of specific environments of deposition. The review paper by Potter emphasizes the complex factors that a geologist must evaluate in interpreting the environments of deposition of ancient sandstones.

With the current emphasis on new sources of energy, petroleum explorationists will be aided by this volume, particularly as they review old petroleum-producing provinces in the search for the remaining subtle traps.

Reprint Series Nos. 7 and 8 were compiled by me. However, the original suggestion for such a compilation was made by K. N. Beckie, Hudson's Bay Oil and Gas Company Limited, Calgary, Alberta, Canada. The papers were assembled by E. M. Tidwell, of the AAPG Editorial Staff.

ROBERT J. WEIMER
Colorado School of Mines
Golden, Colorado
September 5, 1973

TIME-STRATIGRAPHIC ANALYSIS AND PETROLEUM ACCUMULATIONS, PATRICK DRAW FIELD, SWEETWATER COUNTY, WYOMING[1]

ROBERT J. WEIMER[2]
Golden, Colorado

ABSTRACT

The search for new petroleum reserves can be implemented greatly by a more thorough understanding of why petroleum is trapped where it is. The Patrick Draw field, discovered in 1959, started a wave of exploratory effort in the Rocky Mountain area to find additional giant stratigraphic traps in the Upper Cretaceous rocks where porous and permeable sandstone pinches out on structural noses. The lack of success in finding another "Patrick Draw," despite this widespread exploratory effort, means that factors not generally considered must have had a dominant influence in the accumulation. These factors are revealed only by reconstructing the geologic history of the area, beginning with the deposition of the reservoir and source rocks and tracing the structural attitude of these rocks through time.

Although several sandstone reservoirs produce petroleum at Patrick Draw, the principal productive interval consists of two sandstone bars at the top of the Almond Formation (Upper Cretaceous). The spatial dimensions, lithologic character, and stratigraphic framework of these bars suggest that they are barrier-bar sandstone bodies deposited along the margin of the Lewis sea. These porous and permeable linear barrier bars have a general north-south trend and grade updip on the west into impermeable shale and sandstone that were deposited in a swamp and lagoonal environment. A second important productive interval is approximately 40 ft. below the top of the Almond Formation. The areal distribution, lithologic character, and stratigraphic framework of sandstone in this interval suggest that it was deposited as a tidal delta in a lagoon. Each of the three main productive sandstone bodies has a different oil-water contact.

The geologic history of the Patrick Draw area shows that, by the beginning of deposition of the Lance Formation (Upper Cretaceous), conditions were favorable for petroleum accumulation. The reservoir sandstone beds had 1,200 ft. of overburden which had accumulated in the several million years since the reservoir sandstone beds were deposited. An early trap was formed where these sandstone beds were warped over an east-plunging structural nose, and early migration of petroleum produced a large accumulation a few miles south of the present field. When the present Wamsutter arch came into existence in post-early Eocene time, the old trap was opened and the accumulation spilled northward to be trapped at the present location of the Patrick Draw field.

The search for more "Patrick Draws" must include more than an analysis of present structure and potential reservoir rock. The time of formation of the trap, structural modification of the trap through time, and associated origin and migration problems are hidden factors that play the dominant roles in the formation of a large petroleum accumulation. Exploration geologists must learn more about the regional framework of sedimentation—the cause and effect of incipient structural development in depositional areas. They must understand how these factors relate to the geologic history of a region.

INTRODUCTION

The Patrick Draw field, discovered in 1959, is one of the most significant oil discoveries in the Rocky Mountains in the last decade. The field is in southwestern Wyoming approximately 35 mi. east of Rock Springs; it is on the east flank of the Rock Springs uplift and on the south flank of the Wamsutter arch (Figs. 1, 2). The Wasatch Formation (Eocene) is exposed at the surface and production is from the Almond Formation (Upper Cretaceous) at depths ranging from 3,500 to 5,400 ft. More than 25,000,000 bbls. of 40° API gravity oil have been produced in 6 years of production. An estimated 200,000,000–250,000,000 bbls. are in place in the reservoir.

Patrick Draw is unusual for two reasons. (1) The productive strata in the field have a uniform

---

[1] Read before the Rocky Mountain Section of the Association at Billings, Montana, September 29, 1965. An abbreviated form of the paper was the subject, in part, of an A.A.P.G. Distinguished Lecture tour in the spring of 1964. Manuscript received, February 1, 1966; accepted, May 9, 1966.

[2] Professor and head, Department of Geology and Geological Engineering, Colorado School of Mines, Golden Colorado.

The writer appreciates the critical reading of the manuscript by Richard H. DeVoto and Donald P. McGookey. Appreciation is expressed also to W. A. Cobban of the U. S. Geological Survey for making fossil identifications; to D. E. Lawson of the Forest Oil Corporation for making cores and other data from wells in the Patrick Draw field available for study; and to Chester Cassel, consulting geologist, with whom the writer was associated during the original work, for the many hours of thought-provoking discussions about the Cretaceous stratigraphy and associated petroleum occurrences in the Rocky Mountains.

southeast dip of 4° and equivalent strata crop out updip 8 mi. west of the field (Figs. 2, 3). (2) This large oil field is positioned stratigraphically in a part of the section that is noted for gas production rather than oil in the Rocky Mountain area.

Patrick Draw is only one of several oil fields producing from the Almond Formation along the Wamsutter arch, but it is by far the most important (Fig. 2). Because of the importance of these fields to petroleum exploration in the Rocky Mountains, geologists in exploration offices, as well as those in oil-company research laboratories, have attempted to distinguish the stratigraphic changes within the Almond Formation and to determine how these changes have influenced the large petroleum accumulations on the Wamsutter arch. Despite attempts to apply the results of these massive efforts to exploration programs elsewhere in the Rocky Mountains, geologists have not yet found another "Patrick Draw."

During the development drilling of the Wamsutter arch fields in 1961, the Wyoming Geological Association organized a field conference in the area and published a guidebook containing papers describing the stratigraphy and the petroleum occurrences of the area (Wyoming Geol. Assoc., 1961b).

In 1965 the same association organized a symposium on the Cretaceous and Tertiary rocks of the Rock Springs uplift, and a collection of papers was published as the 19th Annual Field Conference Guidebook (Wyoming Geol. Assoc., 1965).

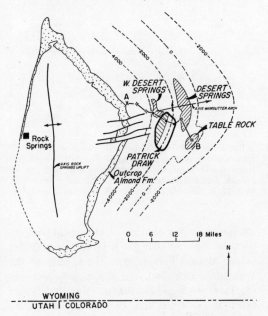

FIG. 2.—Regional structural contour map showing principal petroleum fields producing from Almond Formation on Wamsutter arch. Locations of structural cross section (Fig. 3) and electric-log correlation section A-B (Fig. 4) are indicated. Dashed lines are structural contours in feet; structural datum is top of Almond Formation. Surface elevations in area range from 6,500 to 7,500 ft. Stippled area is outcrop of Almond Formation on flanks of Rock Springs uplift. Diagonal ruling is gas-productive area. Unruled area in Patrick Draw field is oil-productive.

In these guidebooks and in other publications, ideas are presented by many authors relating to the subjects discussed in this paper.

Papers of particular interest on the general stratigraphy of the Upper Cretaceous of the area are by Hale (1950, 1955, 1961), Barlow (1961), Lewis (1961), Smith (1961), Weimer (1961a,b, 1965b), Jacka (1963, 1965), and Ritzma (1963). The Patrick Draw oil field is described by Lawson and Crowson (1961), Burton (1961), Cox (1962), and the Wyoming Geological Association (1961a). The Desert Springs gas field is described by May (1961) and Earl and Dahm (1961). The Table Rock gas field is described by White (1955) and Mees et al. (1961).

The writer first studied the stratigraphy of the Almond Formation and the Lewis Shale from the surface exposures on the east flank of the Rock Springs uplift in 1955–1956. The stratigraphic framework and petroleum potential of these for-

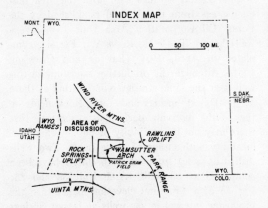

FIG. 1.—Index map showing location of Patrick Draw field and area of discussion in relation to other tectonic elements in southwest Wyoming.

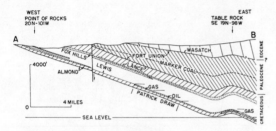

FIG. 3.—West-east structural cross section A-B from surface section of Almond Formation to Table Rock field. Stratigraphic changes and position of Patrick Draw field are indicated. Section is from just west of locality A to locality B, Figures 2 and 5.

mations were the subject of an unpublished report. At the time only six wells had been drilled to the Almond Formation in the vicinity of the Wamsutter arch, and five of the wells were on the Table Rock and Southwest Table Rock structures. It has since been possible to incorporate the results of more than 400 wells into the stratigraphic framework established in the original study. Because most of the development drilling is completed and little additional data are expected, the results of these studies are being presented in this paper. Although several minor petroleum fields have been found on the Wamsutter arch, only the four largest petroleum occurrences are discussed.

PETROLEUM OCCURRENCES—WAMSUTTER ARCH

The Wamsutter arch is one of the classic areas in Wyoming to demonstrate stratigraphically trapped petroleum. The stratigraphic traps have formed on a broad east-plunging structural arch (Fig. 2) where porous and permeable sandstone grades updip into coal-bearing shale, siltstone, and sandstone. Magnitude of dip on the arch is 4°–5° in the Cretaceous section and 1°–2° in the Tertiary section. Only one field, Table Rock, produces from a closed structure. The structure has approximately 350 ft. of closure at the top of the Almond Formation. The principal production on the Wamsutter arch is from the Almond Formation at depths ranging from 3,000 to 6,500 ft.

The first Cretaceous production in the area was established with the completion of The Texas Company Table Rock Unit 5 (sec. 2, T. 18 N., R. 98 W.) in February, 1954, for an open-flow gauge of 8,980 MCFGPD from a depth of 6,500 ft. (electric log, Fig. 4). The 50-ft. productive interval is in sandstone in the upper Almond about 75 ft. below the top of the formation. A Lewis sandstone 420 ft. above the base of the formation yielded commercial amounts of gas on a drill-stem test. A narrow oil ring was found under the gas cap of the Almond reservoir with the deepening of the Unit 2 well in 1957. Ten gas wells and eight oil wells have been completed in the Almond Formation, proving a productive area of about 10 sq. mi. For details on the development of the field, the reader is referred to Mees et al. (1961) and White (1955).

The Desert Springs discovery well was completed by El Paso Natural Gas in January, 1958, from a sandstone in the Almond at a depth of 5,900 ft. The stratigraphic position of the producing sandstone is the same as that of the Table Rock main productive sandstone (bar sequence 1, Figs. 7, 15). The Unit 1 discovery well (sec. 26, T. 21 N., R. 98 W.) flowed 7,700 MCFGPD from 35 net ft. of sandstone in the Almond (May, 1961). A drill-stem test indicated commercial production also from a 35-ft.-thick Lewis sandstone, 650 ft. above the base of the Lewis. Although comprising one producing zone, two sandstone bars in the Almond are productive in the Desert Springs field, each having different gas-water contacts (May, 1961). Nineteen gas wells outline a productive area 13 mi. long and 2 mi. wide. Five of the 19 wells produce from a sandstone in the Lewis ("e") and 16 from the Almond.

Development drilling linked the area of the original Desert Springs Unit with the area producing gas from the Almond established by the Forest Oil Corporation Arch Unit 1 discovery well (sec. 9, T. 19 N., R. 98 W.; Fig. 4), completed in April, 1959 (Lawson and Crowson, 1961). Although the productive sandstone in this well is at the same stratigraphic position as those at the Desert Springs and Table Rock fields on the south, reservoir performance suggests that there may not be continuity of porosity and permeability with either of these fields. Two wells drilled downdip from the Arch Unit 1 have been completed as oil wells (Arch Unit 80 and 83; Fig. 5) at a structural elevation higher than the producing interval in gas wells at Desert Springs and Table Rock. The Arch Unit 77 well (sec. 3, T. 19 N., R. 98 W.) appears to be at the southern, downdip edge of the Desert Springs field, which does not have an oil ring.

The West Desert Springs field was discovered in May, 1959, with the completion of the Texas National UPRR-Rock Springs Grazing Association No. 1 in sec. 17, T. 20 N., R. 99 W. The well flowed 9,700 MCFGPD from a 15-ft.-thick sandstone at a depth of 3,815 ft. The sandstone is at the top of the Almond and is overlain by the Lewis Shale. This production was the first from a sandstone at this stratigraphic position (uppermost Almond) outside of the Table Rock field. Six gas wells have been completed from this sandstone in the West Desert Springs field. Subsequent drilling in West Desert Springs found oil production in a sandstone approximately 30 ft. below the top of the Almond Formation. This sandstone is called the UA-6 sandstone in the field operations (Lawson and Crowson, 1961). Forty-five wells have been completed, or are capable of producing, from the UA-6 sandstone in the West Desert Springs and the northern part of the Patrick Draw fields (Fig. 5). The sandstone interval has an erratic distribution and in places contains more than one thin productive sandstone. Production limits have not been established, but 4–5 sq. mi. is proved (Fig. 5). The thickness ranges from a wedge-edge to 25 ft. but averages approximately 12 ft. Production depth ranges from 3,600 to 4,800 ft. Sea-level datum elevation on the oil-water contact in the UA-6 sandstone is approximately +2,150 ft., and the elevation at the updip edge of the sandstone is approximately +3,650 ft., indicating that there is a 1,500-ft. oil column.

The most important oil discovery in the Wamsutter arch area was made by the El Paso Natural Gas Company at the Patrick Draw Unit 1 in sec. 1, T. 18 N., R. 99 W. The discovery well was completed in November, 1959, flowing 638 b/d of 41° API gravity through a ⅜-in. choke from a productive zone in the upper Almond at a depth of 5,172–5,198 ft. Subsequent discoveries from the same sandstone were made in the Beacon Ridge Unit by Texaco Inc. and in the Arch Unit by the Forest Oil Corporation. All of the original units now are combined as the Patrick Draw field. Details of the field have been summarized in papers by Lawson and Crowson (1961, p. 280), Burton, (1961, p. 276), Cox (1962), and Weimer (1965b, p. 65). Stratigraphic and production details of the productive sandstone are discussed in a subsequent section.

Development drilling has proved the existence of an oil-productive area approximately 10 mi. long and 2 mi. wide with an oil column of 1,050 ft. (Figs. 6, 7). More than 165 oil wells have been drilled on 80-acre spacing. A gas cap in the field covers an area 6 mi. long and 1½ mi. wide with a 600-ft. gas column. The two productive sandstone bars have an average thickness of approximately 20 ft. The field has produced more than 25,000,000 bbls. of oil. Peak production was about 25,000 b/d, but production has declined to a present rate of 9,000 b/d. No gas has been produced from the field, and solution gas is being injected into the gas cap for pressure maintenance. Pilot floods for secondary recovery are in operation.

The total gross value of the producible petroleum in the Wamsutter arch fields discussed above is estimated to be $300,000,000, The value may be considerably above this estimate if the secondary-recovery projects are successful.

FACIES AND FORMATIONS

The formations in the uppermost Cretaceous rocks of the Rock Springs uplift-Wamsutter arch generally correspond to facies that were controlled by environments of deposition. The environments of deposition were within or along the western margin of a seaway. The facies, formations, and postulated environments of the Wamsutter arch are summarized in Table I.

The principal geographic control for the distribution of facies was the western shoreline of the Cretaceous seaway. As the shoreline transgressed or regressed in response to sediment supply and (or) rates of subsidence, the depositional environments were shifted laterally. Because of environment shift, the formations whose dominant lithologic types are described in Table I commonly are intertongued and may contain in a vertical section more than one of the described facies. This relation is illustrated on the restored sections (Figs. 4, 11) and causes obvious problems in placing formation contacts. Moreover, more than one facies were included in some formations, when they were defined originally. The uppermost Almond marine sandstone beds are a distinctive mappable unit and could have been given a separate name. Instead, these sandstone

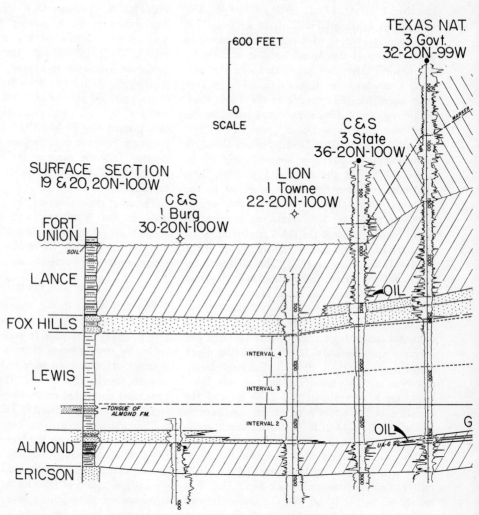

Fig. 4.—West-east electric-log correlation section A-B, showing relation of surface to subsurface stratigruling—non-marine sediments; stippled pattern—marine and transitional sandstone with some siltstone and

beds in the original definition were grouped together with the coal-bearing shale, siltstone, and sandstone as one formation (Schultz, 1909, 1920). Similarly, the sandstone beds described herein as the Fox Hills Formation, a distinct mappable unit, are placed in the basal part of the Lance Formation by some workers.

Principles in Selection of Time-Stratigraphic Intervals, Upper Almond Formation and Lewis Shale

Stratigraphic studies have the most significant meaning to the geologist when time-stratigraphic units are defined accurately and facies changes within them are described. Time-stratigraphic

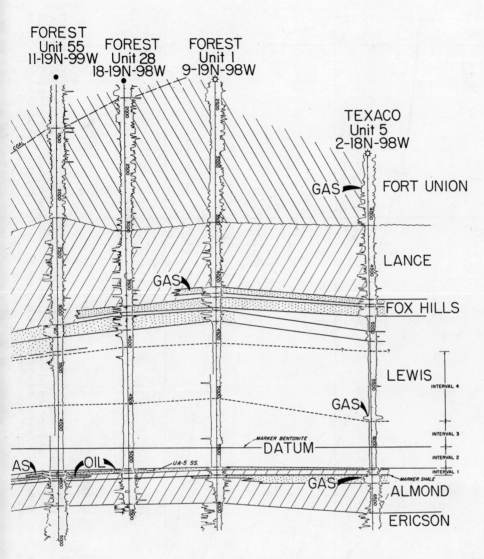

raphy on Wamsutter arch. Stratigraphic position of petroleum-producing sandstone is indicated. Diagonal shale; unmarked area in middle of diagram—marine shale. Location of section shown on Figures 2 and 5.

units are defined as those rocks deposited during an interval of time, the time surfaces bounding the units being determined by organic or inorganic correlation methods. One of the most common methods for determining time surfaces is based on the evolution of life as revealed through a study of fossils. In local basins or small areas of study, inorganic methods, which yield a high degree of precision in correlation, may also be employed. Correlation is the establishment of the mutual time relations between rock units. In this study correlations are based on inorganic methods using electric logs of wells and core descriptions. The method is most akin to that described by

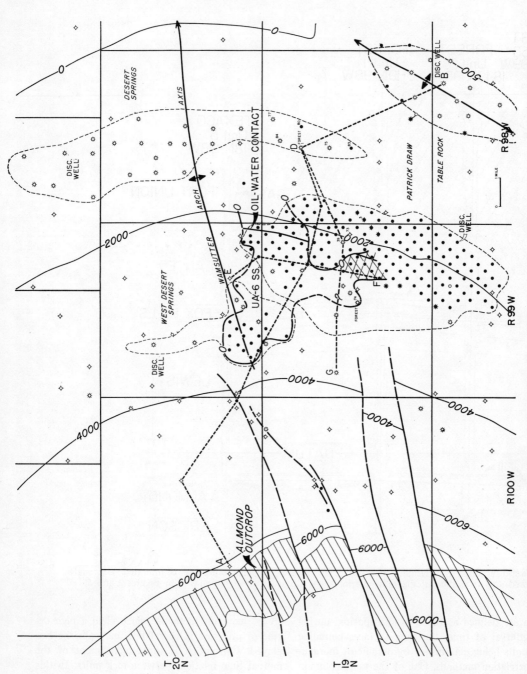

FIG. 5.—Structural contour map on top of Almond Formation with outcrop of formation indicated by diagonal ruling. Important petroleum fields on Wamsutter arch are outlined. Lines of section of Figures 3 and 4 (A-B), 7 (G-D), and 9 (F-E) are indicated. Zero isopachous contour of oil-productive UA-6 sandstone and oil-water contact are shown. Sandstone is productive in north part of Patrick Draw field and in West Desert Springs field. Cross-hatched area in T. 19 N., R. 99 W. is where UA-5 and UA-6 sandstone bodies appear to be in contact (Fig. 9). Contour interval is 2,000 ft. unless otherwise noted.

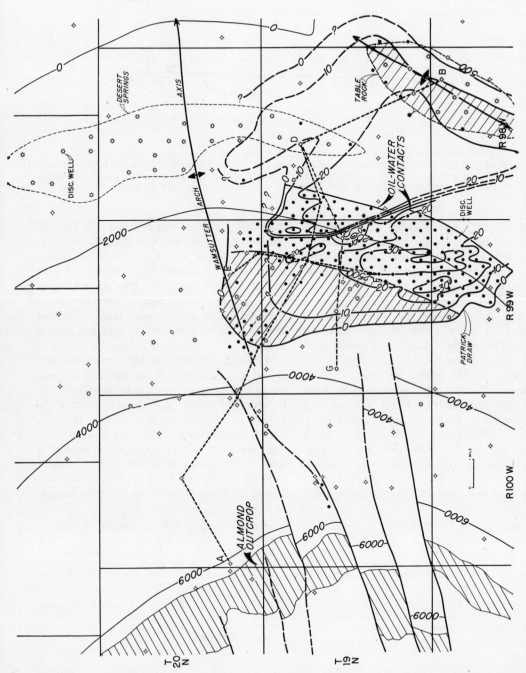

FIG. 6.—Isopachous map of main productive sandstone (UA-5) at Patrick Draw and Table Rock fields superimposed on structural contour map on top of Almond Formation. (Two sandstone bars are isopached and referred to as UA-5.) Isopach interval is 10 ft; structure contour interval is 2,000 ft. unless otherwise noted. Probable gas-productive area from sandstone at top of Almond Formation indicated by diagonal ruling; oil-productive area indicated by oil well symbols in Patrick Draw field. Locations indicated for cross sections on Figures 3 and 4 (A-B), 7 (G-D), and 9 (F-E).

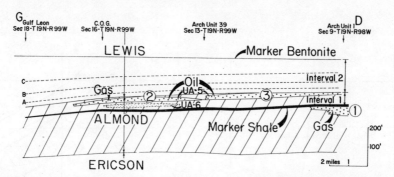

Fig. 7.—West-east restored secton G-D across Patrick Draw field showing stratigraphic position of productive sandstones. Bar 1—main gas-productive sandstone zone, Desert Springs and Table Rock fields. Bars 2 and 3 (UA-5)—oil and gas-productive at Patrick Draw field. Time-stratigraphic Intervals 1 and 2 discussed in text are indicated. Dashed lines labeled A, B, C in Lewis are marker beds. Line of section shown on Figures 5 and 6.

Lowman (1949) and is called "correlation by stratal continuity." Its use in the subsurface by tracing beds among closely spaced wells is similar to the surface-correlation method referred to as "walking the outcrop."

To establish stratal continuity, all logs in a field are studied. Strata that show distinctive electrical or radioactivity characteristics are marked. The lithologic character of each stratum is determined, where possible, by sample or core description. On the Wamsutter arch, bentonite, limestone, siltstone, thin calcareous silty sandstone, thin shale units, and coal beds are used most commonly. Each layer in a vertical stratigraphic sequence records a geologic event which may or may not be synchronous within a large area. By tracing thin distinctive beds from well to well, the continuity of those synchronous events that have been recorded by a single thin layer across a broad area is established with the precision of a few feet. Stratal continuity is established in each field and then is determined among fields by utilizing data from wildcat wells located between the fields. The method works best in marine shale sequences. Bentonite beds in the Lewis Shale are the most easily recognized and have the greatest continuity of any marker beds.

"Phantom horizons" in some cases are employed in correlation (i.e., horizons that are positioned relative to events recorded stratigraphically above or below. A "phantom horizon" is defined as an imaginary time surface to which closely related strata, having only local distribution, are referred. Using a phantom horizon in stratigraphic studies is similar to traversing in plane-table surveying where the elevation of each station is referred to a horizon which, by conven-

TABLE I. FACIES INTERPRETATION OF UPPERMOST CRETACEOUS FORMATIONS OF WAMSUTTER ARCH AREA

Facies	Formation	Postulated Environment
1. Gray shale with thin beds of siltstone, sandstone, limestone, and bentonite	Lewis (600–1,000 ft.)	Marine: offshore neritic (low energy)
2. Gray and white, fine- to medium-grained, cross-stratified, calcareous sandstone with *Ophiomorpha*; sandstone weathers tan and light brown	Fox Hills (150–250 ft.) Uppermost part of Almond (0–400 ft.)	Marine and transitional: barrier island, littoral, shallow neritic (high energy)
3. Gray and tan shale; brown carbonaceous shale; gray, fine-grained, calcareous, lenticular, laminated sandstone, weathers reddish brown; coal beds; layers of the oyster *Ostrea glabra*; exposures have banded appearance	Main part of Almond (200–250 ft.) Lance (700–1,000 ft.)	Non-marine and transitional (coastal plain): lagoon, swamp, fluvial, estuarine(?), tidal flat (generally low energy)

tion, is sea-level. For example, one traceable marker bed among well logs in a field may be 20 ft. above the phantom horizon; another bed may be 30 ft. below the phantom horizon; another bed may correspond locally to the phantom horizon. As the distinctive log characteristics of a traceable bed disappear, another closely related bed may be recognized that can be traced in other parts of the field. By careful work, the stratigraphic position of the phantom horizon can be maintained easily and accurate correlation can be established thereby.

By use of the afore-described methods of maintaining stratal continuity and correlation, four time-stratigraphic intervals have been established in the upper Almond Formation and the Lewis Shale. These intervals are indicated on Figures 4 and 11, and are called time-stratigraphic Intervals 1, 2, 3, and 4.

FACIES CHANGES: UPPER ALMOND FORMATION AND LEWIS SHALE

TIME-STRATIGRAPHIC INTERVAL 1, UPPER ALMOND FORMATION OF PATRICK DRAW FIELD

The oldest time-stratigraphic interval discussed is Interval 1 (70–90 ft. thick), which contains the important oil-productive sandstone bars of the Patrick Draw and West Desert Springs fields (Figs. 4, 7). Interval 1 is in the upper part of the Almond Formation of Late Cretaceous age. The Almond is one of the important petroleum-producing formations of southwestern Wyoming. Until recent years, the formation was noted mainly as the upper coal-bearing formation of the Mesaverde Group in the Rock Springs uplift. During the last 10 years, gas or oil production has been found in sandstone beds throughout the formation, but most important is the upper part which is intertongued with the marine Lewis Shale. Interval 1 is included in this upper part.

The basal bed of Interval 1 rests directly on the gas-productive sandstone beds at Desert Springs and Table Rock. The interval consists of gray, fine- to medium-grained, calcareous sandstone; gray fossiliferous shale; oyster coquina of *Ostrea glabra;* coal beds; and gray laminated shale, siltstone, and sandstone. Lawson and Crowson (1961) gave letter designations to several productive sandstones in the Patrick Draw field. Two important oil-productive sandstone bodies in Interval 1 are the Upper Almond 5 (UA-5) sandstone, the main productive sandstone at Patrick Draw (Fig. 7), and the Upper Almond 6 (UA-6) sandstone, which is productive in the West Desert Springs field and the northern part of Patrick Draw (Figs. 5, 9).

The establishment of boundaries of the time-stratigraphic units is of utmost importance in stratigraphic analysis. Interval 1 is bounded at the base by a 10- to 12-ft., gray, highly fossiliferous, dense claystone or shale. The bed is indicated as the "marker shale" on Figures 4, 7, and 11. As shown by electric logs, the resistivity of this layer is much lower than that of any other shale layer in the upper 100 ft. of the Almond. The western depositional limit of the marker shale is approximately 3 mi. west of Patrick Draw field (Weimer, 1965b, p. 69); the eastern limit is uncertain. Cores of the shale were examined from several wells. The fauna of the shale from 4.191–4,207 ft. in Arch Unit 74 (sec. 22, T. 19 N., R. 99 W., Fig. 5) was identified by W. A. Cobban of the U. S. Geological Survey as having the following forms: *Anomia gryphorhynchus?* Meek, *Volsella (Brachydontes) regularis* White, *Corbula subtrigonalis?* Meek and Hayden, *Corbula undifera* Meek, *Corbicula* sp., *Anodonta parallela* White, and *Ostrea glabra* Meek and Hayden. This fauna is similar to one reported from the Almond Formation in surface exposures (Hale, 1950). The lithologic and electric-log characteristics of the shale are similar to the marine Lewis Shale, but, because of the suggested brackish-water fauna, the marker shale is interpreted as having been deposited in a lagoonal environment.

The top of Interval 1 in the eastern part of the area (Ts. 18, 19, 20 N., Rs. 97, 98, 99 W.) is placed at the base of the Lewis Shale (or the top of the oil-productive sandstone at Patrick Draw). A widespread thin bentonite bed is found in the Lewis Shale about 20 ft. above the base (marker A, Fig. 7). In the Patrick Draw field area, the base of the Lewis is subparallel with this bentonite, suggesting that the base is locally isochronous. West of the pinch-out of the Patrick Draw productive sandstone, the top of the non-marine coal-bearing shale, sandstone, and siltstone of the Almond Formation is the top of Interval 1 (Fig. 7).

Two important facies changes which control petroleum accumulations occur within Interval 1.

The lower facies change is near the middle of the interval where the UA-6 sandstone has porosity and permeability and is oil-productive both in the West Desert Springs field (T. 20 N., R. 99 W.) and in the northern part of the Patrick Draw field (Figs. 5, 7, 9). The sandstone is gray, very fine- to fine-grained, calcareous, and ranges from a wedge-edge to more than 25 ft. thick. The average thickness of the productive sandstones is about 12 ft. Because of the erratic distribution of production (Fig. 5), the fine grain size, and the close association above and below with coal beds and lagoonal shale, the UA-6 sandstone is interpreted as having been deposited as a tidal delta in a shallow-water lagoon. The sandstone could be interpreted as an alluvial deposit but lacks grain-size variation and channeling at the base.

A second facies change is associated with the main oil-productive sandstone (UA-5) at Patrick Draw (Figs. 4, 7). This change occurs near the top of Interval 1. The main productive sandstone is overlain by the marine Lewis Shale (Figs. 7, 9), by oyster-bearing coquina layers, or by 5–10 ft. of carbonaceous shale and impermeable sandstone. The UA-5 sandstone is quartzose, gray, fine- to medium-grained, and calcareous, and contains abundant dark gray to black chert grains and minor amounts of feldspar. Lawson and Crowson (1961, p. 288) reported mineral percentages in the sandstone as follows: quartz 64; chert 32; feldspar (mostly plagioclase) 4; and trace amounts of biotite, muscovite, chlorite, and zircon. Cement is reported to be dominantly silica, but carbonate and clay minerals (kaolinite) also are present. The sandstone ranges in thickness from a wedge-edge to more than 30 ft. (Fig. 6). Coquina layers of *Ostrea glabra* are found near the top of the sandstone in the middle part of Patrick Draw field (Lawson and Crowson, 1961), particularly in the area of thin sandstone between the main development of bars 2 and 3 (Figs. 6, 7). In the northern part of the field, thin laminae of shale and siltstone within the sandstone exhibit angles of cross-lamination as much as 8°, the regional structural dip being about 4°. A few cross-strata dip at higher angles. The UA-5 sandstone rests on a coal bed or associated gray laminated siltstone and carbonaceous shale. Thickness variation within the UA-5 interval results from addition of sandstone both at the base and top of the unit. However, the top of the sandstone shows topographic relief of only 5–10 ft. in relation to marker beds in the overlying Lewis Shale (Fig. 7). The relief at the base of the sandstone is of the same order of magnitude.

The porous and permeable UA-5 sandstone zone occurs through an area at least 10 mi. long and 6–8 mi. wide. The relationship indicated by Figure 10 suggests that the zone can be traced for more than 20 mi. The UA-5 interval in the Patrick Draw field has at least two distinct bars (2 and 3, Figs. 6, 7). Thickness and distribution of these sandstone bars are indicated on Figures 6 and 10. The northern pinch-out of the bars is a gradual tapering to a wedge-edge. Near the northern edge of bar 3, several wells in the northwest corner of T. 19 N., R. 98 W., although in a higher structural position than the main oil production, have a high water cut during production. The cause of this phenomenon is not known.

Although the lithologic character of sandstone bars 2 and 3 is identical, the bars are separate reservoirs having different oil-water contacts. Normally, the two bars are superimposed with the result that production is found despite an area of thin sandstone through the middle of the field (Figs. 6, 7). Bar 2 has a gas cap whereas bar 3 does not. Although drilled in the middle of the field, Arch Unit Well 39 (SE. 1/4, SW. 1/4, sec. 13, T. 19 N., R. 99 W.) and Well 73 NW. 1/4, NE. 1/4, sec. 24, T. 19 N., R. 99 W.) have no productive sandstone (Figs. 5, 7). In place of the productive sandstone in these two wells, oyster coquina layers, carbonaceous shale, and impermeable sandstone were cored.

The UA-5 sandstone pinches out updip at the western margin of Patrick Draw field into impermeable coal-bearing shale and siltstone of Interval 1 (Fig. 7). On the north the oil-productive sandstone grades laterally into impermeable shale, but, in a nearly identical stratigraphic position and in a higher structural position, a water-bearing sandstone is found. The trend and thickness variation of this sandstone are shown by the isopachous contours on Figure 8. Relationship of the sandstone to the Patrick Draw field is shown in Figure 9. Although a separate reservoir, the sandstone is correlated with the UA-5 sandstone because of stratigraphic position, similar lithologic character, and what appears to be the record of a similar geologic event. Based on available sparse well control, the size of this northern

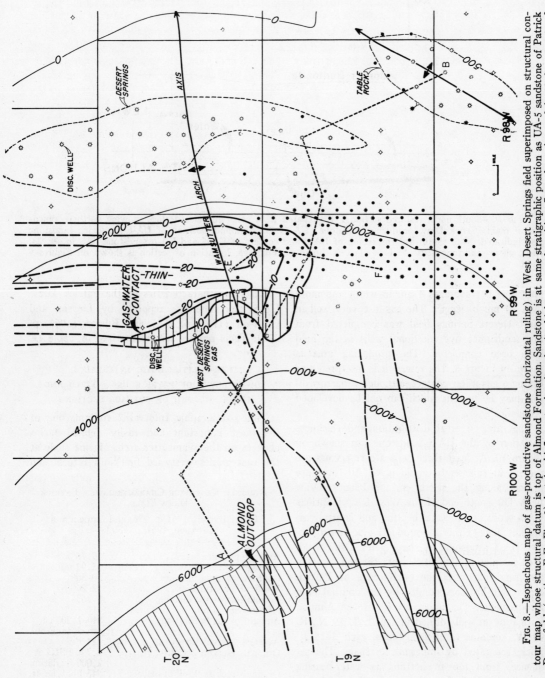

FIG. 8.—Isopachous map of gas-productive sandstone (horizontal ruling) in West Desert Springs field superimposed on structural contour map whose structural datum is top of Almond Formation. Sandstone is at same stratigraphic position as UA-5 sandstone of Patrick Draw field (section F-E, Fig. 9) but is chiefly water-bearing despite high structural position. Isopachous interval is 10 ft.; structural contour interval is 2,000 ft. unless otherwise noted.

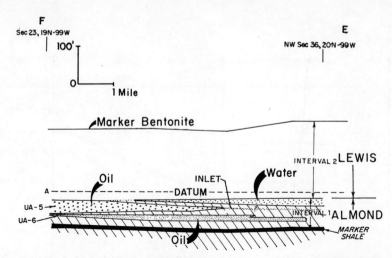

Fig. 9.—South-north restored section F-E prepared from cores and electric logs showing northern pinch-out of main Patrick Draw producing sandstone (UA-5). Stratigraphic position of UA-6 producing sandstone also indicated. Note southward thickening of UA-5 and contact of UA-5 and UA-6 sandstone beds in sec. 23. Area where sandstone beds are in contact is shown on Figure 5. Location of section is shown on Figures 5 and 6.

sandstone bar averages 4 mi. in width and more than 10 mi. in length. The gas-discovery well in West Desert Springs field was completed from this sandstone; five additional wells in the field have been completed. The producing area is shown in Figure 8. The reservoir is interpreted as having a gas-water contact, although a narrow oil ring may be present which can not be developed economically.

The inferred areal distribution and thickness variation of the UA-5 sandstone are shown on Figure 10. At least three separate reservoirs are present at this stratigraphic level. All three are present along the north end of Patrick Draw field. The south end of the West Desert Springs UA-5 sandstone is water-bearing and is stratigraphically above either gas-productive or oil-productive UA-5 sandstone of the Patrick Draw field. Well-completion difficulties from the thin productive sandstone near the north edge of Patrick Draw field are compounded by the unusual fluid arrangements. The upper 50 ft. of the Almond Formation in wells drilled in sec. 2, T. 19 N., R. 99 W. contains three sandstones, each 5–15 ft. thick, separated by impermeable shale. The sequence, from top to bottom, is water-bearing sandstone (UA-5 of north), gas-bearing sandstone (UA-5 of south), and oil-bearing sandstone (UA-6).

Reservoir characteristics of the Patrick Draw UA-5 sandstone are reported by Lawson and Crowson (1961, p. 288) and Burton (1961, p. 276). Their reports are summarized in Table II.

TIME-STRATIGRAPHIC INTERVAL 2,
LOWER LEWIS OF PATRICK DRAW FIELD AND
UPPER ALMOND OF SURFACE SECTIONS

Time-stratigraphic Interval 2 contains one of the most important and easily defined facies changes on the Wamsutter arch. Marine shale at the east grades westward first into littoral and

TABLE II. RESERVOIR CHARACTERISTICS, PATRICK DRAW FIELD

(After Burton, 1961, p. 276; and Lawson and Crowson, 1961, p. 280.)

Average porosity	19.76%
Average permeability	35.92 md.
Average water saturation	30–50%
Residual oil saturation	12–20%
Formation volume factor	1.263
API gravity of oil	41°
Amount of gas in solution	540–1,200 cu. ft./stock-tank bbl.
Average reservoir temperature	121°F.
Original reservoir pressure	1,790 PSIG at +2,000-ft. datum
Original stock-tank oil in place	845 bbls./ac.-ft.
Estimated primary-recovery factor	20%
Approximate average pay thickness	20 ft.
(Gas-solution drive reservoir)	

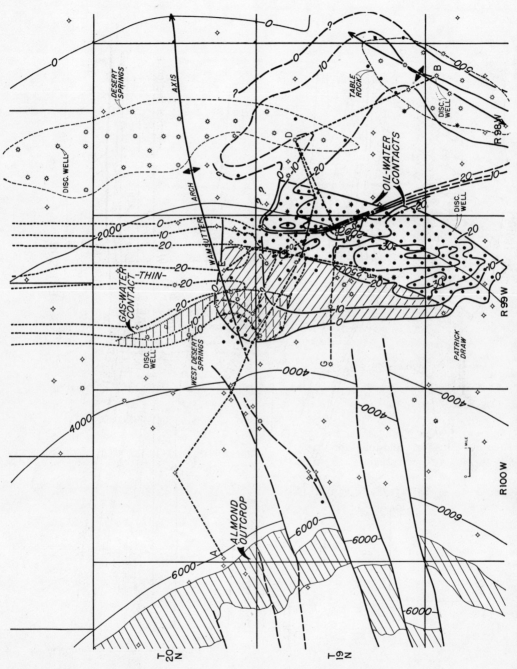

FIG. 10.—Isopachous map of uppermost Almond sandstone bodies (UA-5) in Patrick Draw (solid lines), Table Rock (long dashes), and West Desert Springs (short dashes) area. Isopachous contours are superimposed on structural contour map whose structural datum is top of Almond Formation. Isopachous interval is 10 ft.; structural contour interval is 2,000 ft. unless otherwise noted.

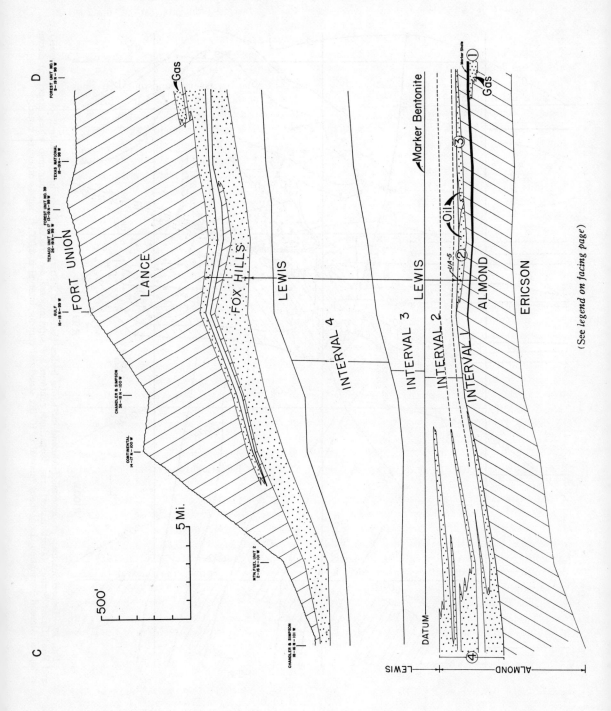

(See legend on facing page)

marine sandstone and second into non-marine coastal-plain sediments. The lower boundary of Interval 2 is the same as the upper boundary of Interval 1, described in the preceding section. The upper boundary of Interval 2 is the most widespread bentonite marker bed in the Lewis Shale (Figs 4, 7). The bed is approximately 3–5 ft. thick and is 10 ft. above a second widely distributed bentonite. The bentonite marker is 20–30 ft. below a widespread calcareous silty sandstone that was used by Barlow (1961) for correlation.

As shown by the sections (Figs. 4, 11) and the isopachous map (Fig. 12), Interval 2 contains 170–200 ft. of gray, non-calcareous shale in the area of the Patrick Draw, Desert Springs, and Table Rock fields. Westward this shale grades into gray, fine- to medium-grained, calcareous, fossiliferous sandstone, approximately 400 ft. thick. Much of this sandstone sequence crops out on the surface on the east flank of the Rock Springs uplift. The sandstone contains marine fossils and is interpreted to have been deposited in marine and barrier-island environments (Jacka, 1965, p. 81). The sandstone intertongues with and grades westward into non-marine, coal-bearing shale, siltstone, and sandstone, approximately 400 ft. thick. These facies changes can be observed from the surface exposures of the Almond Formation both north and south of Point of Rocks (loc. A, Fig. 2). This important change was noted and described by Hale (1950, p. 54), Weimer (1961b, p. 84), Lewis (1961, p. 92), and Jacka (1965, p. 85).

The westward increase in sandstone in Interval 2 is gradual as illustrated by the per-cent sandstone contours on Figure 12. Where exposed on the surface, the sandstone beds contain the fossil *Ophiomorpha*, which is interpreted to be a marine decapod burrow. The fossil also has been referred to as *Halymenites* (Weimer and Hoyt, 1964; Hoyt and Weimer, 1965.). The sandstone is porous and permeable and exhibits sedimentary structures characteristic of barrier-island sand bodies (Jacka, 1965, p. 88). Intertonguing marine and non-marine sediments can be observed at many surface localities.

Porous and permeable sandstone deposited on or associated with barrier islands is described as a barrier bar (Weimer, 1964, p. 381; Jacka, 1965, p. 81). The barrier-bar sandstone is a shoreline deposit characterized by a narrow linear shape, good sorting, porosity and permeability, beach and shallow-neritic sedimentary structures, and littoral and shallow-neritic fauna. Eolian deposits and beach ridges may be preserved but commonly are not. All these features are present in the outcrop of the upper Almond on the east flank of the Rock Springs uplift.

TIME-STRATIGRAPHIC INTERVALS 3 AND 4, MIDDLE AND UPPER LEWIS SHALE

Time-stratigraphic Intervals 3 and 4 are lithologically uniform (Figs. 4, 11) along the outcrop of the Lewis Shale on the east flank of the Rock Springs uplift and near the axis of the Wamsutter arch (T. 20 N., Rs. 97, 98, 99 W.). The intervals are marine and consist of gray non-calcareous shale with thin beds of bentonite, siltstone, calcareous sandstone, and limestone. The boundaries of Intervals 3 and 4 are bentonitic shale beds associated with siltstone and silty sandstone. Intervals 3 and 4 are not productive in the Patrick Draw field, but the gross facies changes within the intervals are discussed briefly.

From Table Rock field to the north end of Desert Springs field (a distance of 15 mi.), Interval 3 thickens from less than 200 ft. to more than 600 ft. Marine shale at Table Rock grades laterally into interbedded sandstone, siltstone, and shale at Desert Springs, where one sandstone in the upper third of the interval is gas-productive (Lewis "e" sand; see May, 1961, p. 292). Northeast of Desert Springs the interval thickens to 1,000 ft. and grades into a mass of siltstone and sandstone. The northeast thickening and facies change are interpreted to be associated with a deltaic center of deposition north of the Wamsut-

FIG. 11.—Restored section C-D showing facies changes in time-stratigraphic intervals in upper Almond Formation and Lewis Shale. Stratigraphic position of inferred barrier-bar sandstone bodies 1, 2, 3, and 4 in upper Almond also shown. Line of section indicated on Figures 12, 14, and 15. Section from electric logs of indicated wells. Diagonal ruling—non-marine sediments; stippled pattern—marine and transitional sandstone with some siltstone and shale; unmarked area in middle of diagram—marine shale.

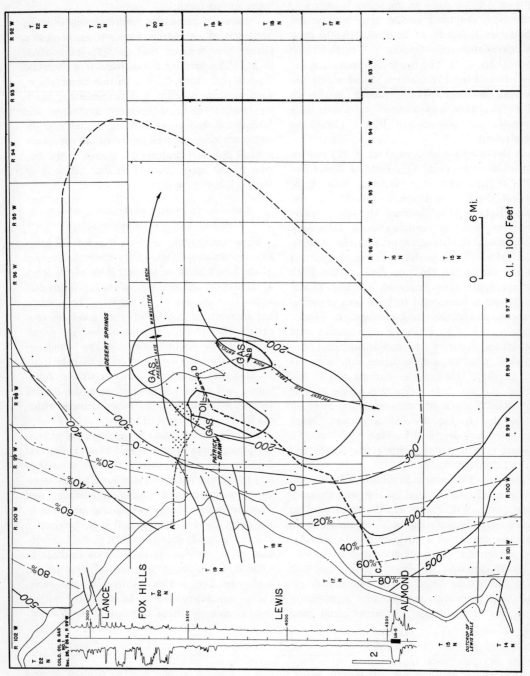

Fig. 12.—Isopachous map of time-stratigraphic Interval 2 (lower Lewis of subsurface; upper Almond and lower Lewis of surface; see Fig. 11 for cross section C-D of interval). Lines of sections A-B and C-D (Figs. 4, 11) show facies changes in interval. Per cent sandstone in interval is indicated by dashed lines. Control points are solid dots; field wells are omitted.

ter arch. Thin bottomset strata at Table Rock grade into thick foreset and topset strata at Desert Springs and toward the northeast.

A similar facies change associated with deltaic sedimentation is seen in Interval 4 southeast of the Wamsutter arch. Interval 4 is 300–400 ft. thick at the surface exposure and in the subsurface of the Patrick Draw field. Southeast across the Washakie basin the interval thickens to 700 ft. at Table Rock and to more than 1,400 ft. on the Dad nose (T. 16 N., R. 93 W.). Accompanying this thickening is a change from a section composed of gray marine shale on the Wamsutter arch (bottomset strata) to a section composed of thick beds of sandstone and siltstone (foreset and topset) on the southeast. This deltaic center of deposition was described by Weimer (1961a, p. 26). Gas production is found in sandstone beds in the lower part of Interval 4 at Table Rock (Fig. 4).

Thus the stratigraphic record shows that during the deposition of the middle and upper Lewis, Wamsutter arch was an area of marine sedimentation between two active centers of deltaic sedimentation. Because of the rapid progradation of the deltaic centers, the facies pattern of Interval 2, which shows a westward increase in thickness and sandstone per cent, is reversed. Interval 3 records an increase in thickness and sandstone content toward the northeast; Interval 4 shows an increase in thickness and sandstone content toward the southeast and northeast.

Youngest Cretaceous and Tertiary Formations

The electric-log characteristics of the Fox Hills Sandstone and Lance Formation (Cretaceous) and the Fort Union Formation (Tertiary) are illustrated on Figure 4.

Fox Hills Sandstone.—The Fox Hills is gray, fine- to medium-grained, fossiliferous sandstone with a thickness ranging from 150 to 250 ft. Many authors include this sandstone in the Lance Formation. A low cuesta is formed by the Fox Hills where it crops out on the east flank of the Rock Springs uplift. The Fox Hills is a regressive marine and transitional sandstone occurring stratigraphically between the marine Lewis Shale and the overlying non-marine Lance Formation. Facies changes between the Lewis, Fox Hills, and Lance are illustrated on Figures 4 and 11. The diagrams and other studies (Weimer, 1961b, p. 82; 1965b, p. 66, 78) suggest that, because of the facies changes, the stratigraphic position of the Fox Hills Sandstone rises approximately 600 ft. from T. 23 N., R. 103 W., to T. 18 N., R. 98 W. (Table Rock field). The Fox Hills is gas-bearing in one well drilled by Texas National in the NE. ¼, NW. ¼ of sec. 8, T. 19 N., R. 98 W. The position of the productive sandstone is shown on Figure 4.

Lance Formation.—The Lance Formation is chiefly gray and tan shale and claystone with thin beds of siltstone, sandstone, and coal. The thickness ranges from a wedge-edge to 1,000 ft., depending on the magnitude of pre-Fort Union truncation of the formation. In the area of the correlation chart (A-B, Fig. 4), the thickness ranges from 600 to 900 ft. The Lance is absent south of Patrick Draw field in T. 17 N., R. 101 W., because of post-depositional erosion. A soil that formed on the Lance during this period of erosion is preserved at the contact with the overlying Fort Union along the east side of the Rock Springs uplift. A sandstone near the base of the Lance (Fig. 4) has yielded small quantities of oil at Red Hill field in a well drilled by Chandler and Simpson (sec. 36, T. 20 N., R. 100 W.).

Fort Union Formation.—The Fort Union Formation consists of gray, fine- to medium-grained, calcareous sandstone interbedded with tan shale and siltstone and light gray shale with numerous coal beds. The thickness ranges from 1,200 to more than 2,000 ft. The formation shows a pronounced westward thinning from the Washakie and Great Divide basins to the Rock Springs uplift. Figure 4 indicates the relation between one marker coal bed in the Fort Union and the base of the formation.

Some workers interpret the thinning of the lower Fort Union as overlap from the basin onto the Rock Springs uplift. Careful tracing of coal beds in West Desert Springs, Patrick Draw, Desert Springs, and Table Rock fields and isopachous mapping of intervals between coal beds indicate that most, if not all, of the westward thinning results from convergence. The per cent of sandstone also decreases toward the outcrop area. This convergence probably is the result of slower rates of sedimentation in the area of the uplift compared with higher rates in basin areas, which were closer to the source of sediments. The Fort

Union yields gas from conglomeratic sandstone on the Table Rock structure (Fig. 4). This productive interval was assigned to the Wasatch Formation by earlier workers (White, 1955, p. 170).

The contact between the Fort Union and overlying Wasatch Formation in the area of the Wamsutter arch is difficult to place. Both formations exhibit similar lithologic character in the outcrop sections and similar electric-log characteristics in the subsurface. The detailed lithology of the formations and the difficulty in placing the contact are described by Davis (1958). Additional work is needed to establish criteria for accurate placement of the boundary between the formations.

EARLY STRUCTURAL DEVELOPMENT AND PETROLEUM ACCUMULATION, PATRICK DRAW AREA

Favorable structural development for petroleum accumulation occurred in the Patrick Draw area soon after deposition of the Almond reservoir rocks. This concept was developed by Barlow (1961, p. 113) and Ritzma (1963, p. 194). The concept is developed more fully in this paper and documented with maps and sections.

An isopachous map of time-stratigraphic Interval 2 (Fig. 12) shows an isopachous thin in the area of the major fields on the Wamsutter arch. Barlow (1961, p. 113) referred to this feature as the "Carlisle high," although it could more appropriately be called the "Carlisle thin." However, if it is assumed that the marker bentonite at the top of time-stratigraphic Interval 2 was deposited essentially horizontally, then the isopachous map (Fig. 12) also may serve as a structural contour map of the top of the Almond Formation. At the time the bentonite bed was deposited, the Almond would have been warped into a regional closed anticline with 100–150 ft. of closure. Depth of burial of the Almond reservoir sandstone bars at this time would have been approximately 160 ft. in the Table Rock area (T. 18 N., R. 98 W.), with increasing depths in all directions.

As depth of burial increased during deposition of time-stratigraphic Intervals 3 and 4 (middle and upper Lewis), the Almond Formation was warped into a broad east-plunging structural nose. Figure 13 is an isopachous map of Intervals 2, 3, and 4 combined, which correspond closely to the total thickness of the Lewis Shale. Although no evidence exists that the marker bed used as the top of Interval 4 was deposited horizontally, it seems to be sub-parallel with coal beds deposited a few hundred feet stratigraphically higher in the lower Lance Formation; these coal beds were deposited horizontally at or near sea-level. Thus, an isopachous map from the base of the Lance coal beds to the top of the Almond Formation has a configuration similar to that of the isopachous trends shown on Figure 13. Accordingly, the isopachous map also may be used as a structural contour map reflecting the structural attitude of the Almond reservoir rocks at the time of the ends of Lewis, Fox Hills, or Lance deposition. The north-south-trending reservoir sandstone beds in the Almond were warped across an east-plunging anticline. The northeast flank of the anticline had a dip of 35 ft./mi. ($\frac{1}{3}°$), whereas the southeast flank had a dip of 65 ft./mi. ($\frac{2}{3}°$). Depth of burial of the reservoir rock ranged from 1,000 to 1,200 ft. An old stratigraphic petroleum accumulation in the Patrick Draw area is believed to have formed on this structural nose in the area indicated by the stippled pattern on Figure 13.

The present structural configuration of the Wamsutter arch (Fig. 14) is different from that near the end of the Cretaceous (Fig. 13). Early Tertiary folding formed the axis of the present arch 10 mi. north of the Cretaceous axis. By mid-Tertiary (post-early Eocene) time, the Almond Formation had developed a 4° dip on the Wamsutter arch (Fig. 14) and the Cretaceous stratigraphic trap (Fig. 13) was opened to the north. Oil and gas moved northward to higher structural positions. The Patrick Draw accumulation is on the south flank of the present arch because the oil and gas could not migrate past the permeability barrier shown on Figures 9 and 10. The uppermost Almond sandstone in the West Desert Springs field (Fig. 8) was in a low structural position (Fig. 13) and therefore water-bearing at the time of original migration (Cretaceous). During Tertiary time this sandstone was elevated to a high structural position, but only a small quantity of gas was available for migration into the trap. An alternate interpretation is that the trap may not have been able to hold a greater gas column than is now present, and large quantities of petroleum may have leaked to the outcrop 8 mi. west.

Several assumptions have been made in developing the hypothesis described above regarding the migration and entrapment of petroleum at Patrick Draw. (1) The oil is assumed to have originated in the marine lower part of the Lewis Shale. (2) The oil migrated and accumulated early in the geologic history of the area, a few million years after the Almond reservoirs were deposited and covered by a few hundred feet of overburden. (3) Two phases of petroleum migration are assumed. The early phase permitted migration of hydrocarbons from shale to the reservoir strata, the shale providing no barrier to migration. After the hydrocarbon accumulated in the reservoir rock, a later phase of migration occurred in which shale acted as a permeability barrier and provided the trapping mechanism now present in the Patrick Draw field.

One can only speculate concerning why the large Patrick Draw oil accumulation is present in an otherwise gas-productive part of the section. Several anomalies are obvious if Patrick Draw field is compared regionally with equivalent strata deposited in the Late Cretaceous basin. The upper Almond and lower Lewis interval is exceptionally thin compared with other areas. The existence of a thin, organic-rich section may have permitted a high concentration of marine organic material favoring oil generation. The location of the Patrick Draw area in a marine embayment between two lower Lewis deltaic centers of deposition (Fig. 16) also may be important.

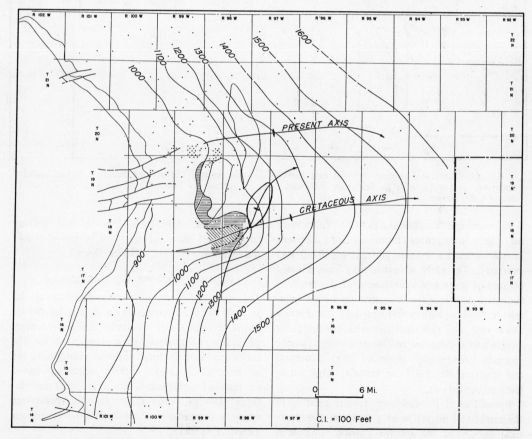

Fig. 13.—Isopachous map of time-stratigraphic Intervals 2, 3, and 4 (approximately equal to Lewis Shale of subsurface). Control points are solid dots; field wells are omitted. Axis of Wamsutter arch during the Cretaceous and present structural axis are indicated. Area of old stratigraphic accumulation during Late Cretaceous shown by horizontal lines; gas area shown by diagonal lines.

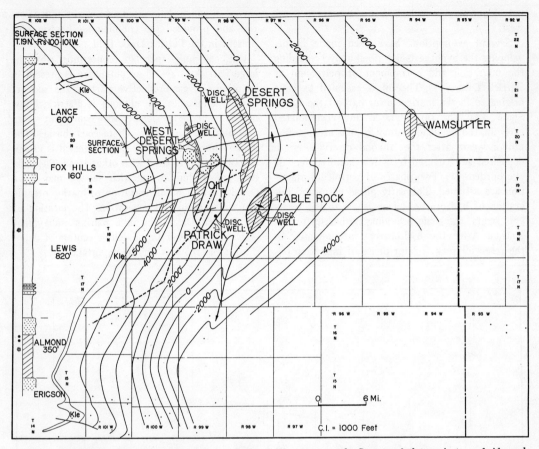

Fig. 14.—Structural contour map of east flank of Wamsutter arch. Structural datum is top of Almond Formation. Discovery wells in fields are indicated. Kle = outcrop of Lewis Shale. Location of Figure 11 (section C-D) shown.

Paleoecologic conditions in this framework may have been more favorable for abundant growth of marine organisms from which oil was generated. The early structure may have formed because of more rapid subsidence of the crust on the north and south in response to deltaic loading. A residual high was formed in the Patrick Draw area, and this, together with intertonguing marine and non-marine sediments having laterally changing permeability, provided ideal structural and stratigraphic conditions favoring large petroleum accumulations.

Regardless of the unknown factors relating to the origin and migration of petroleum, the structural history of the general Patrick Draw area favored entrapment soon after reservoir deposition. Later structural movement, generally interpreted to be related to Laramide orogeny, modified the early traps but did not destroy them. All of these factors must be considered in the search for another "Patrick Draw."

GEOLOGIC HISTORY AND CONCLUSIONS

A geologist describing the geologic history of an area has the opportunity to summarize the interpretative phase of his studies. Subjective interpretations of geologic events recorded by the rocks are likely to differ in detail among different workers. The following interpretations are based on regional surface and subsurface data correlations. Despite weaknesses, the interpretations seem to comprise the best hypothesis to explain all available facts.

The geologic history of the latest Cretaceous rocks in the Wamsutter arch area can be related to shoreline movements along the western margin

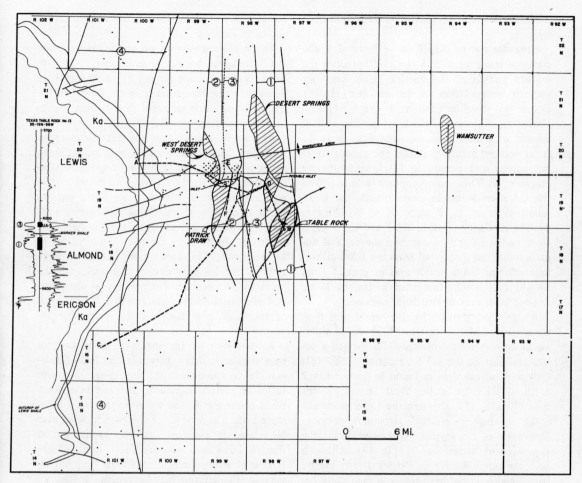

FIG. 15.—Map showing areal distribution of four porous and permeable sandstone zones in upper 100 ft. of Almond Formation. Stratigraphic positions of bar-sandstone bodies 1, 2, 3, and 4 are shown on Figure 11. (Patrick Draw producing sandstone beds referred to in field as UA-5.) Distribution of porosity and permeability in producing interval at Table Rock and Desert Springs and in surface exposures also is shown. Ka = outcrop of Almond Formation; hachures indicate gas-producing areas in each belt of porous sandstone. Impermeable sandstone beds of Intervals 1 and 4 are east of belt of porosity and permeability (refer to Figs. 4, 11). Sandstone bodies are inferred to be barrier-bars. Control wells are solid dots; field wells are not shown. Location of Figure 11 (section C-D) shown.

of the Lewis sea. The first-recognized shoreline deposits are the north-south-trending barrier bars productive of gas at Table Rock and Desert Springs fields (Fig. 15). The 10-ft. marker shale at the base of Interval 1 rests on this barrier-bar zone, which ranges from a wedge-edge to 90 ft. thick. Although this barrier-bar zone was not discussed in detail, it is referred to as shoreline position 1 on Figures 11 and 15. The western updip pinch-out of porous and permeable sandstone into impermeable swamp and lagoonal shale and sandstone is indicated on Figures 7 and 11. This well-defined facies change sets the stratigraphic framework for the overlying time-stratigraphic Intervals 1 and 2.

After this initial transgression, the shoreline regressed to an undetermined position on the east, and 50–70 ft. of coastal-plain sediments accumulated. The first deposition in the Patrick Draw area during or after the regression was the 10–12-ft. lagoonal shale ("marker shale") at the base of Interval 1. Interval 1 contains swamp and lagoonal deposits and an oil-productive sandstone (UA-6) near the middle, interpreted to be a tidal-delta sandstone deposited in a lagoonal environment.

Near the end of deposition of Interval 1, the shoreline transgressed to a new position along the western margin of the Patrick Draw field. By seaward progradation, a barrier bar (UA-5) formed at shoreline position 2 (Figs. 11, 15). During establishment of the beach profile, a small amount of erosion occurred, removing up to 15 ft. of section, before or during the deposition of the porous and permeable UA-5 sandstone by progradation. This erosion appears to have placed the UA-5 sandstone in contact with the UA-6 sandstone in sec. 23, T. 19 N., R. 99 W. (Figs. 5, 9). The similarity of oil in UA-5 and UA-6 sandstone bodies suggests a common source, and migration may have occurred from the UA-5 oil column into the UA-6 in this area of contact (Fig. 5). Oil in the UA-6 migrated northward to be trapped in its present structural position.

A tidal inlet through barrier bars 2 and 3 at the north end of Patrick Draw field caused a permeability seal to be deposited. This formed a petroleum trap in the UA-5 sandstone (Fig. 15). Only a small gas field is found in barrier bar 2 north of the inlet in the West Desert Springs field (Figs. 8, 15), although the UA-5 sandstone is at its highest regional structural position. Regression of the shoreline continued during the deposition of barrier bar 3 (Fig. 15). Although both are oil-productive in Patrick Draw, barrier bars 2 and 3 do not have reservoir continuity because they have different oil-water contacts. The oil-water contact in barrier bar 2 is at +1,500 ft.; the oil-water contact in barrier bar 3 is at + 1,620 ft. Moreover, a thin productive sandstone present through the middle of the Patrick Draw field (Fig. 6) is interpreted as the seaward (eastward) thinning edge of bar 2. As the shoreline prograded seaward (eastward) during the deposition of barrier bars 2 and 3, southward progradation appears to have occurred on the south end of barrier bars 2 and 3 north of the inlet (Fig. 15). This southward growth resulted in an overlap of barrier bar 2 at the north end of Patrick Draw field by a slightly younger water-bearing sandstone, bar 3 (Figs. 8, 9, 10, 15). Approximately 10 ft. of impermeable shale separates the two sandstone bodies and provides the north permeability seal on the UA-5 sandstone at Patrick Draw. Gas production from barrier bar 3 at Table Rock is described by Mees et al. (1961) and is shown on Figure 6.

The writer's interpretation of the origin of the UA-5 sandstone bodies requires some discussion. These bodies are interpreted to be barrier bars deposited along the shoreline of the Lewis sea for the following reasons: areal distribution; north-south isopachous trends; geometry; grain size (fine to medium) and mineral composition (abundant black chert grains); good sorting, indicating high energy; generally low-angle cross-stratification; associated oyster-bearing facies on the west side of bar 3, Patrick Draw field; and the facies association of underlying coal-bearing strata (fresh water) and overlying marine shale. The sandstone pinch-out on the west is interpreted to be the result of a change from barrier-island to swamp and lagoonal environment, whereas the break in permeability and porosity at the north end of Patrick Draw is interpreted as the result of the tidal inlet that cut the barrier-bar sequence.

Paleontology of the sandstone beds has not been studied in detail. However, the fossil *Ophiomorpha*, so common in the barrier-bar sandstone facies in surface exposures, has not been identified positively by the writer in the UA-5 sandstone beds. The absence of this marine fossil may be related to the fact that the barrier bar at Patrick Draw was deposited just south of the tidal inlet where ecologic conditions were not favorable to sustain the burrowing animal.

The slight erosion (10–15 ft. in a distance of 5 mi.) at the base of the productive sandstone southward across the Patrick Draw field might be interpreted as support for a channel-deposit origin (possibly fluvial). However, erosion similar to that which is common at the base of channel deposits also may take place during the formative stages of a shoreline beach or bar deposit, or at the bottoms of channels associated with migrating tidal inlets (Hoyt and Henry, 1965, p. 190).

Transgression of the shoreline approximately 20 mi. west followed deposition of barrier bar 3. During the transgression, an overstep of depositional environments occurred west of Patrick Draw field where offshore neritic shale of the Lewis rests on coastal-plain swamp and lagoonal sediments. During the transgression, the topographically high parts of barrier bars 2 and 3 are inferred to have been beveled and the sand probably was transported to the new shoreline position 4 (Fig. 15). In Patrick Draw field, the depth of beveling may have been controlled by indurated coquina layers of the brackish-water oyster *Ostrea glabra*, which is believed to have been deposited behind the barrier island that formed at shoreline position 3. One coquina layer covered more than 4 sq. mi. near the eastern edge of T. 19 N., R. 99 W. (Lawson and Crowson, 1961, p. 286) (area with diagonal ruling between bars 2 and 3, Fig. 7). Cores from Arch Unit 3, in sec. 19, T. 19 N., R. 98 W. (Fig. 5), show a concentration of burrows and of glauconitic sandstone on top of bar 3 (core depth, 4,959 ft.). This beveled surface marking the base of the transgressive sequence is used as the lower boundary of Interval 2. The Lewis Shale was deposited on the essentially horizontal surface in the area of the Patrick Draw, Table Rock, West Desert Springs, and Desert Springs fields.

Several barrier bars representing shoreline positions of the sea are collectively grouped as shoreline position 4 on Figures 11 and 15. The details of the complex intertonguing of barrier-bar sandstone and swamp and lagoonal deposits can be observed on the surface but are not described here. As indicated on Figure 11, the barrier-bar sandstone of shoreline position 4 embraces most of time-stratigraphic Interval 2.

After deposition of Interval 2, the shoreline of the Lewis sea transgressed farther west, and Intervals 3 and 4 of the Lewis Shale were deposited before the shoreline regressed to the Wamsutter arch area, as recorded by the Fox Hills Sandstone. Following deposition of the Fox Hills, the entire region was covered by non-marine sediments, now included in the Lance Formation. Near the end of the Cretaceous, the Wamsutter arch and Rock Springs uplift were subjected to subaerial erosion which removed part or all of the Lance, Fox Hills, and Lewis. A soil developed during this period of erosion. This soil, which is well exposed on the east flank of the Rock Springs uplift, was preserved when the surface was covered by sediments of the Fort Union Formation during the early Tertiary.

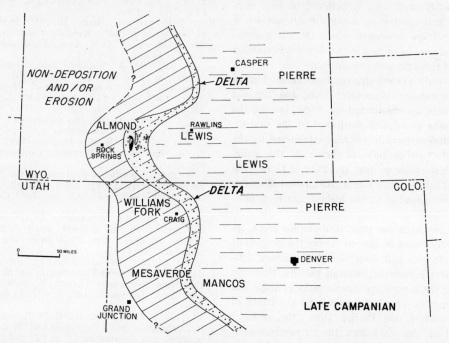

FIG. 16.—Generalized regional lithofacies map showing correlation of upper part of Almond in Wamsutter arch fields and adjacent areas. Environmental interpretation: horizontal ruling = neritic shale and siltstone; diagonal ruling = coastal-plain deposits of impermeable coal-bearing claystone, siltstone, sandstone; mixed stippling and diagonal ruling = zone of intertonguing porous and permeable barrier-bar sandstone beds (shoreline) and coastal-plain deposits (as above). Petroleum fields: 1 = Patrick Draw and West Desert Springs; 2 = Desert Springs; 3 = Table Rock; 4 = Wamsutter.

A generalized regional facies pattern for Intervals 1 and 2 is shown on Figure 16. Locations of the principal fields producing from the sandstone of Interval 1 are indicated. The fields produce from barrier-bar sandstone bodies and associated deposits located regionally between two centers of deltaic sedimentation that are shown on Figure 16 as eastward bulges in the pattern of shoreline deposits. Formation names used for the same stratigraphic interval in northwest Colorado are indicated. Although not a dominant influence on sedimentation in the Wamsutter arch area during deposition of Intervals 1 and 2, the deltaic centers were extremely active and influenced sedimentation strongly during deposition of Intervals 3 and 4. The embayment in the coast line indicated on Figure 16 was called the "Hallville Embayment" by Lewis (1961).

Early structural development of the Wamsutter arch before the end of Lewis Shale deposition played an important role in petroleum accumulation. An east-plunging anticline, whose axis was located approximately 10 mi. south of the present Wamsutter arch axis, is believed to have controlled early petroleum migration. Ancient stratigraphic traps developed where the north-south-oriented sandstone bars in the Almond Formation crossed this east-west-trending anticline.

Post-early Eocene structural movement formed the present-day Wamsutter arch, and the old traps were opened toward the north, causing the petroleum to migrate northward to the first-encountered permeability barrier established by the new structural attitudes of the strata. This structural history shows that the Patrick Draw field formed on the south side of the present Wamsutter arch instead of on the highest structural position.

In the search for large stratigraphic traps, investigators must be able to predict the location of reservoir rock and determine the structural attitude of the reservoir through geologic time. By tracing the history and development of traps and by using the knowledge gained by the study of fields such as the Patrick Draw, the geologist should be able to reduce the tremendous odds which he now faces in finding large stratigraphic traps.

REFERENCES

Barlow, J. A., Jr., 1961, Almond Formation and lower Lewis Shale, east flank of Rock Springs uplift, Sweetwater County, Wyoming: Wyo. Geol. Assoc. Guidebook, 16th Ann. Field Conf., p. 113–115.

Burton, Guy, 1961, Patrick Draw area, Sweetwater Co., Wyoming: Wyo. Geol. Assoc. Guidebook, 16th Ann. Field Conf., p. 276–279.

Cox, J. E., 1962, Patrick Draw field and adjacent areas, Sweetwater County, Wyoming: Billings Geol. Soc. Paper No. 1, p. 1–17.

Davis, R. W., 1958, Stratigraphy of the Wasatch Formation, east flank of the Rock Springs uplift: unpub. Master's thesis, Univ. Wyo.

Douglas, W. B., Jr., and T. R. Blazzard, 1961, Facies relationships of the Blair, Rock Springs and Ericson Formations of the Rock Springs uplift and Washakie basin: Wyo. Geol. Assoc. Guidebook, 16th Ann. Field Conf., p. 81–86.

Earl, J. H., and J. N. Dahm, 1961, Case history—Desert Springs gas field, Sweetwater County, Wyoming: Geophysics, v. 26, no. 6, p. 673–682.

Hale, L. A., 1950, Stratigraphy of the Upper Cretaceous Montana Group in the Rock Springs uplift, Sweetwater County, Wyoming: Wyo. Geol. Assoc. Guidebook, 5th Ann. Field Conf., p. 49–58.

——— 1955, Stratigraphy and facies relationships of the Montanan Group in south central Wyoming, northeastern Utah and northwestern Colorado: Wyo. Geol. Assoc. Guidebook, 10th Ann. Field Conf., p. 89–94.

——— 1961, Late Cretaceous (Montanan) stratigraphy, eastern Washakie basin, Carbon County, Wyoming: Wyo. Geol. Assoc. Guidebook, 16th Ann. Field Conf., p. 129–137.

Hoyt, J. H., and V. J. Henry, Jr., 1965, Significance of inlet sedimentation in recognition of ancient barrier islands: Wyo. Geol. Assoc. Guidebook, 19th Ann. Field Conf., p. 190–194.

——— and R. J. Weimer, 1965, Origin and significance of *Ophiomorpha* (*Halymenites*) in the Cretaceous of the Western Interior: Wyo. Geol. Assoc. Guidebook, 19th Ann. Field Conf., p. 203–208.

Jacka, A. D., 1963, Depositional dynamics of the Almond Formation, Rock Springs uplift, Wyoming (abs.): 13th Ann. Meeting Program, Rocky Mtn. Section, Am. Assoc. of Petroleum Geologists, Casper, Wyo., p. 20.

——— 1965, Depositional dynamics of the Almond Formation, Rock Springs uplift, Wyoming: Wyo. Geol. Assoc. Guidebook, 19th Ann. Field Conf., p. 81–100.

Lawson, D. E., and C. W. Crowson, 1961, Geology of the Arch Unit and adjacent areas, Sweetwater County, Wyoming: Wyo. Geol. Assoc. Guidebook, 16th Ann. Field Conf., p. 280–299.

Lewis, J. L., 1961, The stratigraphy and depositional history of the Almond Formation in the Great Divide basin, Sweetwater County, Wyoming: Wyo. Geol. Assoc. Guidebook, 16th Ann. Field Conf., p. 87–95.

Love, J. D., P. O. McGrew, and H. D. Thomas, 1963, Relationship of latest Cretaceous and Tertiary deposition and deformation to oil and gas in Wyoming: Am. Assoc. Petroleum Geologists Mem. 2, p. 196–208.

Lowman, S. W., 1949, Sedimentary facies in Gulf Coast: Am. Assoc. Petroleum Geologists Bull., v. 33, p. 1939–1997.

May, B. E., 1961, The Desert Springs field: Wyo.

Geol. Assoc. Guidebook, 16th Ann. Field Conf., p. 290–293.

Mees, E. C., J. D. Copen, and J. C. McGee, 1961, Table Rock field, Sweetwater County, Wyoming: Wyo. Geol. Assoc. Guidebook, 16th Ann. Field Conf., p. 294–300.

Ritzma, H. R., 1963, Geology and occurrence of oil and gas, Wamsutter arch, Wyoming: Am. Assoc. Petroleum Geologists Mem. 2, p. 188–195.

Schultz, A. R., 1909, The northern part of the Rock Springs coal field, Sweetwater County, Wyoming: U. S. Geol. Survey Bull. 341, p. 256–282.

―――― 1920, Oil possibilities in and around Baxter basin, in the Rock Springs uplift, Sweetwater County, Wyoming: U. S. Geol. Survey Bull. 702, 107 p.

Smith, J. H., 1961, A summary of stratigraphy and paleontology, upper Colorado and Montanan Groups, south-central Wyoming, northeastern Utah and northwestern Colorado: Wyo. Geol. Assoc. Guidebook, 16th Ann. Field Conf., p. 101–112.

Weimer, R. J., 1961a, Uppermost Cretaceous rocks in central and southern Wyoming, and northwest Colorado: Wyo. Geol. Assoc. Guidebook, 16th Ann. Field Conf., p. 17–28.

―――― 1961b, Spatial dimensions of Upper Cretaceous sandstones, Rocky Mountain area, *in* Geometry of sandstone bodies: Am. Assoc. Petroleum Geologists, p. 82–97.

―――― 1964, Comparison of Recent shoreline sedimentation with the stratigraphy of Upper Cretaceous oil fields, Rocky Mountain area (abs.): Am. Assoc. Petroleum Geologists Bull., v. 48, p. 381.

―――― 1965a, Late Cretaceous deltas, Rocky Mountain region (abs.): Am. Assoc. Petroleum Geologists Bull., v. 49, p. 363.

―――― 1965b, Stratigraphy and petroleum occurrences, Almond and Lewis Formations (Upper Cretaceous), Wamsutter arch, Wyoming: Wyo. Geol. Assoc. Guidebook, 19th Ann. Field Conf., p. 65–81.

―――― and J. H. Hoyt, 1964, Burrows of *Callianassa major* Say, geologic indicators of littoral and shallow neritic environments: Jour. Paleontology, v. 38, p. 761–767.

White, V. L., 1955, Table Rock and Southwest Table Rock gas fields: Wyo. Geol. Assoc. Guidebook, 16th Ann. Field Conf., p. 170–171.

Wyoming Geological Association, 1961a, Supplement to Wyoming oil and gas field symposium, 1957, Map of Patrick Draw-Desert Springs area, p. 542.

―――― 1961b, Symposium on Late Cretaceous rocks, Wyoming and adjacent areas: Wyo. Geol. Assoc. Guidebook, 16th Ann. Field Conf., 351 p.

―――― 1965, Sedimentation of Late Cretaceous and Tertiary outcrops, Rock Springs uplift: Wyo. Geol. Assoc. Guidebook, 19th Ann. Field Conf., 233 p.

SAND BODIES AND SEDIMENTARY ENVIRONMENTS: A REVIEW[1]

PAUL EDWIN POTTER[2]
Bloomington, Indiana

ABSTRACT

Knowledge of sand-body morphology may help to solve many practical exploration problems. Knowledge of the environment of deposition is required to predict sand-body morphology most successfully.

This paper briefly reviews sand-body nomenclature, discusses the problems of representation by cross sections, comments on currently popular concepts of sedimentary environments, and summarizes in detail what is known of sand-body characteristics in six major environments: alluvial, tidal, turbidite, barrier island, shallow-water marine, and desert eolian. The characteristics of alluvial, tidal, and barrier-island sand bodies are better known than those of turbidite, shallow-water marine, and desert-eolian ones. Nonetheless, there is an obvious lack of systematic, quantitative data on the petrology, texture, sedimentary structures, and internal organization of sand bodies for these six major environments.

The problems of sand-body predictions are also briefly summarized. Generally, three types of information are required: depositional environment, regional distribution of sand facies, and paleoslope. Usually prior experience with sandstone bodies of the same facies, preferably in the same basin, is very helpful. Locating a new sandstone body in a basin is usually much more difficult than drilling the first step-out well.

INTRODUCTION

Knowledge of sand-body morphology is the practical means of solving many exploration problems and, if it is well-known, usually is helpful in identifying the dominant transport agent as well. All too commonly, however, whether in exploring for oil and gas, ground water, or predicting the trend of a cutout (washout) of a coal bed, it is the morphology that one desires to predict from either limited subsurface or outcrop data. Knowledge of the environment of deposition of the sand body is one of the most important—if not the most important—factors that makes it possible to determine sand-body morphology.

This paper is a compilation of current knowledge of sedimentary environments in relation to quartzose sand bodies. Only arenites are considered, because of insufficient data relating to carbonate sand bodies. In fact, critical data even for quartzose sand bodies are none too abundant.

PUBLISHED NOMENCLATURE

The terminology of sand-body shape is, like that of sedimentary structures, a mixture of the descriptive and genetic. Moreover, it is probably correct to say that the terminology—or at least the approach it represents—may not be equal to the requirements of most effective prediction. The descriptive terms generally are geometric, whereas the genetic ones—reflecting the close connection between sand-body origin and geomorphology—generally are landform names (Rich, 1938, p. 230–231). Attempts at a purely descriptive, quantitative terminology have been made, but have not been widely followed.

Krynine (1948, p. 146–147) thought that the ratio of width to thickness of a sand body was a good means of estimating its area-volume ratio. Using the ratio of width to thickness, he distinguished four types of sand bodies: (1) blanket (greater than 1,000 to 1), (2) tabular (between 1,000 and 50 to 1), (3) prisms (between 50 and 5 to 1), and (4) shoestring (less than 5 to 1). In practice it has not been easy to impose size limits for such terms. For example, a sandstone body that a reservoir geologist might think of as a blanket is anything but a blanket in the eyes of a basin geologist. McGugan (1965, p. 126) modified and extended this approach when he in-

---

[1] Manuscript received, March 3, 1966; accepted, May 16, 1966.

[2] Geology Department, Indiana University.

It is a pleasure to acknowledge the others who have helped in the preparation of this review. Foremost is Raymond Siever of Harvard University. Several of the paragraphs in the section entitled "Concepts of Environment" were written by him. Francis Pettijohn of The Johns Hopkins University also reviewed the entire manuscript and was most helpful. Others who read it include: Leroy Becker of the Indiana Geological Survey; Alan Horowitz and Albert J. Rudman of Indiana University; Frank Moretti of Imperial Oil, Calgary; W. A. Pryor of the University of Cincinnati; and Ernst Ten Haaf of the University of Utrecht. Kenneth H. Heh and Stephanie V. Hrabar prepared Figure 3.

troduced the persistence factor, which he defined as

$$\frac{\text{areal extent of unit}}{\text{average thickness of unit}}$$

where both numerator and denominator are measured in the same units. He found that some of the widespread sheet and blanket deposits of the neritic and littoral zones of the craton had persistence factors of 396×10^6. Unfortunately, the persistence factor does not differentiate elongate and equant sand bodies of equal area and thickness.

Although the persistence factor and related measures may be useful in some studies, sand thickness is generally of less immediate importance than knowledge of the shape, horizontal dimensions, and orientation of the sand body. Thus workers have introduced the terms blanket and sheet for roughly equidimensional sand bodies and have used descriptive names, such as shoestring (Rich, 1923, p. 103) pod, ribbon, and belt (Potter, 1962, p. 3), for elongate sand bodies, sand bodies whose long dimension is 2 to 100 times greater than width. Beaches, chheniers (a Gulf Coast name for abandoned, raised beaches), barrier bars, as well as alluvial, deltaic, and turbidite channel-fill deposits, and many of the sand bodies of shallow-marine shelves generally provide good examples of elongate sand bodies.

Sand bodies have also been described by their relation to depositional strike. They are called strike sands, if parallel with the depositional strike of the basin, and dip sands, if they extend downdip. The relation of a sand body to an unconformity has prompted such geomorphic names as strike-valley sand bodies (Busch, 1959, p. 2829).

Although a classification of size has been proposed (Krynine, 1948, Table 2), most workers prefer to state simply the typical dimensions and thickness of sand bodies. To this should be added their orientation with respect to depositional strike and appropriate descriptive terms, such as dendritic, anastomosing, bifurcating, branching, straight, and *en échelon*, in order to convey their morphology to the reader. The appropriate genetic name (if it can be correctly assigned), such as washover sand, a sand body that developed transverse to an underlying barrier island, generally contains much of this information.

An additional descriptive term that has proved useful is multistory (Feofilova, 1954, p. 255), referring to the superposition of a sand body of one cycle upon one or more earlier ones, the two or more forming an unusually thick sand section. Although originally proposed for alluvial sand bodies, the name is worthy of general use. Similarly, one might call a laterally coalescent sand body a multilateral sand body. Sullwold (1961, Fig. 2) later used the term bundle for multistory bodies.

Finally, a comment on the term sand body itself. Is it possible to define rigorously a sand body? The answer is probably "no" just as one can not rigorously define most sedimentary structures. But if one feels in need of a definition, one might define a sand body as "a single, interconnected mappable body of sand." Scale enters into this definition. However, the term *mappable* serves to differentiate a small sand body from most single beds.

Terms and Concepts Cited in Text

bar finger	point bar
barrier-island complex	prism
belt	ribbon
blanket	search theory
bundle	sheet
chenier	shoestring
dip sand body	stratigraphic
environment	cross section
environmental sensitivity	strike sand body
facies model	strike-valley
internal organization	sand body
model concept	structural cross
multilateral	section
multistory	tabular
olistostrome	thalweg
persistence factor	washover sand body
pod	

Cross Sections

For a study of sedimentation, cross sections of sand bodies generally are most informative, if they are stratigraphic rather than structural cross sections; *i.e.*, if the sand body is either situated relative to a marker bed or level line above or below the sand body. A series of subparallel, transverse cross sections at right angles to direction of elongation is generally most useful. Longitudinal cross sections also may be valuable, however, especially if one wants to see longitudinally, *e.g.*, downdip, the separation of the base of a sand body from an underlying marker bed (Andresen, 1961, Fig. 9a). For widespread sheet and blanket sand bodies, cross sections either parallel

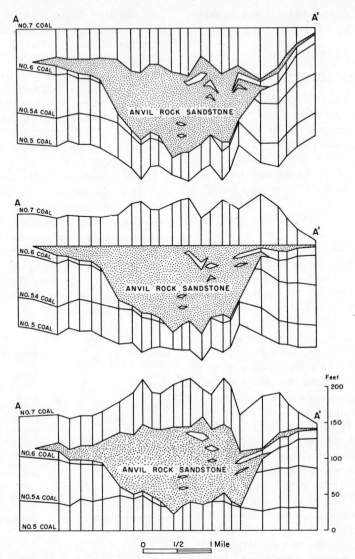

FIG. 1.—Cross sections of Anvil Rock Sandstone (Pennsylvanian) showing effect of different choice of level lines (modified from Potter, 1963, Fig. 37B).

with or at right angles to depositional strike or to transport direction are generally most informative.

Differential compaction of clay and (or) carbonate mud around a sand body distorts reconstruction of its original shape, a fact that is easily seen when one uses first a marker bed above and later a marker bed below an elongate sand body as a level line. If a marker bed above the sand body is assumed level, differential compaction causes the marker beds below the axis of elongation of the sand body to appear to be deformed into a weak syncline. Conversely, use of an underlying marker bed as the level line produces a compactional high over the sand body (Fig. 1).

The effect of differential compaction also depends on the state of consolidation of the underlying sediments. If the underlying sediments are not yet fully consolidated, a channel or valley-fill deposit may subsequently appear as a biconcave lens in the subsurface. Only if previously deposited marker beds were eroded could one safely infer the correct origin. Lack of such proximal beds may, for wider sand bodies, falsely give a channel or valley-fill deposit the appearance of being flat-bottomed and conformable in the subsurface, though in reality the basal contact is erosional and disconformable. If the sand body overlies or occurs not far above the consolidated sediments of an unconformity, however, differential compaction of the underlying sediments will be minimal and original cross-section shape will more likely be largely preserved.

Estimates of the percentage of compaction of clay and sand offer the possibility of reconstructing original shape. Shelton (1962), using contorted sandstone dikes, estimated shale compaction in the Cretaceous of the western United States to be approximately 2.6 times greater than that of the sandstone. In practice such estimates are rarely made, because they are difficult to obtain. Rittenhouse (1961, p. 7–8) discusses some of the interrelations between internal compaction within a sandstone body, topographic relief, and compaction of the enclosing shale. At present there appears to be no really practical satisfactory method for obtaining the original precompaction shape of most sandstone bodies in shale sections. Quartzose sandstone bodies interbedded with carbonates, however, probably reflect more closely their original shape, because carbonates, especially carbonate muds, compact only a little.

A recurring problem in the subsurface is to determine if missing marker beds were eroded at the sand body's edge or if they were not deposited. Usually electric logs alone are insufficient to answer this question and either subsurface lithologic samples or outcrops must be examined.

Concepts of Environment

The word environment is used constantly in sedimentary literature, although all too commonly its precise meaning is not defined. Here various uses are discussed and a definition offered.

Perhaps the most restricted use of the term occurs when one specifies the range of a physical or chemical variable, *e.g.,* pH greater then 7, O_{18}/O_{16} ratios less than some arbitrary quantity, or even "high" or "low" turbulence.

A completely different point of view is the glittering definition of energy in which one refers to "high" or "low" energy, usually meaning kinetic energy. At present, there is no unique way to evaluate this in an ancient sandstone. For example, grain size, grain-mud ratios, and bed thickness have all been used as qualitative measures of current competence, although their exact equivalence is yet to be explored fully.

The "process" point of view is well stated by Krumbein and Sloss (1963, p. 234) who define a sedimentary environment as "the complex of physical, chemical, and biological conditions under which a sediment accumulates."

Shepard and Moore (1955, p. 1488) defined a sedimentary environment as "a spatial unit in which external physical, chemical, and biological conditions and influences affecting the development of a sediment are sufficiently constant to form a characteristic deposit."

The latter definition is reasonably close to the geomorphic concept of environment emphasized by Twenhofel (1950) who, in America, had much to do with shaping attitudes during the 1920s and 1930s. He emphasized a geomorphic unit described by a totality of variables.

The broadest and most meaningful concept of environment, especially for application in ancient sediments, appears to be the geomorphic one defined as follows.

A sedimentary environment is defined by a set of values of physical and chemical variables that correspond to a geomorphic unit of stated size and shape.

Examples are an alluvial fan, a tidal flat, a point bar, and a beach. The word "biological" is omitted in this definition not because it is not important, but because its effect can almost without exception be expressed or measured by physical or chemical variables. Figure 2 diagrams the links

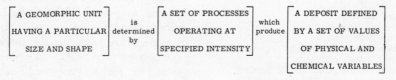

Fig. 2.—Relations between geomorphic unit, processes, and resultant deposit.

between geomorphic unit, sedimentary processes, and resultant deposit and emphasizes environmental reconstruction as paleogeomorphology. This approach emphasizes the idea that in making environmental reconstructions the geologist is in reality a geomorphologist of the past.

The optimistic view of environmental recognition is twofold.

First, one can recognize an ancient environment by inferring a set of variables from their corresponding rock properties. A set or combination is required almost without exception because —if only one variable were used—probably it would not be specific. Turbulence, for example, may be the same in kind and degree behind a sand wave in either a river or on a shallow-marine bank. Thus a combination or set of variables is essential, because it is impossible to define a geomorphic environment by one variable alone.

Choice of environmental variables is determined largely by those that leave a lasting impression on the sand. This is illustrated by the

TABLE I. ENVIRONMENTAL CHECK LIST FOR SAND BODIES

I. STRATIGRAPHIC
  A. Shape
    1. Elongate
      a. curved
        (1) meandering
        (2) dendritic
        (3) bifurcating
      b. straight
      c. relation to depositional strike
    2. Blankets (sheets)
  B. Lateral and vertical associations
    1. Continuous
      a. facies change
      b. oscillatory or cyclical
    2. Discontinuous
      a. unconformity
      b. structural cutoff
  C. Lithologic associations
    1. Range and abundance of rock types
    2. Lithofacies maps

II. PRIMARY SEDIMENTARY STRUCTURES AND BEDDING
  A. Inorganic structures
    1. Cross-bedding
    2. Ripple marks
    3. Parting lineation
    4. Sole markings
    5. Deformational structures
    6. Small-scale channels
    7. Shale chips and clay galls
  B. Burrows, tracks, and trails
  C. Character and organization of bedding
  D. Variability of directional structures
  E. Vertical sequence of bedding and structures

TABLE I—(continued)

III. PETROGRAPHIC
  A. Mineralogy
    1. Primary
    2. Diagenetic
  B. Texture
    1. Grain-size parameters
    2. Shape and roundness
    3. Surface texture
    4. Fabric
      a. grain-to-grain contacts
      b. orientation of framework fraction
    5. Vertical sequence of textures

IV. PALEONTOLOGIC
  A. Autochthonous fossils
  B. Allochthonous fossils
  C. Specialized indicators of marine and non-marine
  D. Ecologic associations

V. GEOCHEMICAL
  A. Primary (Depositional)
    1. Minerals
    2. Isotopes
    3. Organic matter
    4. Trace elements
  B. Diagenetic (Secondary)
    1. Minerals
    2. Isotopes
    3. Organic matter, oil and gas
    4. Formation waters
    5. Structures: concretions, nodules, stylolites, *etc.*

chemical variable pH. Although one can not measure the pH of a sandstone and say it is the pH of an environment, he may, under favorable circumstances, say that certain minerals in a sandstone indicate very high pH values and that their textures indicate that they are primary rather than diagenetic. Thus, pH values are linked with certain restricted alkaline lake environments on the basis of chemistry, modern sediment study, and paleogeomorphic form. In general, the more variables the better, but in practice ordinarily not more than five to ten can be used.

The second point concerns uniqueness and optimistically assumes that a set of processes operating at a specified intensity uniquely produces a *uniquely defined* deposit of stated size and shape. Although in general this is probably true, especially for the larger geomorphic units, one should be aware of the possibility that there may not be invariably an exact one-to-one correspondence of geomorphic unit, processes, and resultant deposit.

Although one may be interested in the environment of a sand body alone, it is necessary to consider the associated lithologic types and fossils, both laterally and vertically, and it is necessary to consider regional as well as local evidence.

Table I gives a check list of properties that can be used in environmental recognition.

CHARACTERISTICS OF SAND BODIES IN DIFFERENT ENVIRONMENTS

Sand bodies of six major origins are recognized: alluvial, tidal, turbidite, barrier island, shallow-water marine, and desert eolian. These types are described in terms of petrology, texture, sedimentary structures, associated fossils and lithologic features, internal organization, and external morphology. By internal organization is meant the vertical sequence of sedimentary structures, bedding, and texture within a sand body. Existing descriptions of sand bodies in these different environments vary in detail, because of the widely differing kinds and amounts of information available. Sand bodies of lacustrine origin are not recognized as a separate class, because they are neither volumetrically important in the geologic record nor do their physical properties differ greatly from those of shallow-water marine origin.

Like many classifications, the foregoing is a mixture of classifying criteria: agents (*e.g.*, streams, tides, winds, and density currents), water depth (*e.g.*, shallow-water marine), and landforms (*e.g.*, barrier islands). Nevertheless, the sands in these six different environments do seem to represent—in the present state of knowledge—naturally distinct and recognizable associations of properties. Many of these environments can be subdivided as, for example, the barrier-island environment, which may include beaches and coastal-dune complexes. The tidal-flat environment provides other good examples: the salt marsh, the higher mud flats, the inner and *Arenicola* sand flats, the lower mud flats, the lower sand flats, and creeks and bordering areas (Evans, 1965). Unfortunately, such detail of discrimination is not as yet generally possible in the majority of ancient sediments. The six groupings of the present study are a compromise between the present state of knowledge, on the one hand, and the distinctions that are deemed necessary for successful prediction in ancient sediments, on the other.

To facilitate presentation, the characteristics of sand bodies in the different environments are presented in two parts: a general summary followed by a documented discussion of specific examples.

ALLUVIAL

Allen (1965a) demonstrated that more is known about the processes and characteristics of alluvial deposition than any other type. Alluvial sand bodies include those deposited on alluvial fans, flood plains, and deltas. Because their characteristics are somewhat different, the sand bodies of alluvial fans and of delta-bar fingers are described separately.

The principal sedimentary structures of alluvial sand bodies are ripple marks, predominantly asymmetrical ones, and cross-bedding, the latter having unimodal orientation in almost all places and usually with small dispersion. Symmetrical ripple marks are very scarce. Commonly, more than 50 per cent of the beds have visible cross-bedding and (or) ripple marks. Cross-bedded units generally do not exceed 3 feet in thickness. Because of the general upward decrease in current velocity that is characteristic of the asymmetrical alluvial cycle of flood-plain deposition, cross-beds 1–3 feet thick in the lower half become thinner upward and are of ripple-mark dimension toward the top, as the sand body grades into clay. Because they are closely correlated, grain size and bed thickness both decrease upward. Lateral migration of the thalweg with resultant lesser stream competence is the cause of this succession. Regardless of size, most beds tend to be lenticular, and small intrasand-body channels are common. Sorting may tend to improve upward partly because of a tendency for basal gravels to form. Collectively, the vertical sequence of sedimentary structures, bed thickness, and grain size in the alluvial sand body has a very definite well-defined internal organization (Allen, 1965b) of which a sandstone body from the Pennsylvanian Mansfield Formation is probably typical (Fig. 3).

Petrologically, alluvial sand bodies are strongly dependent on source and tend to be less mature than their more distal equivalents. Sand bodies of flood plains may have both locally and distally derived gravels. Shale pebbles and shale-pebble gravels are common constituents, and carbonaceous debris is abundant. Cementing agents are a mixture of detrital and chemical material. Modal

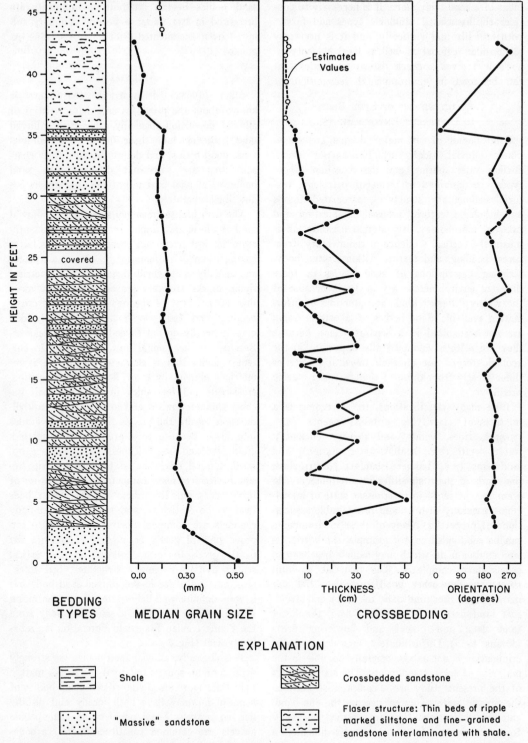

FIG. 3.—Vertical profile of grain size and orientation and thickness of cross-bedding in alluvial sand body of Mansfield Formation (Pennsylvanian). Cut of Southern Railroad, NE. 1/4, SW. 1/4, NW. 1/4, sec. 25, T. 25 N., R. 2 W., Perry County, Indiana.

diameter tends to be related to sand-body or channel dimension. Alluvial sands tend to have relatively high percentages of fine-end tails, especially in the silt range, and are generally not well sorted. Variability of textural parameters between samples is high.

Fragmental plant remains, rootlets, and a few logs may be present. Most alluvial sands are faunally barren with only scattered fresh- or brackish-water pelecypods and gastropods present, commonly as channel-bottom concentrates. In the Mesozoic and later, however, concentrations of vertebrate remains are not uncommon.

Alluvial sand bodies may range from 100 feet wide or less to lateral composite bodies as wide as 30 miles. The latter size is the anastomosing complex of a valley-fill deposit of a large alluvial river in its lower reaches. Alluvial sand bodies may range in thickness from less than 10 feet to more than 100 feet; few of them probably are more than 200 feet thick. Almost all alluvial sand bodies consist of valley-fill deposits, have a disconformity at their base, and are elongate downdip with map patterns that range from relatively straight to bifurcating, markedly meandering and anastomosing with distributaries on deltas. Along a major unconformity with a sharply entrenched second-cycle valley system—usually dendritic but possibly trellis—basal alluvial sand-body morphology may be controlled almost completely by paleotopography.

The associated lithologic types are variable, and variability depends on the proximity of the shoreline facies. If the sand body is in a non-marine sequence, the associated sediments will consist of silt and silty clay and possibly a few lenticular coal or peat beds, or coal beds may be widespread as in many coal-measure sequences. Silt is nearly always present. Generally it tends to cap the sand body, but it may also be prominent—especially in channels—as a lateral equivalent. Other sand bodies may be superimposed vertically to form multistory sand bodies or joined laterally to form multilateral sand bodies, such as are characteristic of many large alluvial flood plains. A tabular or sheet sand body of alluvial origin is nearly always a coalescent sand body, which is formed by the coalescence of many elongate ones. Such an alluvial, coalescent sheet, however, normally is more heterogeneous than a comparable marine-shelf sheet sand. Correlation is difficult in a non-marine sequence, because good marker beds—unless they are coal or peat—are lacking. If marine shale or thin limestone is present, the sandstone body will pass upward into the marine elements of the cycle or may be superimposed on an underlying marine section. Where marine units are present, correlation of the sand bodies of a particular cycle by means of the marine units is then possible. In a mixed marine-non-marine sequence, the alluvial sand bodies are seen clearly to be its most disruptive, non-layer-cake elements, because they are elongate, because there is commonly a truncated section at their base, and because of differential compaction of the clay around the sand body.

Table II gives the principal characteristics of alluvial sand bodies.

TABLE II. CHARACTERISTICS OF ALLUVIAL SAND BODIES

(Alluvial-fan and bar-finger deposits not included)

PETROLOGY

Detrital. Abundant shale pebbles and shale-pebble conglomerates. Generally carbonaceous debris. Petrographically immature to moderately mature. Pebbles and cobbles, if present, may be both local and distal. Detrital plus chemical cements. Faunal content low to absent.

TEXTURE

Poor to moderate sorting and moderate to low grain-matrix ratio. Abundant silt in fine-end tail. Tendency to poor rounding. High variability.

SEDIMENTARY STRUCTURES

Asymmetrical ripple marks and abundant well-oriented cross-bedding, commonly unimodal. Parting lineation and deformational structures are common minor accessories. Beds tend to be lenticular with erosional scour. Some tracks and trails.

INTERNAL ORGANIZATION

Strong asymmetry. Upward decrease in grain size and bed thickness, possibly with conglomerate near base. Larger channel-fill sandstone bodies tend to be coarser-grained than smaller ones.

SIZE, SHAPE, AND ORIENTATION

Commonly very elongate. Width ranges from a few tens of feet to composites of 30 miles. Dendritic as well as anastomosing and bifurcating patterns. Elongate downdip. Excellent correlation of internal directional structures and elongation.

ASSOCIATED LITHOLOGIC TYPES

Vertical: overlying silty shales, commonly of alluvial origin. Possible peat and coal. Basal contact commonly sharply disconformable. Multistory sandstone bodies. Marine units in mixed sections. Lateral: silty shale and siltstone commonly with abundant carbonaceous material as well as roots, leaves, and stems. Multilateral sandstone bodies. Correlation generally difficult.

A related type of sand body—one that shares many alluvial characteristics but perhaps could be described separately—is the bar-finger sand body described by Fisk (1961). Examples from the Mississippi delta indicate that these sand bodies are elongate downdip, have unidirectional cross-bedding parallel with the long axis, and are fine-grained. They are as thick as 250 feet, as wide as 5 miles, and as long as 30 miles. They thin and narrow upstream. Fauna generally is absent. These sand bodies differ from alluvial ones, according to Fisk (1961, p. 45), in that their basal contact is conformable and transitional with the underlying marine, prodelta silt and clay, and they coarsen upward unlike the typical alluvial sand body. In an alluvial-delta complex, such sand bodies may be intercalated with ordinary alluvial ones of flood-plain origin. Downdip bifurcation or branching, which can also occur on a smaller scale in sand bodies of flood-plain origin, is characteristic of bar-finger sand bodies, whose interbranch width increases seaward.

Sand bodies of alluvial-fan origin are coarser-grained than those previously described. Poorly sorted gravel and even boulders are typical, sand usually being subordinate. Petrologically, these materials are very immature, composition usually being closely linked to nearby bedrock. Size gradients parallel cross-bedding orientation, from fan head to fan base, and are strong. Lenticular bedding, cut and fill, and channeling are characteristic. Textural variation between samples is very high. Sand bodies of alluvial fans characteristically form wedges that thin in the direction of transport. In plan view, these wedges are approximately semicircular. Lateral coalescence to form composite bodies many miles long parallel with a fault line is not uncommon. Thickness of such composites may be many hundreds and even reach thousands of feet and is controlled by the subsidence of the downthrown block.

Descriptions of sand bodies in alluvial and deltaic deposits are moderately abundant. Generally, these descriptions pertain to the lower-river course and for most ancient deposits include deltaic deposits as well.[3]

Shanster (1951) described modern alluvial deposits with special reference to European Russia and emphasized the origin and characteristics

[3] An excellent source of information was discovered while this paper was in proof; this source is Shirley (1966).

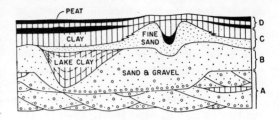

A. Braided river, with probable channel bottom deposits
B. Large meandering river
C. Small meandering river, strongly branched
D. Swamp and peat formation

FIG. 4—Idealized sequence and structure of alluvial Pleistocene deposits in the Netherlands (modified from Zagwijn, 1963, Fig. 8).

of the alluvial sequence. Other later Russian studies include (Lopatin, 1952; Shamov, 1959; and Kartashov, 1961—all cited in Allen, 1965a, p. 185–189). Fisk et al. (1954) and Fisk (1959, 1961) described the bar-finger sands of the Mississippi delta, and Doeglas (1962) and Harms et al. (1963) have reported on the sedimentary structures of alluvial point bars, the sand accumulation of meander bends. Zagwijn (1963) shows in a diagrammatic cross section the alluvial cycle as inferred from Pliocene and Pleistocene deposits in the Netherlands (Fig. 4). With slight modification, this alluvial cycle doubtlessly can be duplicated in the alluvium of the Ohio River at Owensboro (Fig. 5). Blissenback (1954) described, mostly from outcrop, alluvial deposits in arid regions. Leopold et al. (1964, p. 198–332) discussed channel form and processes at length.

Studies of sand bodies of inferred alluvial origin in ancient sediments are much more numerous. Good description of petrography, textures, and sedimentary structures of Devonian sandstone bodies in England is given by Allen (1962, 1964). Examples of Carboniferous alluvial sandstone bodies are especially numerous. Lee et al. (1938, p. 12) early recognized multistory sandstone bodies of alluvial origin in the Pennsylvanian of north Texas. Pepper et al. (1954) described a series of channel-fill sandstones in the Mississippian Bedford Shale and Berea Sandstone of Ohio. The longest of these systems have been traced 60 miles. Zhemchuzhnikov (1954) and Zhemchuzhnikov et al. (1959, 1960) have described numerous examples in the Carboniferous of the Donets coal basin in Russia. These studies contain much regional information and many in-

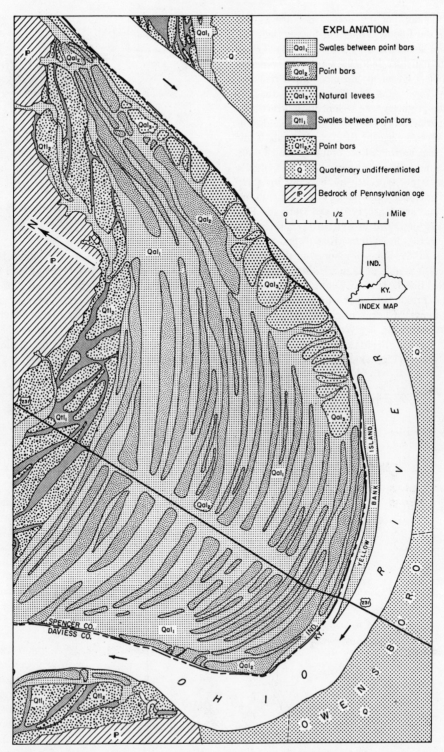

Fig. 5.—Surface expression of Cary and post-Cary point bars and natural levees in alluvial deposits of Ohio River near Owensboro, Kentucky (modified from Ray, 1965, Pl. 1).

teresting details, but are not so detailed as the studies of the Illinois Geological Survey, most of which were summarized by Potter (1963). Friedman (1960) gives essentially similar information for Indiana. Excellent documentation is also given by Hewitt and Morgan (1965), who described the geometry, the sedimentary structures, and the permeability of the Robinson Sandstone of Pennsylvanian age in the Illinois basin. They observed that horizontal permeability subparallels the direction of sand-body elongation. The Illinois basin studies are especially noteworthy for their details of sand-body morphology, both in subsurface and outcrop. Lins (1950) and Doty and Hubert (1962) described in outcrop Pennsylvanian sandstone bodies in western Missouri and in Kansas. Busch (1959, Fig. 12) mapped in the subsurface a good example of a small distributary system of the Booch delta of Pennsylvanian age in Oklahoma. Presumably, many of these sandstone bodies are of bar-finger origin. Excellent documentation in outcrop is given by Wurster (1964) for the Triassic Schilfsandstein in Germany. Wurster shows (Fig. 6) an anastomosing-distri-

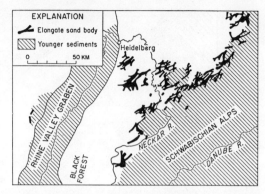

Fig. 6—Regional pattern of elongate sand bodies in Triassic Schilfsandstein (modified from Wurster, 1964, Fig. 25).

butary deltaic pattern that is very similar to that described earlier by Stokes (1961, Fig. 14) for the Saltwash Member of the Jurassic Morrison Formation in Arizona. Stokes, however, emphasized deposition by streams on a large alluvial fan (p. 173). Wurster also shows (Fig. 7) the good correlation between cross-bedding orientation and sand-body elongation. Schlee and Moench (1961)

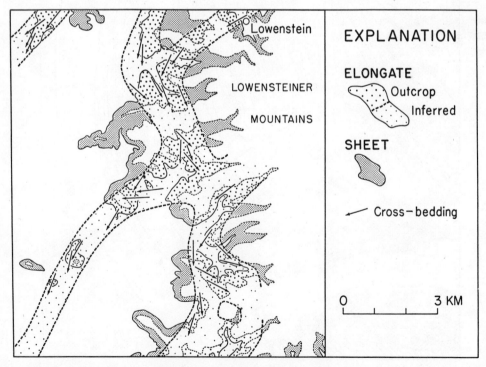

Fig. 7.—Detail of Figure 6. Note correlation between cross-bedding orientation and elongation of sand bodies (modified from Wurster, 1964, Fig. 49).

give a good analysis of channel morphology, sedimentary petrology, and directional structures of the Jurassic "Jackpile sandstone" of New Mexico. The U. S. Geological Survey has published other useful descriptions of channel morphology in the Colorado Plateau (Witkind, 1956, p. 235; Finch, 1959, p. 134–145; Witkind and Thaden, 1963, p. 69–82). Nanz (1954) studied an Oligocene sandstone body in the subsurface of the Gulf Coast. Dondanville (1963) described deltaic and marine sandstones in the Cretaceous Fall River Formation from outcrop. He recognized channel sandstones as thick as 80 feet that have typical deltaic patterns. Of particular interest is his finding that the detrital mineralogy of the channel sandstones differs from that of the associated marine sandstones. This suggests that careful heavy-mineral study of a deltaic-marine sequence may be very rewarding in identifying the type of sand body and predicting its lateral extent. Such study would probably be rewarding, however, only if the heavy-mineral suite were relatively immature. Potentially, detailed study of the Pleistocene alluvial fill of preglacial valleys could contribute much to the understanding of paleochannels and occurrence of sand bodies in such valleys. Dury (1964a, b) gives many examples of point bars, meander patterns, and the fill of modern valleys, which may be particulary useful in ground-water exploration in alluvial deposits.

TIDAL

Sand bodies of the tidal environment appear to be of two principal types: larger ones in estuaries and passes between barrier islands and smaller ones in creeks and inlets that transgress the intertidal zone along the strand. Available information suggests that both types share many of the characteristics of alluvial sand bodies. At present, unfortunately, it is difficult to distinguish clearly between many tidally influenced sand bodies of shallow-marine shelves and those of tidal estuaries.

Most tidal sand bodies, like alluvial ones, are elongate.

Orientation of tidal sand bodies which develop in creeks and inlets of the intertidal zone is at right angles to the strand line. Such sand bodies usually are thin, perhaps exceeding 10 feet in a few places; width is small, probably exceeding 50–100 feet in very few places. Thin but abundant cross-bedding and asymmetrical ripple marks of sand-wave origin usually parallel channel axis and dip seaward. However, inclined bedding of lateral accretion origin also is present and is both thicker and more variable in orientation.

Elongate tidal sand bodies in the estuary itself probably parallel closely the axis of the estuary. Cross-bedding and ripple marks predominate with inclination parallel with the long dimension. Thickness may exceed 20–25 feet and width more than 500–1,000, although little information is available.

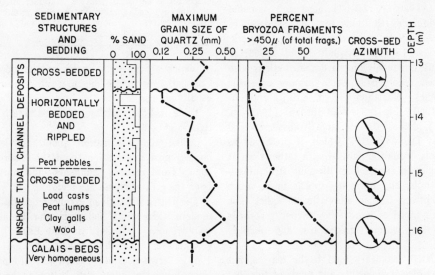

Fig. 8.—Vertical profile of inshore tidal sand body in Rhine-Maas estuary (modified from Noorthoorn van der Kruijft and Lagaaij, 1960, Figs. 3, 5).

In both types, beds are lenticular and small channels and washouts are common. The basal contact is disconformable and commonly marked by shelly gravel. Internal organization is strong; grain size decreases upward, commonly from a basal gravel (Fig. 8). Bed thickness shows a similar decline upward. Thus the sequence of structure and grain size in a tidal sand body has a cycle very similar to the alluvial one. Tidal sands have both chemical and detrital cements. Petrographically, tidal sands are comparable with those of most alluvial sands, except in their more advanced maturity. Grain size probably can be related to size of the sand body, that is, size of the tidal channel.

There are several striking contrasts between alluvial and tidal sand bodies, however—ones that should easily serve to distinguish them, if data are available.

Skeletal debris and glauconite are relatively common in modern tidal sands. The basal gravel, if present, consists mostly of locally derived materials in which shells are prominent. Depending on the predominance of the flood (landward) or ebb (seaward) tides, cross-bedding in axial estuarine tidal sand bodies may dip predominantly landward, seaward, or be bimodal, if flood and ebb currents are of approximately equal strength. Bimodal cross-bedding and marine fossil debris are probably most important in distinguishing a sand body that developed in response to tidally influenced currents from alluvial ones. Zhemchuzhnikov (1926, p. 60–63) suggested bimodal cross-bedding distributions for tidal sand bodies and these may be most characteristic (Reineck, 1963, Figs. 15–19; Terwindt *et al.*, 1963, p. 256), but unimodal landward distributions also have been recorded in some inshore estuarine sand bodies (Noorthoorn v. d. Kruijft and Lagaaij, 1960, Figs. 3, 4). Jordan (1962, p. 940–942) described the large transverse sand waves of Delaware Bay, which, if preserved as individuals, would be parallel with rather than at right angles to strand line. Biogenic structures—tracks, trails, and burrows—appear to be much more abundant than in alluvial sand bodies. The associated fauna of Recent tidal deposits consists of mollusks, worms, crustaceans, and possibly algae.

Associated sediments tend to be marine, but they may also include silts and muds of tidal-flat origin, as well as barrier-island sand bodies, most

TABLE III. CHARACTERISTICS OF TIDAL SAND BODIES

PETROLOGY

Detrital. Argillaceous rock fragments, skeletal debris and collophane plus argillaceous material. Possibly some glauconite and authigenic feldspar. Detrital and chemical cements.

TEXTURE

Fair sorting, moderate to high grain-matrix ratio. Possibly shell conglomerate at base. Very similar to texture of alluvial sandstone. Peat, clay galls, and wood common.

SEDIMENTARY STRUCTURES

Asymmetrical and symmetrical ripple marks common. Abundant cross-bedding of variable thickness; commonly bimodal but may dip seaward or landward. Lenticular bedding dominant. Tracks, trails, and burrows abundant. Channels and washouts.

INTERNAL ORGANIZATION

Good asymmetry. Strong vertical decrease in grain size and bedding thickness, possibly with some conglomerate at base.

SIZE, SHAPE, AND ORIENTATION

A few tens of feet to more than 1,000 feet wide, mostly very elongate. Long axis at right angles to shoreline or parallel with estuarine axis. Straight to moderately meandering, dendritic patterns, the latter as tidal inlets. Also lunate bars in passes between barrier islands. Cross-bedding parallel with elongation; principal mode may point seaward as well as landward in estuaries.

ASSOCIATED LITHOLOGIC TYPES

Vertical: variable according to regressive or transgressive origin. Associated with interlaminated shale and sandstone of tidal origin and marine sediments. Disconformable basal contact. Lateral: interbedded siltstone and shale of tidal-flat origin, commonly with mollusks, worms, crustaceans, and possibly algae as well as marine sediments.

of which are at right angles to the tidal ones. Correlation of sand bodies in a littoral complex is usually not easy.

Table III summarizes the characteristics of tidal sand bodies.

Unlike the information about alluvial sand bodies, most information about tidal sand bodies comes from modern sediments—and most of it from the greater North Sea region.

The Dutch have given excellent descriptions of modern tidal sand bodies in the Rhine delta (Noorthoorn v. d. Kruijft and Lagaaij, 1960; Oomkens and Terwindt, 1960; Terwindt *et al.*, 1963). Terwindt *et al.* (1963, Fig. 7) found a distinct break in grain size between the tidally influenced estuarine sands and the alluvially

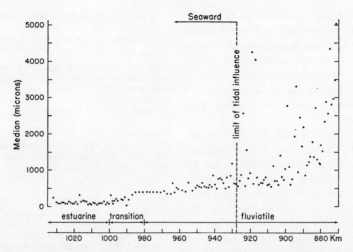

Fig. 9.—Median grain size in inshore estuarine, transition, and fluviatile sands of Rhine and Rhine-Maas estuary (modified from Terwindt et al., 1963, Fig. 7).

transported ones, which are much coarser (Fig. 9). Useful information also is given by van Straaten (1954a) on a tidal-flat environment, although tidal sand bodies are not stressed. Klein (1963) emphasizes sedimentary structures in his study of Bay of Fundy tidal deposits. Evans (1965) described the intertidal sediments of the Wash in Great Britain, also emphasizing sedimentary structures. Off (1963) called attention to and summarized the morphological features and occurrence of tidal sand bodies on marine shelves, where tides are strong. The sand bodies described by him were typically elongate at right angles to the strand line and had dimensions of 2–3 by 10–20 miles. Reineck (1963) described tidal sand bodies of the North Sea Watt, citing most of the relevant previous literature.

Ancient well-documented examples of inferred tidal sandstone bodies have not been widely described in the published literature, although the tidal-flat environment is mentioned fairly frequently (cf., van Straaten, 1954b).

TURBIDITES

Documented examples of both modern and ancient sand bodies of turbidite origin are not numerous. Lack of good markers in most turbidite sequences restricts mapping of individual sand bodies and, because most turbidites are pre-orogenic in the tectonic cycle, later complex structural deformation can also be a limiting factor. In contrast, sedimentary structures have been extensively studied and fairly adequate petrologic information is available. Reviews that have been very helpful in obtaining the characteristics listed in Table IV include those by Emery (1964) and Kuenen (1964), who compared and contrasted modern and ancient turbidites as well as descriptions of turbidite basins off southern California by Gorsline and Emery (1959), Shepard and Einsele (1962), and Hand and Emery (1964).

Petrologically, most turbidite sands are immature and have a framework fraction that ranges from subgraywacke to graywacke, but they may also include carbonate and volcanic-rich sands as well as sands of orthoquartzite composition. Rock fragments and immature minerals generally are abundant as is marine skeletal debris, especially in Cretaceous and younger turbidites. The skeletal debris commonly is a mixture of both shallow- and deep-water origin. Glauconite may be present, and carbonaceous material is common. Gravel tends to be derived locally and most clasts are shale. Clay matrix is the principal cementing agent in ancient turbidites, but it is conspicuously absent in most modern ones. Thus the grain-matrix ratio is much lower in ancient than in modern turbidites.

Graded bedding, abundant sole marks, convoluted lamination, asymmetrical ripple marks, small-scale cross-bedding, and long, even beds of sandstone interlaminated with shale are the most striking sedimentary structures of turbidite sandstone bodies. This combination occurs in abundance only in turbidites.

Turbidite sandstone bodies range from those

that are almost entirely sandstone to those that are rhythmically interbedded with much shale. Because thin, uniform sandstone beds interbedded with shale are one of the characteristics of the turbidite facies, a turbidite sand body may require a definition in terms of some arbitrary sandstone-shale ratio, such as a ratio greater than one.

The sparse data indicate that elongate turbidite sand bodies in the Recent occur as fairly straight channel-fill deposits. But both dendritic and bifurcating patterns also exist (Hand and Emery, 1964, Figs. 3, 8). The dendritic patterns are proximal to the "feeding ground" of the shelf; the bifurcating patterns are downdip on the fan and on the abyssal plain. Width ranges from a few hundreds of feet to possibly more than several miles. As in alluvial sand bodies, length can be many times width. Elongate sandstone bodies of turbidite origin occur in channels as deep as 1,200 feet in the Miocene of California (Martin, 1963, Figs. 5, 6). But individual sandstone bodies in such deep channels may be no more than 200–400 feet thick. Such channel-fill sandstone deposits are differentiated from surrounding lithologic types principally by their higher sandstone-shale ratio and separated from them by a disconformable basal contact. Associated lithologic types are other more shaly turbidites. Presumably, grain size may be related to channel dimension and may, together with bed thickness, decrease upward as in alluvial and tidal sand bodies, although no published data are available. Elongation is downdip. Grain size may be expected to decline in the downdip direction (Hand and Emery, 1964, Fig. 7). Correlation between long dimension and directional structures, which are unimodal with small variability, is excellent. Multistory sand bodies probably are common.

Sullwold (1960, 1961) and Martin (1963) described elongate turbidite sandstone bodies in the Tertiary of southern California.

How extensive are elongate turbidite sand bodies? It is difficult to answer this question with certainty, but elongate turbidite sand bodies may well be the exception rather than the rule. One can surmise that most of the turbidites preserved in ancient basins represent deposition on the submarine fans at the base of their "continental shelves." Sandstone bodies within these fans are dominantly sheet and blanket-like, because here even single beds can cover wide areas.

TABLE IV. CHARACTERISTICS OF TURBIDITE SAND BODIES

PETROLOGY
Detrital. Rock fragments and immature minerals abundant as well as marine skeletals of both deep-water and shallow-water origin. Carbonaceous material generally present. Shale-pebble conglomerate. Cements are very largely detrital.

TEXTURE
Very poor to fair sorting and low grain-matrix ratio. Rhythmic alternation of beds produces abrupt juxtaposition of shale and sandstone. Rounding almost all inherited.

SEDIMENTARY STRUCTURES
Graded beds rhythmically interbedded with shale. Absence of large-scale cross-beds. Beds notable for their long lateral persistence. Sole marks, asymmetrical ripple marks, and laminated and convoluted beds are common. Trails and tracks are generally present.

INTERNAL ORGANIZATION
Few data. Grain size may be related to sand-body dimensions.

SIZE, SHAPE, AND ORIENTATION
Elongate sandstone bodies up to several miles; fairly straight but dendritic and bifurcating possible. Extend downdip into basin. Excellent correlation of directional structure and shape. But sheet and blanket-like deposits probably predominate. Large olistostromes not uncommon.

ASSOCIATED LITHOLOGIC TYPES
Vertical: commonly other marine shale and turbidite sandstone. Multistory sandstone bodies possible. Lateral: except for lower sandstone-shale ratio, notably little lithologic contrast. Mixed benthonic and pelagic faunas in shale and possibly reworked shelf faunas in sandstone.

Another feature of turbidite sand bodies that should be mentioned is the large, mass slump sheet—olistostromes as such sheets have been called; these occur in turbidite sequences, and have been reported in southern Italy and in Sicily (Flores, 1959). Olistostromes not only disrupt the normal stratigraphic section but, through rotation and crumpling during sliding, can complicate vastly the interpretation of sandstone bodies in turbidite sequences.

BARRIER ISLAND

The term barrier-island complex (Hayes and Scott, 1964, p. 238) includes sand bodies that formed from offshore bars (shoreface barriers), from beaches, and from dunes, the whole constituting what Shepard (1960) called Gulf Coast barriers. The essential feature of a Gulf Coast barrier is an elongate sand body (representing a mixture of environments) that parallels the strand line and separates marine from non-marine environments. Separation of inner-shelf sands from the barrier-island sand bodies generally is more difficult than separation from continental sand bodies. Because it is generally not possible to distinguish the different subenvironments of a barrier-island complex (redundant in ancient sandstone bodies), they are grouped together, though future studies may lead to more effective differentiation.

Petrologically, the sands of most barrier-island complexes are mature. Good sorting in the surf zone and dunes eliminates much clay, and strong abrasion promotes roundness and eliminates many rock fragments. Folk and Robles (1964, p. 290) suggest that the $\sigma_I = 0.30$ to 0.60ϕ represents the limiting value of sorting for beach sands. Variability of textural parameters between samples is low. Fossils and skeletal debris are common in beach and offshore sands. Heavy-mineral concentrates and laminae in the beach sands also reflect good sorting. Cements are mostly chemical.

Sedimentary structures differ between the dune and water-deposited sands of the barrier-island complex. In beach sands, lamination and lineation are conspicuous. Weakly inclined bedding dipping seaward and cross-bedding, which probably tends to have bimodal orientation, are present. The landward mode represents the dip of backshore sand and the seaward mode the foreshore sand (Logvinenko and Remizov, 1964, Fig. 4). Cross-bedding dipping roughly parallel with or obliquely toward the shoreline, formed by obliquely moving sand waves, commonly is present (van Straaten, 1953, Fig. 7; Seibold, 1963, Fig. 14). If the barrier-complex sand body includes a dune complex, as it commonly does, cross-beds in the dune complex are very abundant and probably will have a large variance. Foreset dip of the eolian cross-bedding may be somewhat higher than that of most water-deposited sands; several investigators report dips in excess of 35°. Ripple marks are abundant and of many varieties, especially on the beach, but varieties are more restricted in the eolian sands. Burrows and trails are abundant in the water-transported sands, and the fauna is chiefly the more robust forms.

Available data indicate a vertical increase in grain size, especially in regressive sequences, that distinguishes the barrier-island complex from elongate sand bodies of alluvial origin (Fig. 10).

Barrier-island sand bodies separate marine deposits from lagoonal or alluvial deposits. Thus there is very strong lithologic and biotic contrast on opposite sides of the elongate barrier sand body—marine fossils and possibly glauconite seaward in contrast to a brackish or continental biota landward of the beach-dune barrier. Laterally, barrier sand bodies may coalesce, a condition depending on stability of the shoreline. Crescent-shaped sand bodies of tidal origin may be present in passes, more commonly on the landward side, between barrier sand bodies. Vertical sequences will change according to regression or transgression: the barrier-sand body may be overlapped by continental deposits, if it is part of a regressive cycle, or by marine sediment, if it is part of a transgressive cycle. Evidence suggests that the basal contact is fairly even and probably transitional. Multistory sand bodies probably are relatively common and may be of mixed origin—alluvial or tidal sand bodies superimposed at right angles on barrier-island sand bodies.

The Recent deposits of the Gulf Coast indicate that thickness ranges from 20 to 60 feet. Width ranges from a few hundred feet to as much as 2 miles. On the Gulf Coast lengths may be many miles, possibly 20–60, and some ancient examples of comparable length also have been reported. Shape generally is fairly straight to gently curving, and there are some weakly branching patterns, but sinuosity is much less than in alluvial sands. Elongation is parallel with the strand line. Cross-bedding, if of eolian origin, has no fixed relation to sand-body elongation. If subaqueous, however, it may be bimodal, the two modes being at right angles to the long dimension. Grain fabric of the surface beach sands is at right angles to sand-body elongation.

Table V summarizes these characteristics.

Thompson (1937) described the sedimentary structures of beaches, as did McKee (1957) later. McBride and Hayes (1962), Land (1964), and Logvinenko and Remizov (1964) measured cross-bedding orientation of dune and subaqueous origin in barrier sand bodies. Byrne *et al.* (1959) and Gould and McFarlan (1959) describe and illustrate the stratigraphic relations and give cross sections of cheniers and their lithologic associations in the Gulf Coast. Another valuable description is given by Corbeille (1962) for the

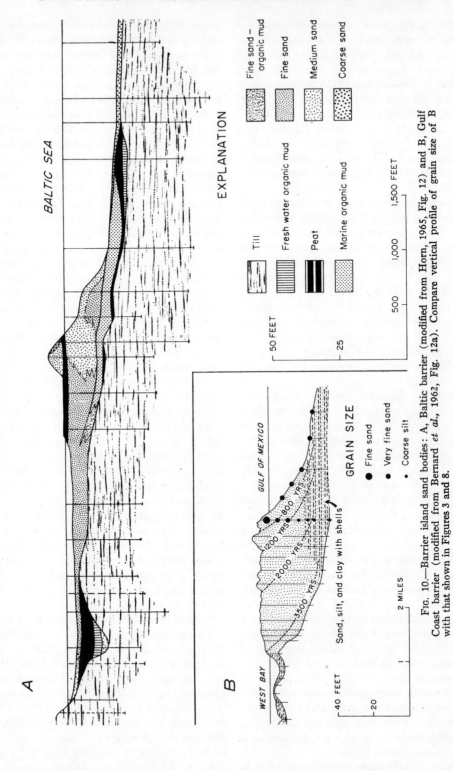

FIG. 10.—Barrier island sand bodies: A, Baltic barrier (modified from Horn, 1965, Fig. 12) and B, Gulf Coast barrier (modified from Bernard *et al.*, 1962, Fig. 12a). Compare vertical profile of grain size of B with that shown in Figures 3 and 8.

TABLE V. CHARACTERISTICS OF BARRIER-ISLAND SAND BODIES

PETROLOGY
Detrital. Heavy-mineral concentrates common. Skeletal debris, collophane, and minor glauconite. Conglomerates, if present, generally locally derived and may be mostly shells. Fauna are generally the more robust species. Mostly chemical cements.

TEXTURE
Commonly excellent sorting and very high grain-matrix ratio. Low variability of textural parameters between samples. Commonly good rounding.

SEDIMENTARY STRUCTURES
Asymmetrical ripple marks. Abundant gently inclined bedding. Lamination and lineation conspicuous on beach. Cross-bedding moderately abundant and may be eolian as well as water-laid. Variability of cross-bedding orientation, moderate to large. Bimodal distributions may occur on the beach. Burrows and channeling common.

INTERNAL ORGANIZATION
Sparse data indicate vertical increase in grain size and bed thickness, especially in regressive sequences.

SIZE, SHAPE, AND ORIENTATION
Widths from a few hundreds of feet to more than several miles. Thickness 20-60 feet. Very elongate, parallel with strand line. Sandstone bodies generally straight to gently curved. Grain fabrics and cross-bedding can be variable, especially if eolian transport important.

ASSOCIATED LITHOLOGIC TYPES
Vertical: variable according to regressive or transgressive origin. Basal contact generally fairly even and may be transitional. Lateral: separates marine from lagoonal or terrestrial deposits giving maximum lithologic contrast. Multilateral sandstone bodies common.

New Orleans barrier island, a shallowly buried sand body consisting of a clean, well-sorted sand that coarsens upward. It is 35 feet thick, 3 miles wide, and at least 10 miles long. Shepard (1960) presents much descriptive material. Bernard *et al.* (1962, p. 202–205) and Curray and Moore (1964) show how a coalescent series of barrier sand bodies develops. Hayes and Scott (1964) give a short but informative summary of the environments and processes that produce barrier sand bodies. Seibold (1963) summarizes the results of granulometric, petrologic, and sedimentary structure study of the bars and barriers along parts of the Baltic and North Sea coasts. Horn (1965) gives excellent documentation and cross sections for a "Baltic barrier" of late Pleistocene and Holocene age.

Ancient examples with good documentation are exceedingly scarce. Hollenshead and Pritchard (1961) show examples of barrier-island sandstone bodies in the Cretaceous Mesaverde Group of the San Juan basin in Colorado and New Mexico. Sabins (1963a), integrating petrology, grain size, and electric-log cross sections, gives good documentation for the Cretaceous Gallup Sandstone in New Mexico, where he recognized forebar, bar, backbar, and beach-sand bodies. Generalized descriptions of two composite barrier-island systems, the Terryville Sandstone (Jurassic) and the Frio Formation (Oligocene), have been made in the Gulf Coast (Thomas and Mann, 1966; Boyd and Dyer, 1966). Both sandstone bodies are thick (600–1,400 feet), regressive, multistory, composite sandstone bodies 80–160 miles long and 10–30 miles wide. Zimmerle (1964) relied on petrography and sedimentary structures to identify in outcrop a Tertiary beach deposit in Bavaria. Shelton (1965) gives convincing evidence for the barrier island origin of the Cretaceous Eagle Sandstone in Montana: low-angle inclined bedding, mottled structure, upward increase in grain size, transitional basal contact, and long dimension parallel with depositional strike.

SHALLOW-WATER MARINE

The sand bodies of shallow-water marine shelves have many properties—especially petrographic and textural ones—that can be found in sand bodies of both tidal and barrier-island sand bodies. Because of Pleistocene changes in sea-level, many modern shelf sands may actually be relics of tidal and barrier-island sand bodies. Ancient examples indicate that shallow marine-shelf sand bodies are distinguished from strand-line ones primarily by their dimensions, shape, and lithologic associations and by presumed distance from the strand line. In analogy with modern shelves, most of the ancient marine-shelf sandstones are believed to have formed in less than several hundred feet of water. Although a common feature of most sedimentary basins, good descriptions of both ancient or modern marine-shelf arenites are scarce, and thus one is forced to surmise and synthesize many of their properties from few and scattered data.

The sand of marine shelves tends to be mineralogically mature, especially in ancient sediments, presumably because it has passed through the shoreline complex, especially on stable cratons.

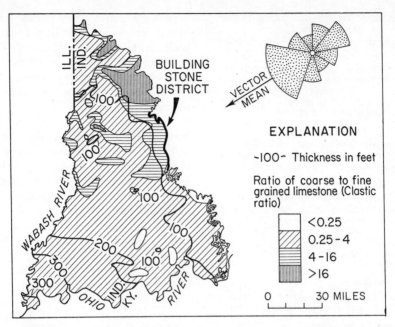

FIG. 11.—Cross-bedding orientation, thickness, and ratio of coarse- to fine-grained limestone in Salem Limestone (Mississippian). Cross-bedding obtained from building-stone district only. Note strong bimodality of current system with principal mode pointing downdip (modified from Pinsak, 1957, Pl. 3; and Sedimentation Seminar, 1966, Fig. 6).

Thus argillaceous rock fragments and unstable minerals generally are not very abundant. Glauconite, detrital carbonate, skeletal debris, marine fossils, and collophane usually are present. Relic fauna and anomalously coarse grains are common in the upper few feet of some shelf sands. Gravel or conglomerate may be derived locally. Cementing agents are mostly chemical, but may include appreciable detrital material proximal to clay or shale transitions. Sorting generally is good to excellent, and there is perhaps a tendency for the development of better-than-average rounding. The variability of textural parameters between samples is generally very low.

Ripple marks, both symmetrical and asymmetrical, generally are abundant, although varietal types are probably less diverse than in the barrier-island complex. Cross-bedding commonly is conspicuous and as a rule more variable in orientation than in alluvial and turbidite sand. Zhemchuzhnikov (1926, p. 60–63) early suggested this variability. In one marine-shelf sandstone, Wermund (1965) found a random orientation of cross-bedding. Where tidal currents are a factor, bimodal distributions also may occur. Figure 11 shows an example of the strong bimodality of cross-bedding orientation in the Mississippian Salem Limestone, a skeletal limestone of shallow marine-shelf origin. Changeable wind-driven waves also probably contribute to variability. Trails and burrows may be abundant, during periods when sedimentation is slow. A few minor channels and washouts—commonly of smaller magnitude than in alluvial, turbidite, and tidal sand bodies—also occur.

Vertical trends in grain size of marine-shelf sand bodies are known best from the widespread sand sheets of the past, especially in the Cretaceous of western North America, where grain size declines upward in transgressive sandstones and increases upward in regressive ones. Similar relations may prevail for smaller isolated marine-shelf sand bodies, if they are part of a regressive or transgressive sequence—or vertical trends may be but weakly developed, although little actual information is available. Because grain size and bed thickness correlate, bed thickness generally will increase as grain size increases. Pronounced vertical trends may not always be present; however, some thin sheet-like marine units—in the Illinois basin (Potter, 1963, p. 25)—have no pronounced vertical size trend.

TABLE VI. CHARACTERISTICS OF MARINE-SHELF
SAND BODIES

PETROLOGY
Detrital. Commonly few argillaceous fragments and micas. Conglomerates, if present, are locally derived. Authigenic feldspar and glauconite and some detrital carbonate may be present. Mostly chemical cements. Relict faunas are common at tops of sand.

TEXTURE
Good to excellent sorting and high grain-matrix ratio. Variability of textural parameters between samples is very low. Tendency toward good rounding.

SEDIMENTARY STRUCTURES
Symmetrical and asymmetrical ripple marks. Crossbeds may be abundant; orientation commonly variable and may be bimodal. Trails and burrows. Minor channels and washouts.

INTERNAL ORGANIZATION
Data sparse, but vertical trends in grain size and bedding thickness either may be irregular or may increase or decrease, according to possible transgressive or regressive origin.

SIZE, SHAPE, AND ORIENTATION
Size and shape, highly variable, ranging from irregular, small pods through elongate ribbons to widespread sheets of many miles. Bifurcating and dendritic patterns absent. Elongate bodies have variable orientation with respect to depositional strike: parallel, perpendicular, and random. Relation of cross-bedding orientation to elongation not well-known, but probably variable.

ASSOCIATED LITHOLOGIC TYPES
Vertical: variable according to regressive or transgressive origin, but mostly marine shale and (or) carbonates. Basal contact may be disconformable, but generally not of great magnitude. Lateral: generally relatively uniform silt-free shales or carbonates, commonly with rich and varied marine fauna.

Marine shale—usually relatively silt-free—usually with a rich and varied marine fauna, is a common lateral equivalent of discrete, isolated sandstone bodies. In some cases, carbonates may be the dominant associated lithologic type. In ancient basins, laterally persistent marine marker beds, if present, facilitate correlation.

Size, shape, and orientation of marine-shelf sand bodies pose challenging problems, principally because of the diversity of sand-body types that can occur. Widespread sheets—both thick and thin—have been interpreted to have a shallow marine-shelf origin. Better evidence generally is available for the thin rather than thick sheets, which may be, in reality, complex composite bodies and may include some barrier-island sand bodies. They also may be the complex record of transgressive strand-line deposits. Most of these sheets have a relatively simple geometry, forming either relatively uniform blankets or wedges. Textural gradients tend to be slight and uniform. Elongate sand bodies have variable orientation with respect to depositional strike: parallel, perpendicular, and random. Relation of elongation to cross-bedding orientation is not well-known, but it probably is variable. Less is known about the isolated elongate shallow-water marine sand bodies than probably any other type.

Table VI gives the principal characteristics.

Development of acoustic sounders and interest in microrelief of the sea floor have contributed important information. Stride (1963) shows in detail the microrelief of the sea floor on the English Channel and parts of the Irish Sea. Here tidal currents are strong, and sediment transport largely parallels the strand line. Jordan (1962, Fig. 1) mapped the complex pattern of sand-wave crests on Georges Bank. The sand waves of the tidally influenced sand bodies of the North Sea described by Reineck (1963) also may be characteristic of much shallow-shelf sedimentation. Most of these studies probably are more of interest for their contribution to mechanism of sand transport on shallow waves, however, than for their documentation of marine-shelf sand bodies that are preserved in sedimentary basins. Modern equivalents of the widespread sheet and blanket sands of the past have not been described, perhaps because of abnormality of present continental shelves. That is, many shelf sands may be relics from lower Pleistocene sea-levels rather than the result of present-day processes. If so, where should one look for a modern analogue of a shelf sand? One possibility might be a desert-dune complex, which might provide useful comparative insight to sand sedimentation on a shallow-marine shelf.

Because integrated descriptions—those that combine detailed subsurface morphology with petrography and sedimentary structures—are hard to find, most investigators have concentrated on only one or two aspects. Although old, Dake's (1921) classic study of the St. Peter Sandstone is still useful. Cannon (1966) also gives a good example.

DESERT EOLIAN

The geologic literature indicates that eolian sand bodies of desert origin are volumetrically important in the fill of some sedimentary basins. Nevertheless, systematic tabulations of the petrology, sedimentary structures, and geometry of individual sand bodies of eolian origin are scarce—scarce enough to tempt one not even to include their discussion! One of the difficulties is the uncertainty about identifying ancient eolianites. Abundant and thick cross-beds, the absence of fossils, and a clay-deficient sand are the chief criteria for an eolian origin. In modern deserts, frosted grains also are very characteristic. Bigarella and Salamuni (1961, p. 1092) reported that ventifacts are common in some beds of the Botucatú Sandstone of Brazil.

Petrologically, comparatively little is known of the characteristics of eolian sands except that the grain-matrix ratio probably is high, because much clay has been winnowed out. Depending on the source rocks, the framework fraction may be immature. Sorting of the framework fraction generally is considered to be fair to good; micaceous minerals generally are absent. Rounding commonly is observed to be good. Cementing agents are probably mostly chemical. Gravel generally is rare, although shale chips can be abundant in some eolian sands.

The principal sedimentary structure is cross-bedding, which may be thick; individual beds are as thick as 100 feet or more. Variability of cross-bedding is small to moderate, although larger variances have been reported in the cross-bedding of some modern coastal dunes, presumably because of variable onshore and offshore winds or because of seasonal changes in wind direction. McKee and Tibbits (1964) show that modern barchans have relatively uniform unimodal internal cross-bedding, but in a sief dune—a longitudinal dune—cross-bedding is bimodal. Shotten (1956, Fig. 3) had earlier deduced the same conclusion based on Bagnold's work. Nonetheless, the variability of cross-bedding shown by most of the thick sheet sands of the Colorado Plateau that are believed to be largely eolian—the Webber, Nugget, and Navajo (Poole, 1964, Figs. 2, 7)—is comparable with that of most alluvial sands. This fact suggests that sief dunes are not a prominent part of these deposits. Ripple marks, generally a characteristic asymmetrical type with low amplitude and high index, are common in modern desert sands. A good reference is McKee (1966).

Fossils, tracks, and trails generally are scarce.

No information is available for the internal organization of eolian sand bodies, but strong vertical trends seem doubtful.

Associated lithologic types are difficult to generalize, because vertically the bounding units tend to be separated by unconformity, especially the basal contact. However, laterally a sheet eolian deposit may grade into either marine or continental deposits. Near such transitions, environmental discrimination is especially difficult. Sand itself, however, is almost the only sediment present. Perhaps because repositories are not present, size sorting—other than clay removal—is minimal, and the typical thick eolian sheet sand probably will have only weak textural gradients. Thus an eolian blanket sand may be the most uniform of all sand deposits. Jobin (1962, Table 6) contrasts the characteristics of marine and eolian sandstones with sandstones of alluvial origin in the Colorado Plateau region.

Although individual dunes—but doubtlessly composite ones—may reach enormous sizes in deserts, where heights as great as 600 feet have been reported, such individuals have yet to be identified in ancient sandstone deposits. Most ancient eolianites consist of widespread thick sheets of cross-bedded sandstone. Conceivably, eolian sands may contain the largest of all contiguous sand bodies.

The literature of ancient eolianites is relatively extensive, but, unfortunately, comparatively little of it is relevant to the purpose of this paper. The descriptions by Stokes (1961, p. 155–161), Baars (1961), and Poole (1964) of Pennsylvanian and Permian eolianites in the American Southwest are among the most useful. Shotten's study (1956) of the New Red Sandstone in Great Britain also is valuable.

PROBLEMS OF SAND-BODY PREDICTION

Prediction has two major aspects—where to look for a new sandstone body and how to outline efficiently its areal extent with a minimum of wells once it is found. Both problems are among the most challenging that a geologist can be asked to solve. The location problem generally is much more difficult than the extension problem.

Generally, a combination of three kinds of knowledge is required: (1) depositional environment of the sandstone body, (2) regional distribution of facies in the basin combined with knowledge of paleoslope, and (3) prior experience with similar sandstone bodies of the same facies, preferably but not necessarily in the same basin.

EXTENSION PROBLEM

The extension problem can be stated as follows: given an isolated well that intercepts one or more sandstone bodies—assumed to be elongate—what is their orientation and probable lateral extent? The ideal solution for prediction of elongation direction requires knowledge of (1) the environment of deposition, (2) paleoslope of the basin, and (3) orientation of internal directional structures.

Because they are a building block of sand-rich facies, types of sandstone bodies generally are related closely to the regional facies and the associated lithologic types in which they occur. As a result, one usually is not faced with the problem of differentiating all six major environments, but only several.

Perceptive contouring of sandstone bodies requires a model, which corresponds with the probable shape of a sandstone body of a particular origin. To identify the correct model, one uses a combination of internal and external characteristics. The principal internal characteristics are environmentally sensitive minerals and fossils, sedimentary structures, internal organization, and possibly grain size.

The principal external characteristics are the associated lithologic types, the nature of the basal contact, and paleoslope (regional depositional strike). The associated lithologic types and paleoslope have the principal predictive significance. Visher (1965) has described the vertical sequences of six environments—regressive and transgressive marine, alluvial, deltaic, bathyal-abyssal, and lacustrine—as predictive guides to exploration of sandstone bodies. Visher (p. 41) stresses Walther's Law of Facies, which he states as follows: ". . . where there are no time breaks in a stratigraphic section, those sediments which were areally adjacent must succeed each other vertically. . . ." Although not emphasized by Visher, knowledge of paleoslope is essential for maximum predictive success with this approach, because paleoslope commonly controls the direction of facies migration.

Once experience with a sandy facies in a basin has been acquired, it may be possible to develop a relatively small group of criteria—perhaps largely empirical—by which one can distinguish not only the major environments described previously, but possibly subenvironments within them. For example, Sabins (1963b) showed how a combination of fossils, mineral evidence, and information on median grain size could be analyzed in terms of a "yes" or "no" response in a computer-like flow diagram to distinguish seven subfacies of barrier-island and marine-shelf sand bodies. With present knowledge, it is generally not possible to transfer indiscriminately a particular set of such criteria from a sand facies of one basin to a like facies elsewhere.

Once the environment is identified, how successfully can one predict probable direction of elongation and lateral extent of a sand body?

Direction of elongation can be attacked in one of three ways. Probably the simplest is by analogy with orientation of similar sand bodies elsewhere in the basin. From what is known, this tends to yield good results simply because paleoslope in many basins was stable through appreciable periods of time and, as a result, patterns of sand-body orientation in successive cycles change comparatively little. The foregoing empirical approach depends on the relation between orientation of a particular kind of elongate sand body and paleoslope. Almost all tidal, barrier-island, turbidite, and alluvial sand bodies have a close relation to paleoslope. Marine-shelf sand bodies also may have a close relation, but so little is known about them that generalization is difficult. Finally, orientation of internal directional structures is good for predicting the elongation of alluvial, tidal, and turbidite sand bodies. Cross-bedding orientation is the best structure to measure in alluvial and tidal deposits, where it correlates well with long dimension. It is also the best structure to measure in marine-shelf sands, but its relation to long dimension is not generally established. Nor in barrier-island sandstone bodies will cross-bedding generally have a simple relation to elongation direction. Whether or not orientation of cross-bedding 0.5–3.0 feet thick can be determined in the well wall by a remote sensing device, such as a dipmeter, is a current active subject of research (Jizba et al., 1964).

The paleotopography of an unconformity also can be a valuable guide to elongation of the sand bodies that lie just above it. It appears to be useful because—for a particular amount of well control—one can usually outline paleodrainage and paleovalleys much more accurately than one can outline the sand bodies that later may have partly buried the topography. Paleotopography probably is most helpful when one is working with a youthfully dissected unconformity—an unconformity with entrenched second-cycle valleys—of either dendritic or trellis origin. Under these conditions of sharply entrenched valleys and alluvial fill, there is good correlation between sand body and valley elongation. Topographic control on basal sedimentation is likely to be less strong where an unconformity is in maturity or late maturity, especially if the basal sands are largely of marine or barrier-island origin.

Where location of elongate sand bodies is believed to have been influenced by structure, structure maps also have been used to predict sand-body orientation.

More difficult than the direction of elongation of a sand body is prediction of its lateral extent. In a sandstone-shale sequence a dipmeter may effectively indicate an edge of a sandstone body by the steeper dip of either the top or basal contact as it thins near an edge. Although published information is scant, there is probably a relation between thickness, grain size, and bedding type for most sandstone bodies that might be helpful, if cores are available.

Geophysical techniques also, under certain conditions, may be rewarding, although few published case histories are available. Anstey (1960, p. 245) has suggested that synthetic seismograms in connection with seismic reflection surveys can be useful in locating the edge of a sandstone body between two wells, one with and the other without sandstone.

Proper contouring also can contribute to location of the margins of a sandstone body. Contouring sand thickness on the basis of arithmetic spacing between control points generally yields poor estimates of sand-body width. More realistic is the determination of the width of a sand body and of the rate of change of thickness along its margins in areas of good well control and the extrapolation of these values into areas of less dense control.

LOCATION PROBLEM

The techniques of locating a new sandstone body are not as good as those for outlining its areal extent after discovery. Not much is known about locating isolated or apparently isolated sandstone bodies in the subsurface.

The pessimistic view of this problem is that elongate sandstone bodies—at least some types of them—occur more or less randomly within a particular facies in a basin. A large amount of evidence in some well-drilled basins tends to support this view.

Geologic solutions to the location problem may be either regional or local. A regional solution attempts at an early stage to outline areas—mostly groups of townships—of greater and lesser probability of sand-body occurrence. One way to do this is to outline roughly the sand-dispersal system of the basin by determining sand occurrence and directional structures in outcrop and supplementing this with the available subsurface data. Both structure and paleotopography have been used in some basins in the belief that these features, which have substantial continuity, exercise a control on sand occurrence. Because narrow elongate sandstone bodies in a shale section usually produce a compactional distortion in the adjacent shales, small density contrasts between the shale and sandstone may be sufficient to permit identification at shallow depths of the position of a sandstone body in much the same manner as gravity has been successfully used in exploration for buried valleys of Pleistocene age (Hall and Hajnal, 1962).

Geologically, the most difficult prediction problem is the new basin with isolated marine-shelf sand bodies and no outcrop. Comparatively little can be done at an early stage except to drill structures. In contrast, best success can be expected in basins where an outcrop is nearby and the sand bodies are principally of alluvial origin. With such conditions there is the possibility of making downdip predictions by using directional structures and sandstone thickness measured in outcrop or, if a structure has been drilled, to use paleoslope and the downdip orientation of alluvial sand bodies to locate the first step-out well. Probably the biggest effort at downdip prediction with directional structures in an alluvial sequence has been made in the Triassic and Jurassic uraniferous sandstone bodies of the Colorado Plateau.

The foregoing assumes that sandstone bodies of a basin have a rational arrangement related to the regional sand-dispersal system, basin geometry, and distribution of sources of supply. This rational arrangement has been termed a facies model (Potter, 1959). The concept of sand-body distribution in a basin has been developed best primarily for alluvial sand bodies (Allen, 1965b, Figs. 35, 36) but, as knowledge increases, more models for other environments can be expected. Computer simulation of sandstone distribution in a basin, as has been done by Harbaugh (1966) for marine carbonates, appears to be the next logical step in the development of the facies-model concept.

The facies-model concept with its emphasis on the existence of relatively few recurring models represents cause-and-effect "deterministic geology"—an approach that attempts to relate distribution and orientation of sand bodies in a basin to measurable, causal factors. Future research will doubtlessly incorporate a random, probabilistic element as well in such models so that they will cease to be fully deterministic. Incorporation of such a probabilistic element is recognition of the fact that, for a stated amount of information, wholly deterministic prediction may be impossible.

Another approach is search theory, developed from operations research. Here one outlines a drilling campaign on the assumption that sand bodies of assumed dimensions and orientation are essentially randomly distributed either within an entire basin, or more likely, within a small part of it, perhaps several townships. Applied first to ore bodies, applications to sandstone-body search are obvious (Celasun, 1964; Griffiths and Drew, 1964; Marshall, 1964).

INTERNAL CHARACTERISTICS

Internal characteristics have two distinct aspects: vertical sequence and lateral heterogeneity, both being the response to such primary characteristics as grain size, clay content, sedimentary structures, and bedding as well as secondary ones, principally chemical cementation.

The vertical sequence for alluvial and tidal sand bodies is well-established. These sand bodies usually have good vertical zonation—medium- to thick-bedded, coarse-grained sands, usually crossbedded, overlain by thin-bedded and finer-grained ripple-marked and parallel-laminated sands and silts. Because of both fine grain size and the common association of ripple marks with very thin clay partings, permeability and porosity usually decline upward. Thus alluvial and tidal sand bodies commonly show, in the absence of strong cementation, good vertical zonation of horizontal permeability. Such zonation usually will be even more pronounced if these sand bodies are of multistory origin. Figure 12 shows the vertical zonation by grain size, sedimentary structures, and permeability of the Robinson Sandstone (Pennsylvanian) of the Illinois basin. Permeability gradients can be either reduced or even totally obliterated by secondary development of calcite, kaolinite, and silica cements.

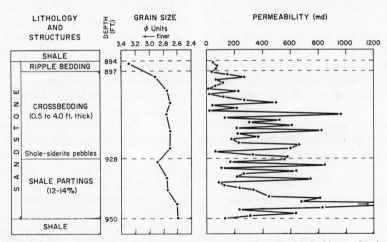

FIG. 12.—Permeability, grain size, and sedimentary structures in alluvial Robinson Sandstone (Pennsylvanian) of Illinois basin. Contrasts in grain size and bedding at 928 feet may indicate multistory sand body (modified from Hewitt and Morgan, 1965, Fig. 7).

Published descriptions of lateral heterogeneity in sand bodies are few, especially the relation between rock properties and fluid flow. Partly this reflects the difficulty of both data collection as well as the lack of an effective means of analysis.

Geologically two approaches have been followed: the petrographic approach based on grain fabric, cementing agents, and size distribution (*cf.* Griffiths, 1964) versus the sedimentary structure-bedding approach (*cf.* Montadert, 1963, and Zeito, 1965). Johnson and Greenkorn (1963) incorporate to an extent both approaches. Scale is perhaps the essential difference between the two. When is fluid flow more dependent on rock properties that can be specified in a thin section than on characteristics such as cross-bedding and interbedded shale breaks that are of 10^4 to 10^6 greater dimension? Although no clear answer is available, bedding features may be more important—at least before secondary cementation is far advanced. Moreover, because bedding features are products of primary sedimentation, they may be more regularly distributed within a sand body as well as more rationally related to its external shape than the distribution of cementing agents. Obviously, directional structures should be considered in predicting reservoir heterogeneities of primary origin. Although the reason is not understood fully, horizontal permeability parallels cross-bedding orientation in the few cases where the two have been studied (Hewitt and Morgan, 1965, Fig. 7). Possibly internal shale breaks, whose primary dip could be recorded by a dipmeter, also can be related to current system.

A study of the fluid flow characteristics of individual sand bodies, such as made from a regional point of view by Jobin (1962, Table 6) for the Colorado Plateau, would be most useful. Such a classification should include mean permeability and porosity, their within-well variability, permeability gradients within the sand body, and a consideration of the internal organization and external geometry of the sandstone body.

Where We Stand

The preceding discussion is perhaps most useful not for its details, but for the opportunity it gives to look back and broadly appraise the entire subject.

The environment that is best known—in terms of both sand-body characteristics and processes—is the alluvial one. Here, although additional profitable work can be done, knowledge approaches being equal to the task of reasonably predicting sand-body occurrence and characteristics. Sand bodies of tidal origin are probably the second best documented and understood in the Recent, although few have been recognized in ancient sediments. Moreover, distinguishing some types of shallow-shelf marine sand bodies, sand bodies that may have accumulated on tidally influenced marine banks, for example, from those that formed in large estuaries, usually is not easy. Barrier-island sand bodies probably are as well-understood as tidal ones, if not better, especially because orientation with respect to depositional strike is much more rigid. On the other hand, knowledge of sand bodies in the turbidite, shallow-water marine, and desert-eolian environments is very inadequate. Certainly, with reference to sand-body morphology, only the barest essentials are known about the turbidite and desert-eolian environments. Each may prove to have a rather small number of morphologic types, however, and thus contrast with shallow marine-shelf sand bodies, which appear to have the most variable morphologic characteristics of all. Regularity of orientation with respect to depositional strike appears to be a major problem for marine-shelf sand bodies.

The principal difficulty—both in preparing this review and in recognizing sand bodies of different origin in ancient sediments—is an obvious one. It is the lack of systematic, quantitative data on the petrology, texture, sedimentary structures, and internal organization of sand bodies. No amount of insight into sedimentary processes can substitute for such a systematic quantitative study, because it is by finding the best match between the characteristics of a particular ancient sandstone body and those from a modern environment that an environmental interpretation of an ancient sandstone body is made. This lack of information is all the more disturbing, when one realizes that new, "far out" research is not required, but rather only the systematic application of modern sedimentary petrology combined with appropriate field studies, studies that emphsize bedding, sedimentary structures, internal organization, and relation of the sand body to bounding lithologic types. Systematic petrographic study alone could be rewarding. More study of Füchtbauer's (1963) suggestion that the color of biotite and possibly even tourmaline in sandstones is environmentally sensitive is one example of work that might be done with profit. The environmental significance of cementing agents is another potentially rewarding area. In this age of proposals for $300,000,000 particle accelerators in physics and of vast sums for space research, is it too much to expect a coordinated effort by geologists, using standardized techniques, to obtain the basic data for sandstone bodies? By incorporating the recent emphasis on internal organization of sandstone bodies and their relation to the vertical sequence, such systematic study would be most rewarding. The planners of such a program might profitably remember that a total description in either the modern or ancient is neither possible nor desirable, even if it were possible. Rather, what is needed is an investigation that yields descriptions relevant to a philosophy that is applicable to and workable for ancient sandstone bodies.

The above proposal is, of course, largely within the classic framework of finding the best match between an ancient unknown and its presumed modern equivalent. Because environmental recognition has been one of geology's more difficult tasks, one is tempted to ask, "Is more of the same what is really required?" Until the data that the classic approach requires are available, however, it seems less than fair to prejudge the method. Nevertheless, every geologist who has ever toiled at length with a sandstone body, particularly one whose environment of deposition has proved elusive, must have been troubled at some time or other by the question, "Isn't there a better or easier way to do this?" This immediately suggests a second question: "Have we really been asking the right questions?"

REFERENCES CITED

Allen, J. R. L., 1962, Petrology, origin and deposition of the highest lower Old Red Sandstone of Shropshire, England: Jour. Sed. Petrology, v. 32, p. 657–697.

——— 1964, Studies in alluvial sedimentation: six cyclothems from the Lower Old Red Sandstone, Anglo-Welsh basin: Sedimentology, v. 3, p. 163–198.

——— 1965a, A review of the origin and characteristics of Recent alluvial sediments: Sedimentology (Special issue), v. 5, 191 p.

——— 1965b, Finding upwards cycles in alluvial successions: Liverpool and Manchester Geol. Jour., v. 4, p. 229–246.

Andresen, M. J., 1961, Geology and petrology of the Trivoli Sandstone in the Illinois basin: Ill. Geol. Survey Circ. 316, 31 p.

Anstey, N. A., 1960, Attacking the problems of the synthetic seismogram: Geophysical Prospecting, v. 8, p. 242–259.

Baars, D. L., 1961, Permian blanket sandstones of Colorado Plateau, in Geometry of sandstone bodies: Am. Assoc. Petroleum Geologists, p. 179–207.

Bersier, A., 1959, Séquences détritiques et divagatious (fluviales): Ecologae Geol. Helvetiae, v. 51, p. 854–893.

Bernard, H. A., and others, 1962, Recent and Pleistocene geology of southwest Texas, in Geology of Gulf Coast and central Texas, Guidebook of excursions: Houston Geol. Soc., p. 175–224.

Bigarella, J. J., and R. Salamuni, 1961, Early Mesozoic wind patterns as suggested by dune bedding in the Botucatú Sandstone of Brazil and Uruguay: Geol. Soc. America Bull., v. 72, p. 1089–1106.

Blissenback, Erich, 1954, Geology of alluvial fans in semi-arid regions: Geol. Soc. America Bull., v. 65, p. 175–189.

Boyd, Don R., and Byron F. Dyer, 1966, Frio barrier bar system of south Texas: Am. Assoc. Petroleum Geologists Bull., v. 50, p. 170–178.

Busch, D. A., 1959, Prospecting for stratigraphic traps: Am. Assoc. Petroleum Geologists Bull., v. 43, p. 2829–2843.

Byrne, J. V., and others, 1959, The chenier plain and its stratigraphy, south-western Louisiana: Gulf Coast Assoc. Geol. Soc. Trans., v. 9, p. 1–23.

Cannon, James L., 1966, Outcrop examination and interpretation of paleocurrent patterns of the Blackleaf Formation near Great Falls, Montana: Billings Geol. Soc., 17th ann. field conf., p. 71–111.

Celasun, M., 1964, The allocation of funds to reconnaissance drilling projects, in International symposium, Applications of statistics, operations research, and computers in the mineral industry, part A: Colo. School Mines Quart., v. 59, p. 169–185.

Corbeille, R. L., 1962, New Orleans barrier-island: Gulf Coast Assoc. Geol. Soc. Trans., v. 12, p. 223–230.

Curray, J. R., and D. G. Moore, 1964, Holocene regressive littoral sands, Costa de Nayarit, Mexico, in Deltaic and shallow marine deposits, Developments in sedimentology, v. 1: Amsterdam, Elsevier Pub. Co., p. 76–82.

Dake, C. L., 1921, The problem of the St. Peter Sandstone: Mo. Univ. School Mines and Metallurgy Bull., Tech. Ser., v. 6, 228 p.

Doeglas, D. J., 1962, The structure of sedimentary deposits of braided rivers: Sedimentology, v. 1, p. 167–190.

Dondanville, R. F., 1963, The Fall River Formation, northwestern Black Hills: lithology and geologic history: Guidebook, First Joint Field Conf., Wyo. Geol. Assoc. and Billings Geol. Soc., Aug. 8–10, 1963, p. 87–99.

Doty, R. W., and J. F. Hubert, 1962, Petrology and paleogeography of the Warrensburg channel sandstone, western Missouri: Sedimentology, v. 1, p. 7–39.

Dury, G. H., 1964a, Principles of underfit streams: U. S. Geol. Survey Prof. Paper 452-A, 67 p.

——— 1964b, Subsurface exploration and chronology of underfit streams: U. S. Geol. Survey Prof. Paper 452-B, 56 p.

Emery, K. O., 1964, Turbidites—Precambrian to present, *in* Studies in oceanography: Tokyo Univ., p. 486–495.

Evans, Graham, 1965, Intertidal flat sediments and their environments of deposition in the Wash: Quart. Jour. Geol. Soc. London, v. 121, p. 209–245.

Ewing, M., D. B. Ericson, and B. C. Heezen, 1958, Sediments and topography of the Gulf of Mexico, *in* Habitat of oil: Am. Assoc. Petroleum Geologists, p. 994–1053.

Feofilova, A. P., 1954, The place of alluvium in sedimentation cycles of various orders and time of its formation: Akad. Nauk S.S.S.R., Geol. Inst., Trans. 151, Coal ser. 5, p. 241–272 (in Russian).

Finch, W. I., 1959, Geology of uranium deposits in Triassic rocks of the Colorado Plateau region: U. S. Geol. Survey Bull. 1074-D, 164 p.

Fisk, H. N., 1959, Sand facies of Recent Mississippi delta deposits: Fourth World Petroleum Cong. Proc., Rome, Sec. I/C, Rept. 3, p. 337–398.

——— 1961, Bar-finger sands of Mississippi delta, *in* Geometry of sandstone bodies: Am. Assoc. Petroleum Geologists, p. 29–52.

——— E. McFarlan, Jr., C. R. Kolb, and L. J. Wilbert, Jr., 1954, Sedimentary framework of the modern Mississippi delta: Jour. Sed. Petrology, v. 24, p. 76–99.

Flores, G., 1959, Evidence of slump phenomena (olistostromes) in areas of hydrocarbon exploration in Sicily: Fifth World Petroleum Cong. Proc., New York, Sec. 1, Paper 13, p. 259–275.

Folk, Robert L., and R. Robles, 1964, Carbonate sands of Isla Perez, Alacran reef complex, Yucatán: Jour. Geology, v. 72, p. 255–292.

Friedman, S. A., 1960, Channel-fill sandstones in the Middle Pennsylvanian rocks of Indiana: Ind. Geol. Survey Rept. Prog. 23, 59 p.

Füchtbauer, Hans, 1963, Zum Einfluss des Ablagerungsmilieus auf die Farbe von Biotiten und Turmalinen, *in* Unterscheidungs-möglichkeiten mariner und nichtmariner Sedimente: Fortschr. Geol. Rheinld. u. Westf., v. 10, p. 331–336.

Gorsline, D. S., and K. O. Emery, 1959, Turbidity-current deposits in San Pedro and Santa Monica basins off southern California: Geol. Soc. America Bull., v. 70, p. 279–290.

Gould, H. D., and E. McFarlan, Jr., 1959, Geologic history of the chenier plain, southwestern Louisiana: Gulf Coast Assoc. Geol. Soc. Trans., v. 9, p. 1–10.

Griffiths, J. C., 1964, Statistical approach to the study of potential oil reservoir sandstones, *in* International symposium, Applications of statistics, operations research and computers in the mineral industry: Colo. School Mines Quart., v. 59, p. 637–668.

——— and L. J. Drew, 1964, Simulation of exploration programs for natural resources by models, *in* International symposium, Applications of statistics, operations research, and computers in the mineral industry, part A: Colo. School Mines Quart., v. 59, p. 187–206.

Hall, D. H., and Z. Hajnal, 1962, The gravimeter in studies of buried valleys: Geophysics, v. 27, pt. 2, p. 939–951.

Hand, B. M., and K. O. Emery, 1964, Turbidites and topography of north end of San Diego trough, California: Jour. Geology, v. 72, p. 526–542.

Harbaugh, J. W., 1966, Mathematical simulation of marine sedimentation with IBM 7090/7094 computers: Kans. Geol. Survey, Computer Contr. 1, 51 p.

Harms, J. C., D. B. MacKenzie, and D. G. McCubbin, 1963, Stratification in modern sands of the Red River, Louisiana: Jour. Geology, v. 71, p. 566–580.

Hayes, M. O., and A. J. Scott, 1964. Environmental complexes, south Texas Coast: Gulf Coast Assoc. Geol. Soc. Trans., v. 14, p. 237–240.

Hewitt, C. H., and J. T. Morgan, 1965, The Fry in situ combustion test—reservoir characteristics: Jour. Petroleum Technology, v. 17, p. 337–353.

Hollenshead, C. T., and R. L. Pritchard, 1961, Geometry of producing Mesaverde sandstones, San Juan basin, *in* Geometry of sandstone bodies: Am. Assoc. Petroleum Geologists, p. 98–118.

Horn, Dietrich, 1965, Zur geologischen Entwicklung der südlichen Schleimündung im Holozän: Meyniana, v. 15, p. 42–58.

Jizba, Z. V., W. C. Campbell, and T. W. Todd, 1964, Core resistivity profiles and their bearing on dipmeter survey interpretation: Am. Assoc. Petroleum Geologists Bull., v. 48, p. 1804–1809.

Jobin, D. A., 1962, Relation of the transmissive character of the sedimentary rocks of the Colorado Plateau to the distribution of uranium deposits: U. S. Geol. Survey Bull. 1124, 151 p.

Johnson, C. R., and R. A. Greenkorn, 1963, Description of gross reservoir heterogeneity by correlation of lithologic and fluid properties from core samples: Internatl. Assoc. Sci. Hydrology Bull., v. 8, p. 52–63.

Jordan, G. F., 1962, Large submarine sand waves: Science, v. 136, p. 939–948.

Klein, G. deVries, 1963, Bay of Fundy intertidal zone sediments: Jour. Sed. Petrology, v. 33, p. 844–854.

Krumbein, W. C., and L. L. Sloss, 1963, Stratigraphy and sedimentation: 2nd ed., San Francisco, W. H. Freeman and Co., 660 p.

Krynine, Paul D., 1948, The megascopic study and classification of sedimentary rocks: Jour. Geology, v. 56, p. 130–165.

Kuenen, Ph. H., 1964, Deep sea sands and ancient turbidites, *in* Turbidites, Developments in sedimentology, v. 3: Amsterdam, Elsevier Pub. Co., p. 3–33.

Land, L. S., 1964, Eolian crossbedding in the beach dune environment, Sapelo Island, Georgia: Jour. Sed. Petrology, v. 34, p. 389–394.

Lee, Wallace, and others, 1938, Stratigraphic and paleontologic studies of the Pennsylvanian and Permian rocks in north-central Texas: Univ. Tex. Pub., Bull. 3801, 252 p.

Leopold, L. B., and others, 1964, Alluvial processes in geomorphology: San Francisco, W. H. Freeman and Co., 522 p.

Lins, T. W., 1950, Origin and environment of the Tonganoxie Sandstone in northeastern Kansas: Kans. Geol. Survey Bull. 86, p. 107–140.

Logvinenko, N. V., and I. N. Remizov, 1964, Sedimentology of beaches on the north coast of the Sea of Azov, *in* Deltaic and shallow marine deposits, Developments in sedimentology, v. 1: Amsterdam, Elsevier Pub. Co., p. 244–252.

Marshall, K. T., 1964, A preliminary model for determining the optimum drilling pattern in locating and evaluating an ore body, *in* International symposium, Applications of statistics, operations research and computers in the mineral industry: Colo. School Mines Quart., v. 59, p. 223–236.

Martin, B. O., 1963, Rosedale channel—evidence for late Miocene submarine erosion in Great Valley, California: Am. Assoc. Petroleum Geologists Bull., v. 47, p. 441–456.

McBride, E. F., and M. O. Hayes, 1962, Dune cross-bedding on Mustang Island, Texas: Am. Assoc. Petroleum Geologists Bull., v. 46, p. 546–551.

McGugan, Alan, 1965, Occurrence and persistence of thin shelf deposits of uniform lithology: Geol. Soc. America Bull., v. 76, p. 125–130.

McKee, E. D., 1957, Primary structures in some Recent sediments: Am. Assoc. Petroleum Geologists Bull., v. 41, p. 1704–1747.

——— 1966, Structures of dunes at White Sands National Monument, New Mexico (and a comparison with structures of dunes from selected other areas): Sedimentology (special issue), v. 7, 69 p.

——— and G. C. Tibbetts, Jr., 1964, Primary structures of a seif dune and associated deposits in Libya: Jour. Sed. Petrology, v. 34, p. 5–17.

Montadert, L., 1963, La sédimentologie et l'étude détaillée des hétérogénétiés d'un réservoir: application au gisement d'Hassi-Messaoud: Inst. français Pétrole Rev., v. 18, p. 241–257.

Nanz, R. H., Jr., 1954, Genesis of Oligocene sandstone reservoirs, Seeligson field, Jim Wells, and Kleberg Counties, Texas: Am. Assoc. Petroleum Geologists Bull., v. 38, p. 97–117.

Noorthoorn v. d. Kruijft, J. L., and R. Lagaaij, 1960, Displaced faunas from inshore estuarine sediments in the Haringvliet (Netherlands): Geologie en Mijnbouw, v. 39 (n. s. 22), p. 711–723.

Off, Theodore, 1963, Rhythmic linear sand bodies caused by tidal currents: Am. Assoc. Petroleum Geologists Bull., v. 47, p. 324–341.

Oomkens, E., and J. H. J. Terwindt, 1960, Inshore estuarine sediments in the Haringvleit (Netherlands): Geologie en Mijnbouw, v. 39 (n. s. 22), p. 701–710.

Pepper, J. F., and others, 1954, Geology of the Bedford Shale and Berea Sandstone in the Appalachian basin: U. S. Geol. Survey Prof. Paper 259, 111 p.

Pinsak, Arthur, 1957, Subsurface stratigraphy of the Salem Limestone and associated formations in Indiana: Ind. Geol. Survey, v. 11, 62 p.

Poole, F. G., 1964, Paleowinds in the western United States, in Problems in paleoclimatology, in N.A.T.O. Paleoclimates Conf. Proc., Univ. Newcastle upon Tyne, 1963: London, Interscience Publisher, p. 394–405.

Potter, Paul Edwin, 1959, Facies model conference: Science, v. 129, p. 1292–1294.

——— 1962, Late Mississippian sandstones of Illinois: Ill. Geol. Survey Circ. 340, 36 p.

——— 1963, Late Paleozoic sandstones of the Illinois basin: Ill. Geol. Survey Rept. Inv. 217, 92 p.

Ray, Louis, L., 1965, Geomorphology and Quaternary geology of the Owensboro quadrangle, Indiana and Kentucky: U. S. Geol. Survey Prof. Paper 488, 72 p.

Reineck, H. E., 1963, Sedimentgefüge im Bereich der südlichen Nordsee: Senckenberg. naturf. Gesell. Abh., no. 504, 64 p.

Rich, J. L., 1923, Shoestring sands of eastern Kansas: Am. Assoc. Petroleum Geologists Bull., v. 7, p. 103–113.

——— 1938, Shorelines and lenticular sands as factors in oil accumulation, in The science of petroleum, v. 1: London, Oxford Univ. Press, p. 230–239.

Rittenhouse, Gordon, 1961, Problems and principles of sandstone-body classification, in Geometry of sandstone bodies: Am. Assoc. Petroleum Geologists, p. 3–12.

Sabins, F F., Jr., 1963a, Anatomy of stratigraphic trap, Bisti field, New Mexico: Am. Assoc. Petroleum Geologists Bull., v. 47, p. 193–228.

——— 1963b, Computer flow diagram in facies analysis: Am. Assoc. Petroleum Geologists Bull., v. 47, p. 2045–2047.

Schlee, J. S., and R. H. Moench, 1961, Properties and genesis of "Jackpile" sandstone, Laguna, New Mexico, in Geometry of sandstone bodies: Am. Assoc. Petroleum Geologists, p. 134–150.

Sedimentation Seminar, 1966, Cross bedding in the Salem Limestone of central Indiana: Sedimentology, v. 6, p. 95–114.

Seibold, Eugene, 1963, Geological investigation of nearshore sand-transport-examples of methods and problems from Baltic and North Seas, in Progress in oceanography, v. 1: New York, The Macmillan Co., p. 1–70.

Shantser, E. V., 1951, Alluvium of the river plains of the temperate zone and its significance for the knowledge of regularities of structure and formation of alluvial deposits: Akad. Nauk S.S.S.R., Geol. Inst., Trans. v. 135, Geol. ser. 55, 274 p. (in Russian).

Shelton, J. W., 1962, Shale compaction in a section of Cretaceous Dakota Sandstone, northwestern North Dakota: Jour. Sed. Petrology, v. 32, p. 874–877.

——— 1965, Trend and genesis of lowermost sandstone unit of Eagle Sandstone at Billings, Montana: Am. Assoc. Petroleum Geologists Bull., v 49, p. 1385–1397.

Shepard, F. P., 1960, Gulf Coast barriers, in Recent sediments, Northwest Gulf of Mexico: Am. Assoc. Petroleum Geologists, p. 197–220.

——— and Gerhard Einsele, 1962, Sedimentation in San Diego trough and contributing submarine canyons: Sedimentology, v. 1, p. 81–133.

——— and D. G. Moore, 1955, Central Texas coast sedimentation: characteristics of sedimentary environment, recent history, and diagenesis: Am. Assoc. Petroleum Geologists Bull., v. 39, p. 1463–1593.

Shirley, Martha Lou, ed., 1966, Deltas in their geologic framework: Houston Geol. Soc., 251 p.

Shotten, F. W., 1956, Some aspects of the New Red desert in Britain: Liverpool and Manchester Geol. Jour., v. 1, p. 450–465.

Stokes, W. L., 1961, Alluvial and eolian sandstone bodies in Colorado Plateau, in Geometry of sandstone bodies: Am. Assoc. Petroleum Geologists, p. 151–178.

Straaten, L.M.J.U. van, 1953, Megaripples in the Dutch Wadden Sea and the basin of Arcachon (France): Geologie en Mijnbouw, v. 15, p. 1–11.

——— 1954a, Composition and structure of Recent marine sediments in the Netherlands: Leidse Geol. Mededel., v. 19, 11 p.

——— 1954b, Sedimentology of Recent tidal flat deposits and the psammites du Condroz (Devonian): Geologie en Mijnbouw, v. 16, p. 25–47.

Stride, A. H., 1963, Current-swept sea floors near the southern half of Great Britain: Quart. Jour. Geol. Soc. London, v. 119, p. 175–199.

Sullwold, H. H., Jr., 1960, Tarzana fan, deep submarine fan of late Miocene age, Los Angeles County, California: Am. Assoc. Petroleum Geologists Bull., v. 44, p. 433–457.

——— 1961, Turbidites in oil exploration, in Geometry of sandstone bodies: Am. Assoc. Petroleum Geologists, p. 63–81.

Terwindt, J. H. J., and others, 1963, Sediment movement and sediment properties in the tidal area of the Lower Rhine (Rotterdam Waterway): Koninkl. Nederlands Geol.-Mijnb. Genoot. Verh., Geol. ser., v. 21-2, p. 243-258.

Thompson, W. O., 1937, Original structures of beaches, bars, and dunes: Geol. Soc. America Bull., v. 48, p. 723-752.

Thomas, William A., and C. John Mann, 1966, Late Jurassic depositional environments, Louisiana and Arkansas: Am. Assoc. Petroleum Geologists Bull., v. 50, p. 178-182.

Twenhofel, W. H., 1950, Principles of sedimentation: 2d ed., New York, McGraw-Hill Book Co., Inc., 673 p.

Visher, G. S., 1965, Use of vertical profile in environmental reconstruction: Am. Assoc. Petroleum Geologists Bull., v. 49, p. 41-61.

Wermund, E. G., 1965, Cross-bedding in the Meridian Sand: Sedimentology, v. 5, p. 69-79.

Witkind, I. J., 1956, Channels and related swales at the base of the Shinarump Conglomerate, Monument Valley, Arizona: U. S. Geol. Survey Prof. Paper 300, p. 233-237.

──── and R. E. Thaden, 1963, Geology and uranium-vanadium deposits of the Monument Valley area, Apache and Navajo Counties, Arizona: U. S. Geol. Survey Bull. 1103, 171 p.

Wurster, Paul, 1964, Geologie des Schilfsandsteins: Hamburg Geol. Staatsinst. Mitt., v. 33, 140 p.

Zagwijn, W. H., 1963, Pleistocene stratigraphy in the Netherlands, based on changes in vegetation and climate, *in* Trans. Jubilee Convention, Pt. 2: Koninkl. Nederlands Geol.-Mijnb. Genoot. Verh., Geol. ser., v. 21-2, p. 173-196.

Zeito, George A., 1965, Interbedding of shale breaks and reservoir heterogeneities: Jour. Petroleum Technology, v. 17, p. 1223-1228.

Zhemchuzhnikov, Yu A., 1926, Type of cross-bedding as criteria of origin of sediments: Ann. Inst. Mines, Leningrad, v. 7, p. 35-69 (in Russian).

──── 1954, Alluvial deposits in coal measures in the middle Carboniferous of the Donets basin: Akad. Nauk S.S.S.R., Geol. Inst., Trans. 154, Coal ser., 296 p. (in Russian).

──── and others, 1959, Structure and accumulation conditions of the principal coal-bearing formations and coal beds of the middle Carboniferous of the Donets basin, pt. 1: Akad. Nauk S.S.S.R., Geol. Inst., Trans. 15, 346 p. (in Russian).

──── and others, 1960, Structure and accumulation conditions of the principal coal-bearing formations and coal beds of the middle Carboniferous of the Donets basin, pt. II; Akad. Nauk S.S.S.R., Geol. Inst., Trans. 15, 346 p. (in Russian).

Zimmerle, Winfried, 1964, Sedimentology of a Tertiary beach sand in the subalpine molasse trough, *in* Deltaic and shallow marine deposits, Recent developments in sedimentology, v. 1: Amsterdam, Elsevier Pub. Co., p. 447-457.

COMPARISON OF MARINE-BAR WITH VALLEY-FILL STRATIGRAPHIC TRAPS, WESTERN NEBRASKA[1]

F. A. EXUM[2] AND J. C. HARMS[2]
Littleton, Colorado 80121

ABSTRACT

Marine-bar and valley-fill stratigraphic traps in the Cretaceous "J" sandstone in Cheyenne and Banner Counties, Nebraska, illustrate control of reservoir shape, size, and characteristics by depositional environment.

Reservoirs deposited as shallow-marine bars are elliptical lenses 2-5 mi long, 0.5-1.5 mi wide, and less than 25 ft thick. Sandstone grades laterally into marine mudrock. There are two generations of bars in the area, closely spaced stratigraphically, but with different directions of elongation. These lenses now are tilted with a regional southwest dip. Entrapment is independent of structural closure. Most bar bodies are entirely oil filled.

Reservoirs deposited as a valley fill are within a prism of sandstone more than 20 mi long, 2,000 ft wide, and 50-80 ft thick. The boundaries of this body are erosional. Oil is trapped only where the valley-fill trend crosses plunging anticlines. The valley fill interconnects all pools as a single aquifer system.

Exploration and production efforts are guided by several considerations. Position of marine-bar reservoirs can be predicted by techniques which map gradients in sandstone-shale proportions, such as those based on mechanical logs. Bars in the study area are scattered and not in chains; orientation is varied. Structure is unimportant. In contrast, valley-fill reservoirs are separated by erosional boundaries from enclosing rocks; hence, they cannot be detected by examination of the enclosing facies. Where present, however, the valley fill has great continuity and persistence of trend. Structure is essential. Valley-fill reservoirs have water drive and high primary recovery, whereas marine-bar reservoirs have only solution-gas energy.

Environmental interpretation of these reservoirs is based on fossils, sedimentary structures, textures, facies relations, and geometry. A single core commonly allows correct interpretation. Exploration and production programs are guided profitably by use of environmental concepts at an early stage.

INTRODUCTION

Two types of sandstone reservoir, marine bars and a valley fill, are compared. Although they are within the same area and stratigraphic interval, these two types of reservoir sandstone are different in origin, and therefore have distinctly different sizes, shapes, distribution patterns, and internal features. These characteristics profoundly affect exploitation of oil reservoirs. The primary aim of this paper is to illustrate the way in which recognition of the origin of these reservoirs can be applied to exploration and production practices.

The writers believe that sandstone bodies similar to the ones described form reservoirs in other parts of the Denver basin and probably in other areas as well. Therefore, they can be used beyond provincial boundaries as models to guide exploitation. A secondary aim is to describe and compare smaller scale characteristics in detail, so that similar sandstone bodies might be recognized elsewhere, even if core material is limited.

The marine-bar reservoirs are described in detail. Changes from sandstone to shale, governed by depositional controls, are solely responsible for oil trapping. Reservoirs in the valley fill have been reported earlier (Harms, 1966), but their important features are reviewed briefly for comparison. These traps involve a combination of stratigraphic and structural conditions.

LOCATION AND GEOLOGIC SETTING

The reservoirs compared and contrasted are in the Denver basin in western Nebraska. Locations of the areas of marine-bar reservoirs and of valley-fill reservoirs are shown in Figure 1. The axis of the basin trends north-south. The basin is asymmetric, and the steeply dipping west flank is

---

[1] Given at the 41st Annual Meeting of the Society of Economic Paleontologists and Mineralogists, Los Angeles, California, April 12, 1967, and at the 17th Annual Meeting of the Rocky Mountain Section of the Association, Casper, Wyoming, October 11, 1967. This paper received both the 1967 A. I. Levorsen Memorial Award for the Rocky Mountain Section and the SEPM Outstanding Paper Award for 1967. Manuscript received, November 15, 1967; accepted, February 8, 1968.

[2] Marathon Oil Company.

Published with permission of Marathon Oil Company. Many Marathon geologists provided data essential for this study. Although those who have given assistance are too numerous to mention individually, the writers wish to acknowledge the help of S. E. Drum, who made the map on which Figure 1 is based. D. B. MacKenzie, R. Dana Russell, F. D. Spindle, Jr., and L. D. Traupe read the text and suggested improvements.

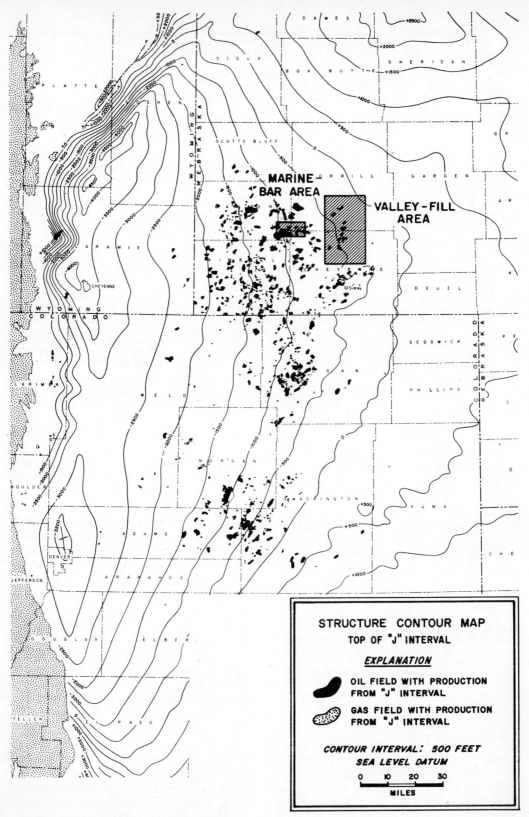

FIG. 1.—Structural contour map of Denver basin. Datum is top of Dakota (Cretaceous) "J" interval. Depth datum is sea level. Cross-hatched blocks show marine-bar and valley-fill study areas. CI = 500 ft.

bounded by the Precambrian of the Front Range. The east flank dips only about 30 ft/mi (0°21′) and includes most of the productive oil fields of the region.

The marine-bar study area is approximately 72 sq mi (186 km^2) in eastern Banner County and western Morrill and Cheyenne Counties. The valley-fill study area, 20 mi (32 km) east, is about 250 sq mi (648 km^2) in the central part of Morrill and Cheyenne Counties. Structural datum for the contours in Figure 1 is the top of the Cretaceous Dakota "J" sandstone, the major producing zone in both areas. Surface elevations range from 4,300 to 4,700 ft (1,300 to 1,450 m); the "J" is at subsurface depths of 4,300 to 5,400 ft (1,300 to 1,750 m). Locations of fields in the Denver basin that produce from this unit also are shown in Figure 1.

A generalized stratigraphic column for the two study areas is shown in Figure 2, together with common names of units used by many geologists. The nomenclature for the Dakota section is used informally, but in the same way as outlined by Harms (1966, p. 2121).

The Dakota section is underlain by nonmarine Jurassic rocks and overlain by marine Late Cretaceous rocks. Because transgressions and regressions by Dakota seas were common in the area, both marine and continental rocks are represented. Stratigraphic complexity related to transgression and regression is seen even within the relatively thin "J" unit (Fig. 3).

In both the marine-bar and valley-fill study areas, sedimentary strata record two asymmetric, regressive cycles. The older cycle began with deposition of the marine Skull Creek Shale, continued with deposition of increasingly coarser "J$_2$" sandstone near or along a shoreline, and concluded with the deposition of coal and the development of a widespread root zone in subaerial environments. The "J$_1$" unit was deposited during a repetition of the cycle, but with certain variations. Foraminifer-bearing marine shale was deposited at the base, deposition of sandstone as nonemergent marine bars followed, and emergence marked by local areas of root-zone development and valley cutting concluded the cycle. This second emergence apparently was caused by withdrawal of the sea, rather than shoreline progradation, because no beach deposits are recognized. After this emergence the Huntsman sea trans-

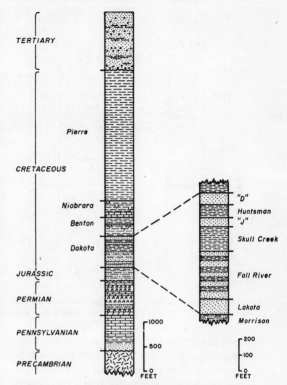

Fig. 2.—Generalized stratigraphic column, western Nebraska.

gressed a broad area in western Nebraska.

The most productive oil reservoirs are in the upper part of the "J" unit. Marine-bar reservoirs are scattered sandstone bodies in two distinct intervals within the "J$_1$," and the valley-fill reservoirs are in the prism of nonmarine deposits laid down by a stream at the end of "J$_1$" deposition.

Marine-Bar Reservoirs

Lenticular sandstone bodies, interpreted to be shallow-marine bars, form the important oil reservoirs in the western area outlined on Figure 1. These bodies are in two distinct intervals in the upper and lower parts of the "J$_1$," as indicated on Figure 3. The bars are elongate and elliptical in map view, with dimensions ranging from 1.5 by 5 mi (2.4 by 8.1 km) for the largest to 0.3 by 1.5 mi (0.5 by 2.4 km) for the smallest (Fig. 4). Their central areas have a maximum thickness of 25 ft (7.7 m) of sandstone; these sandstone bodies merge by lateral interfingering with shaly intervals only about 5 ft (1.5 m) thick. Contacts

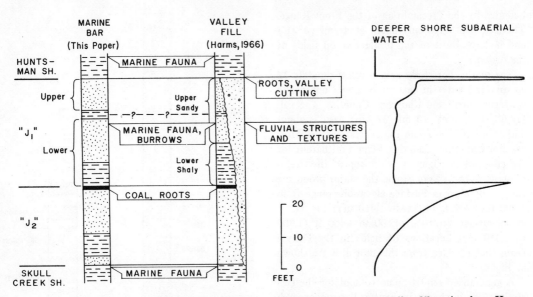

Fig. 3.—Columnar section of "J" unit in marine-bar and valley-fill study areas. Valley-fill section from Harms (1966). Graph shows generalized environmental interpretation.

between the sandy central parts and overlying and underlying shale layers are commonly sharp, or transitional through very small intervals.

Long axes of the bars in the upper part of the "J_1" trend NW-SE (Fig. 4). In the lower part of the "J_1" axes trend NE-SW (Fig. 5). Although well oriented, the bars are not linked in chainlike patterns, but rather are scattered. Study of 20 cores and 270 mechanical logs shows that the elliptical sandstone lenses are concentrated in two distinct stratigraphic intervals, within each of which orientation is good, but alignment is poor.

The writers interpret these bodies to be marine bars formed a considerable distance from shore-

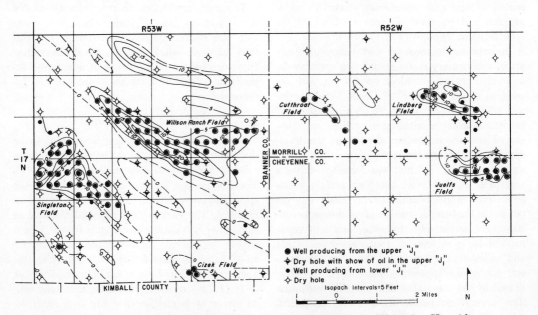

Fig. 4.—Isopach map of net sandstone in upper "J_1," marine-bar area, Nebraska. CI = 5 ft.

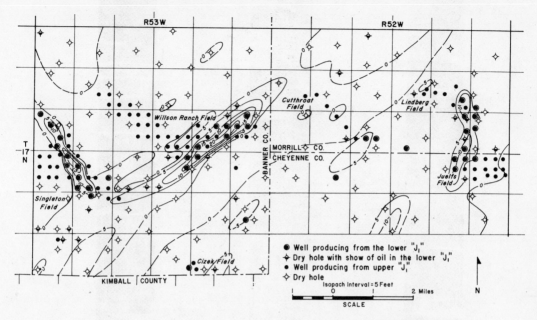

Fig. 5.—Isopach map of net sandstone in lower "J_1," marine-bar area, Nebraska. CI = 5 ft.

line. Interbedded and laterally equivalent shales contain foraminifers and are burrowed. Sandstone grades into thin shale units. In places shale beds are only one-fifth as thick as the sandstone; thus, sand areas probably were topographic highs even before compaction. There is no evidence that the bars ever were emergent or that they effectively separated lagoons from the Cretaceous sea. These considerations suggest that the term "marine bar" is appropriate, although the writers cannot provide well-documented modern analogs.

"J_1" LITHOLOGY

The "J_1" member is composed largely of shale, but also contains at two levels discontinuous, elongate, lenticular sandstone bodies which interfinger with their enclosing facies. The upper "J_1" unit ranges in thickness from 2 to 24 ft (0.6 to 7.3 m) and the lower "J_1" from 6 to 29 ft (1.8 to 8.8 m). The thickest sections are in areas where the sandstone bodies are of maximum development. Figure 4 is an isopachous map showing the net sandstone in the upper part of the "J_1" member. Net sandstone was estimated from intervals of separation between curves of contact resistivity logs, and addition to this figure of the high resistivity "spikes" caused by cemented sandstone beds. Figure 5 is an isopachous map of net sandstone in the lower part of the "J_1." The isopachous maps show that the productive sandstone bodies are discrete, elongate pods enclosed by mudrock.

The lateral change from sandstone to shale is transitional and interfingering. This observation is important to exploration for two major reasons. First, it suggests that the sandstone bodies were bathymetric prominences rather than channeled lows, and the distribution pattern should reflect this depositional process. Second, the nature of the change may allow recognition of proximity indicators by which present control could be used to predict the location of other potential reservoir bodies.

The transition from sandstone to mudrock in the "J_1" member is photographically represented by short core segments shown in Figure 6. The transition is by gradual changes through three typical facies.

Central-bar facies.—The central-bar facies of "J_1" member bars is illustrated by the oil reservoir at Willson Ranch field (Fig. 4). The two most prominent characteristics of this facies are the sandy texture, interrupted by relatively few shale partings, and the abundance of low-angle cross-stratification.

The sandstone is composed of very fine to

Fig. 6.—Major facies of marine bar. Central-bar facies is mostly porous sandstone with low-angle cross-stratification. Bar-margin facies is severely burrowed, interbedded sandstone and shale. Interbar facies is parallel-stratified claystone and shale with a few thin, tightly cemented sandstone beds.

fine, rounded, well-sorted grains. Median grain diameter of about 0.125 mm and a maximum diameter of 0.350 mm were estimated by microscopic examination. No systematic lateral or vertical variation in maximum grain diameter was observed within this facies. Clay-size material is present both as matrix and as thin laminae. The matrix clay is less than 5 percent of the rock volume, as determined from point counts of thin sections. The clay partings or laminae, conspicuous because of their dark color, are very thin and constitute only a small part of the total unit. In many wells, the content of clay-size material is greatest near the base of the sandstone facies and decreases upward. Mineralogic composition of the sand-size fraction (Fig. 7) indicates an eastern provenance (MacKenzie and Poole, 1962). The clay-size fraction (Fig. 8) is shown by X-ray analysis to contain abundant kaolinite and illite.

Granule- or pebble-size material is nearly absent in the clean sandstone of the central-bar facies. Phosphatic fragments are found in the thin sandstone bed marking the top of the "J" section. A few siderite or clay pebbles are present near the base of the upper "J_1" bar at Willson Ranch field in a few cores where the central-bar facies is developed best. These pebbles are volumetrically insignificant, but may serve to identify areas where limited scour accompanied deposition.

Sedimentary structures in the central-bar facies are limited to two scales of cross-stratification. About half the central-bar facies is cross-stratified in sets greater than a few centimeters thick, and the rest is cross-stratified in very small sets, interpreted as products of ripples. Deformational structures and structures caused by organic activity have not been recognized within the central-bar facies. The characteristic low angle of cross-stratification in the larger sets is illustrated in Figure 9, which contains histograms showing the dip of cross-stratification as a function of frequency in percent. Approximately 79 percent of the dips are less than 10°. The average of 32 dips is 7° in the upper "J_1" bar, and is 8° for 14 dips in the lower "J_1." The cross-laminae are planar, as seen in the restricted view provided by a 3½-in. (9-cm) core. The boundary between each set of cross-strata truncates the laminae of the underlying set along an apparently planar surface dipping at low angle. The laminae at the base of each set parallel the underlying surface of truncation. The maximum thickness of sets cannot be

	No. of Samples	No. of Points		Quartz	Chert	Potash Feldspar	Plagioclase Feldspar	Muscovite	Other	Quartz/Chert Ratio	Total Feldspar
Upper "J_1"	6	1800	Average Percent	70.3	1.8	0.1	2.1	0.1	25.6	39.0	2.2
			Range of Percents	61.6–81.0	0.3–3.3	0.0–0.4	0.4–5.7	0.0–0.4	16.0–35.6	20.0–25.0	0.0–5.7
Lower "J_1"	8	2400	Average Percent	68.9	3.1	0.1	1.3	0.1	26.5	26.1	1.4
			Range of Percents	51.3–79.0	1.4–5.3	0.0–1.0	0.3–2.6	0.0–0.4	17.3–46.1	10.2–39.2	0.0–2.6
"J_2" member	7	3145	Average Percent	61.9	7.0	0.0	0.7	0.1	30.3	11.8	0.7
			Range of Percents	50.2–84.2	2.5–11.3	0.0–0.0	0.0–2.7	0.0–0.4	24.0–47.2	4.9–20.1	0.0–2.7

FIG. 7.—Mineralogic composition of sand- and silt-size grains in marine bars determined from point counts using standard thin sections.

	No. of Samples		Kaolinite	Illite-Sericite	Na Montmorillonite	Chlorite	Quartz	Pyrite	Others
Huntsman Shale	4	Average Percent	10	12	29	0	43	3	6
		Range of Percents	5-15	5-20	10-55	0	25-50	tr-5	0-15
Upper "J$_1$"	2	Average Percent	25	35	8	3	22	5	2
		Range of Percents	none	none	0-15	tr-5	5-40	*-10	0-5
Lower "J$_1$"	8	Average Percent	23	28	0	3	41	tr	4
		Range of Percents	10-35	15-40	none	tr-15	20-65	0-tr	0-5
"J$_2$" member	8	Average Percent	44	16	0	*	31	tr	9
		Range of Percents	15-80	0-25	none	0-*	15-50	0-tr	0-10

*possible occurrence, less than trace amounts

FIG. 8.—Mineralogic composition of fraction finer than 0.016 mm in marine bars, as determined by X-ray diffraction analysis.

established because the cores have been sampled at close intervals for core analysis and both the top and base of a set are not seen in a single piece of core. However, sets must be at least 1 ft thick, perhaps much more.

The ripple forms are difficult to characterize from core material. Complete forms rarely are preserved. Where ripples are fairly well preserved, the forms are symmetrical, rounded, and have slopes less than 25°, typical of wave ripples. Reconstruction based on fragmentary evidence suggests that the ripples are less than 1 cm

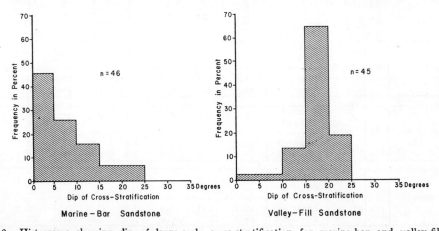

FIG. 9.—Histograms showing dip of large-scale cross-stratification for marine-bar and valley-fill rocks. Although range of dips is same, more than 80 percent of dips in marine bar are less than 15° and more than 80 percent of dips in valley fill are greater than 15°.

high and spaced at intervals of a few centimeters. More commonly, series of laminae mark successive positions of migrating ripples. The tops of the laminae are truncated by the next overlying set, and no estimate can be made of the maximum height of such ripples. The distribution of large- and small-scale cross-stratification does not appear to be systematic within the central-bar facies.

The clean sandstone in the central-bar facies forms the important petroleum reservoirs in the study area. As reported by Boardman (1959, p. 141), the average porosity from cores of two wells in the Willson Ranch field is 20.2 percent for the upper part of "J_1" and 22.1 percent for the lower part. No values are reported for permeability, but an average of 150 md was estimated from production figures.

Bar-margin facies.—The bar-margin facies consists predominantly of a mottled intermixture of fine- to very fine-grained sandstone and dark-gray to black shale. Burrowing obscures most of the primary sedimentary structures. The texture of the sandstone blebs is nearly identical to that of the sandstone in the central-bar facies. The shale is composed primarily of clay-size material and a small admixture of silt- and very fine sand-size grains. This facies is illustrated in Figure 6.

A variety of burrow forms is recognized, the most common of which is shown in Figure 10A. Large sand-filled tubes, elliptical in cross section, dip at angles of 0° to 45°. The tubes are commonly 1–2 cm in diameter and several centimeters long. A core through this type of structure produces a peculiar and distinctive *augen* appearance (Fig. 10A). The tubes are filled with fine- to very fine-grained, light-gray sandstone which contrasts with the darker mudrock surrounding each burrow. In many places the sandstone filling is horizontally transected by very thin laminae of dark-gray to black shale. A pattern to the distribution of this type of burrowing has been recognized.

Cross-stratification like that of the central-bar facies is common in beds which have not been severely churned or reworked by burrowing organisms. Parallel stratification is common in the mudrock, together with a few small-scale deformational or slump features.

Interbar facies.—The interbar facies is primarily shale with scattered, thin layers of quartzitic sandstone. The sandstone is identical in texture and structures to that in the central-bar facies. It differs only in that the cement—authigenic quartz overgrowths—is pervasive and has greatly reduced the permeability.

The shale is composed of clay-size material with abundant disseminated silt- and sand-size grains. The stratification of the shale is mostly parallel and the units average about 5 cm in thickness. These units are slightly different in color and texture. Some of the claystone has very fine laminae. A very few small-scale slump structures were observed. Only a small percentage of the mudrock contains burrows and the stratification is only slightly disturbed. The burrows are small and generally are in zones less than 5 cm thick. One type is predominantly horizontally oriented, less than 1 mm in diameter, and filled with light-gray siltstone or very fine-grained sandstone (Fig. 10B). The burrow filling is possibly a concentration of the silt- and sand-size material normally found disseminated throughout the mudrock. Another burrow type consists of vertically oriented tubes with bulbous pouches or pockets at the base (Fig. 10B). The tubes are about 8 cm long, 5 mm wide at the top, and as much as 2 cm wide at the base. They commonly are filled with claystone, siltstone, and very fine-grained sandstone. The sand-filled parts of the tube are primarily in the lower broader pouch.

Facies relations.—Several kinds of evidence indicate that the change from predominantly one facies to another is by interfingering. The strongest evidence for interfingering is provided by cores and is illustrated in Figure 6. The cores are arranged in order of increasing distance from the axis of the reservoir. The core on the left consists almost entirely of clean sandstone and represents the central-bar facies. The core in the middle contains some clean, unburrowed sandstone like the central-bar facies, but most of it consists of burrowed sandstone and shale closely mixed. This core is typical of the bar-margin facies. The core at the right is typical of the nonproductive interbar facies and consists of dark mudrock and a few thin beds of quartzitic sandstone. The three cores, seen together, provide evidence of a progression from reservoir rock to nonreservoir rock by interfingering.

Other evidence for interfingering of reservoir and nonreservoir rock types is based on the

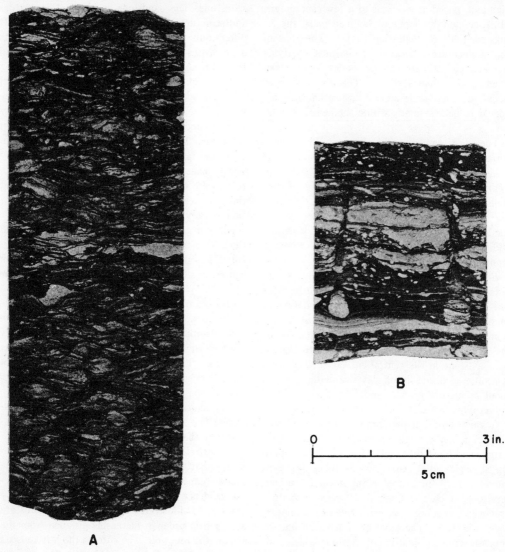

Fig. 10.—Burrow types in the "J₁" member, marine-bar area. Sample A shows *augen* burrows which are best developed along periphery of central-bar facies. Sample B shows large pouch burrows and very small tube burrows (white specks) seen in interbar facies.

progressive change in the character of the electric logs. For example, Figure 11 is an east-west section across the Willson Ranch field with the top of the "J" unit as datum. The section approximately parallels the long axis of the upper "J₁" bar and is transverse to the lower "J₁" bar. The thicknesses, net sandstone contents, and average spontaneous potential amplitudes change progressively within each interval. There are no marked discontinuities.

The bar-margin facies is peripheral to the central-bar facies, forming a halo around it. This facies is developed best in areas where net sandstone thickness is between 5 and 10 ft (1.5 and 3.0 m). Within this facies, the degree of burrowing increases from a minimum near the mudrock to a maximum adjacent to the central-bar facies, where burrowing apparently ends abruptly. It is in the most severely burrowed area next to the central-bar facies that the *augen* form of burrow predominates.

The progressive change in clay content at bar

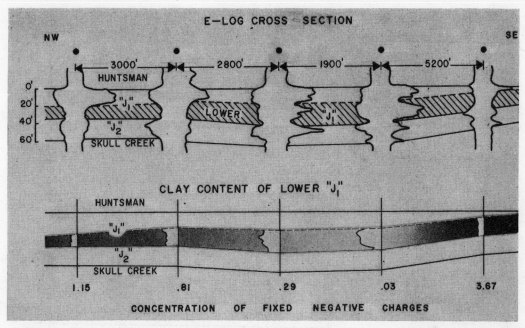

Fig. 11.—Electric-log cross section through Willson Ranch field, Banner County, Nebraska, showing marine bars in both upper and lower "J_1." Line of section is indicated on Figure 12. Lower "J_1" cross section traverses complete bar from interbar facies at the left, through bar-margin, central-bar, and bar-margin facies, to interbar facies at right. Lower cross section shows same lower "J_1" unit. Concentration of fixed negative charges (gm equiv. per l), calculated from self-potential curve, is proportional to amount of clay present. Intensity of shading within lower "J_1" also is proportional to amount of clay present. Lower clay, higher sandstone region of marine-bar reservoir appears in center of section.

peripheries can be demonstrated semiquantitatively by calculating clay content from spontaneous potential logs. The lower, shaded cross section in Figure 11 is drawn through the same wells as the upper cross section. Clay content, expressed as the concentration of fixed negative charges, has been calculated according to techniques suggested by De Witte (1955) and Slack and Otte (1960). The values computed for the lower part of the "J_1" are shown below each well. In addition, the lower "J_1" has been shaded on the basis of these numbers; darker shading represents greater clay content. Clay content increases with distance from the bar axis.

The same technique can be used to map the distribution of clay, and Figure 12 shows such a map prepared for the lower part of the "J_1" at the Willson Ranch field. The gradients in clay content can be detected for distances of 1–2 mi beyond field boundaries and can be useful in guiding exploration toward low-clay (high-sand) areas.

All lines of evidence indicate systematic and progressive changes in the proportions of sandstone and mudrock within both the upper and lower parts of the "J_1" and indicate facies changes by interfingering. Presumably the bar-margin facies was deposited originally as thin interbeds of sandstone and mudrock near the edges of the main sandstone bodies. Postdepositional modification by burrowing organisms churned these beds into mottled mixtures of sandstone and shale. The thin quartzitic sandstone stringers within mudrock are identical to the central-bar facies except in degree of cementation, and probably represent the outermost extensions of the sandstone bodies.

Distribution of facies.—The net sandstone isopachous maps for the upper and lower "J_1" show a distinct depositional "grain" or trend. The sandstone bodies in the upper part are elongated NW-SE (Fig. 4). Similar deposits in the lower part generally are aligned NE-SW (Fig. 5). Hence the upper and lower "J_1," which are lithologically very similar, have sandstone trends differing in orientation by nearly 90°.

The depositional patterns are apparent only if the two units are differentiated effectively by

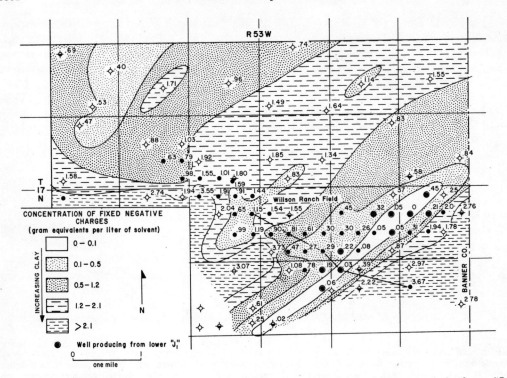

Fig. 12.—Contour map showing concentration of fixed negative charges (clay content) for lower "J_1" in part of Willson Ranch field, Banner County, Nebraska. Gradients toward field can be recognized beyond productive limits of reservoir. NW-SE-trending cross section line is Figure 11.

proper correlation. Correlation in this case is based on a thin, shaly bed between the upper and lower sandy parts that is present in most of the area. The unit is a medium- to dark-gray clay shale with a few very thin quartzitic sandstone stringers. It is expressed characteristically on electric logs by the lowest spontaneous potential and resistivity values found in the "J_1" member (Fig. 11). Where core material is available, this shale separates overlying sandstone beds that contain trace amounts of green glauconite from underlying sandstone beds that contain tan clay aggregates of unknown mineralogy and very small amounts of glauconite. This contrast in glauconite and clay-aggregate content is the single known mineralogic difference within the "J_1," but it supports the stratigraphic value of the intervening shale bed. In the Willson Ranch field, the shale unit has an average thickness of 4 ft (1.2 m) but is locally thinner, perhaps because of erosion. The exact relation of this shale marker bed to beds 20 mi (32 km) east in the valley-fill area is uncertain, but the probable correlation is suggested in Figure 3.

Several of the sandstone bodies are 3–5 mi (4.8–8.0 km) long and 0.5–1.0 mi (0.8–1.6 km) wide, although there is a range of sizes. Their present cross-sectional shape is lenticular, but it is not possible to determine their exact shape at the time of deposition and before the compaction of the surrounding mudrock. In map view, the long axes of the sandstone bodies generally do not fall along a single line. Rather, the bodies are scattered at lateral spacings of about 2–3 mi (3.2–4.8 km).

"J_1" PALEONTOLOGY

Fossil remains found in the "J_1" member include foraminifers, radiolarians, sponge spicules, palynomorphs, brachiopods, teeth and bones of presumed fish origin, and plant fragments. Only the foraminifers were studied in detail. Most foraminifer species recovered are arenaceous, although calcareous forms in small numbers were found in two localities. Of 40 samples in the "J₁" member, 15 were barren, 16 were moderately fossilferous, and 9 had abundant specimens. The types of foraminifers identified from the "J" unit and Huntsman Shale are summarized in Figure 13.

Foraminifers were studied from the lower "J₁" in off-bar areas around the Willson Ranch reservoir, in the search for faunal differences that might suggest environmental differences between the two sides of the sandstone body. All samples contained a similar fauna, which is indicative of similar environmental conditions on both sides of the body. This observation suggests that the bar did not effectively separate a sea and lagoon, although core material to test this possibility is scant. There is no evidence, such as root zones or partly weathered glauconite, that the sandstone bodies ever became emergent.

Other fossil material found in the "J" unit includes fragments of the phosphatic brachiopod *Lingula*. These fragments are a common component of the phosphatic sandstone at the "J"-Huntsman contact, but also are found scattered throughout the sandstone and mudrock of the "J₁" member. In places, the large size and angularity of the fragments indicate that the shells have not been transported appreciable distances. Sponge spicules, fish teeth and bones, spores, pollen, and other plant debris are present within the mudrock of the "J₁." Plant debris also is seen as imprints showing vascular structure on carbonaceous or shaly partings within the central-bar facies. Radiolarians are common within the Huntsman Shale, but only one specimen was found in the "J₁" member.

STRUCTURE

A structure map, with the top of the "J" unit as datum, is in Figure 14. A block diagram of the same area is shown in Figure 15. The contour horizon on Figure 14 is the contact of the thin phosphatic sandstone with the overlying Huntsman Shale, a marker easily selected from spontaneous potential and resistivity logs. Regional dip is southwest or west at a rate of about 25 ft/mi (4.8 m/km). This regional dip is modified locally by weak west-plunging anticlinal noses south of the Singleton field and through the Juelfs field. A small northeast-trending "high" coincides with the lower "J₁" reservoir in the Willson Ranch field. However, structural closure is only about 25 ft, less than half the height of the oil column in the reservoir. This "high" probably reflects the super-

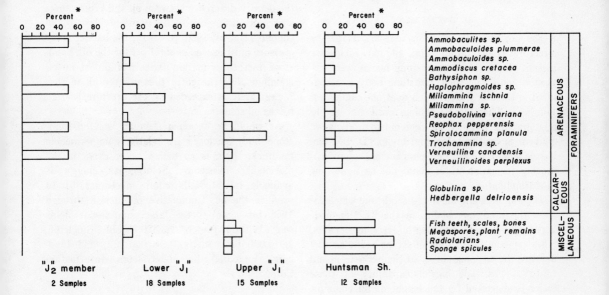

* Percent of productive samples in which species is present.

FIG. 13.—Microfossils of "J" unit, marine-bar study area. Identifications made by C. H. Ellis.

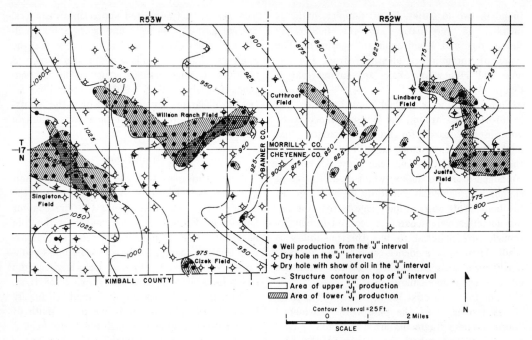

Fig. 14.—Subsea structure contours, datum top of "J" unit, marine-bar area, Nebraska. Depth datum is sea level. Upper and lower "J₁" productive areas are shown by hatched patterns. Prdouction clearly is not related to structural closure. CI = 25 ft.

position of upper on lower "J_1" bars and their influence on compaction.

OIL ENTRAPMENT

The positions of oil fields are not related to structural configuration (Figs. 14, 15). Fields are present where thicker sandstone bodies were deposited; only one such sandstone body is recognized in the study area which does not contain oil (Fig. 4). Therefore, trapping must be largely stratigraphic. Because the regional hydrodynamic gradient is relatively low and updip (D. B. MacKenzie and H. K. van Poollen, personal commun.), moving formation water could not be important in entrapment.

The mudrock and burrowed sandstone which enclose the sandstone bodies in the "J_1" member apparently serve as effective oil-trapping barriers. Silica cementation of thin sandstone beds also may improve the trapping ability of the bar-margin and off-bar facies. The barrier capability of these facies is suggested by the height of oil columns. The top of the upper "J_1" reservoir sandstone at the east end of the Willson Ranch field is at −943 ft (−288 m) and the oil column extends westward through the tilted body to −1,017 ft (−310 m), or a vertical distance of at least 74 ft (22.6 m). If the density difference between oil and brine causes a pressure difference of 0.095 psi per foot of oil column (Harms, 1966, p. 2141), the barrier entry pressure must exceed about 7 psi for the oil. Sandstone in the bar-margin facies has almost no oil staining; thus the entry pressure for oil in that facies rarely is exceeded by the pressure exerted by the oil column.

As would be anticipated in an isolated sandstone body enclosed by relatively impermeable mudrock, there is no active water drive in the marine-bar reservoirs. Solution-gas energy is available, but all wells require mechanical lift to produce the oil. Cumulative production through 1966 for some of the larger marine-bar fields was: Willson Ranch, 7,370,000 bbl; Singleton, 7,940,000 bbl; Juelfs, 4,280,000 bbl; and Lindberg, 1,426,000 bbl (Petroleum Information Corp., 1967).

VALLEY-FILL RESERVOIRS

The valley-fill area, 20 mi (32 km) east of the marine-bar area, is shown on Figure 1. The stra-

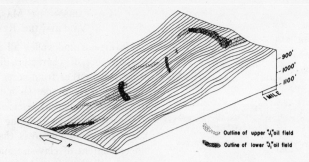

Fig. 15.—Isometric block diagram of marine-bar study area with productive limits outlined for upper and lower "J_1" reservoirs. Structural relief is represented by lines formed by intersection of top of "J" unit with regularly spaced east-west vertical planes. Oil accumulations are independent of structural closure.

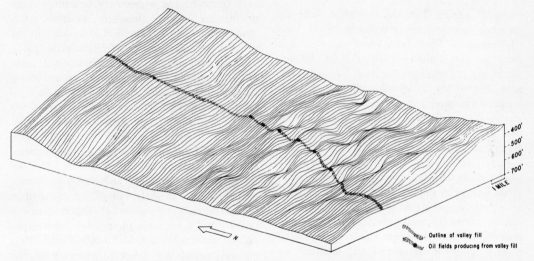

Fig. 16.—Isometric block diagram of valley-fill study area. Structural relief is represented by lines formed by intersection of top of "J" unit with regularly spaced east-west vertical planes. Oil accumulations are present where ribbon of valley-fill rocks crosses northwest-plunging anticlinal noses.

tigraphy and oil fields of the "J" unit in this area were described by Harms (1966). Only the major conclusions are reviewed briefly for comparison of marine-bar and valley-fill reservoirs.

The "J" unit in most of the valley-fill study area closely resembles that of the marine-bar area; it can be divided conveniently into "J_1" and "J_2" members which are exact equivalents of the "J_1" and "J_2" members in the marine-bar area. Each member is predominantly marine, relatively thin, and widespread, and is comparable with the corresponding member in the marine-bar area (Fig. 3). There are two notable differences: (1) after deposition of the marine "J_1," the area emerged and a stream cut a narrow valley which was filled mainly with sandstone, and (2) no marine sandstone bodies in the "J_1" member of the valley-fill area are known to be as thick or as large areally as those in the marine-bar area.

The major oil reservoirs in the valley-fill area are within the stream-deposited sandstone. The valley fill can be traced as a narrow ribbon for nearly 20 mi (32 km) along a north-south trend (Fig. 16). Within this trend, the marine "J_1" and "J_2" members have been totally or partly removed by erosion. Figure 16 illustrates the distribution and geometry of the valley-fill rocks. The valley fill is at least 20 mi (32 km) long, about 2,000 ft (610 m) wide for most of its length, and ranges in thickness from 40 to 80 ft (12 to 24 m).

The sedimentary rocks of the valley fill are distinctly different from those of the "J_1" member in many important respects, although they are similar in others. Mechanical logs of typical

ROCK PROPERTIES

	MARINE BAR	VALLEY FILL
MINERALOGY	GLAUCONITE PRESENT	GLAUCONITE ABSENT
TEXTURES	CLAY CONTENT DECREASES UPWARD PEBBLES RARE	CLAY CONTENT INCREASES UPWARD PEBBLES COMMON
STRUCTURES	LOW-ANGLE CROSS STRATA SYMMETRIC RIPPLES BURROWED	HIGH-ANGLE CROSS STRATA ASYMMETRIC RIPPLES DEFORMED
PALEONTOLOGY	MARINE MICROFOSSILS, POLLEN, SPORES	SPORES, POLLEN

FIG. 17.—Comparison of rock properties of marine bars and valley fill.

valley-fill sections lack the consistent form and sequence of deflections recorded in the "J_1" and "J_2" members. Shaly beds are most common in the upper part of the valley fill. Perhaps the most significant and readily recognized lithologic characteristics of the valley-fill rocks are the sedimentary structures and textures. High-angle cross-stratification, with an average dip of 20°, is common in all valley-fill sandstone (Fig. 9). These larger cross-stratified sets range in thickness from 3 in. (7.5 cm) to 2 ft (61 cm). Cross-laminae in places grade into underlying small-scale cross-strata formed by ripples. The form of these small-scale sets indicates that the ripples were asymmetric and had downstream faces sloping more than 32°, typical of current ripples. Where sandstone is closely interstratified with mudrock, deformational structures are common but organic structures are very scarce.

The valley-fill sandstone commonly bears siderite or mudrock pebbles, some an inch or more in diameter, although in texture and mineralogic character the sand-size material is very much like "J_1" sandstone. No glauconite has been observed; no foraminifers or macrofossils have been recovered from the valley fill.

Entrapment of oil in the valley fill requires a suitable combination of stratigraphy and structure. Oil accumulations are present in the valley trend where it crosses the axes of northwest-plunging anticlinal noses (Fig. 16). Shale and sandstone of the "J_1" and "J_2" members provide updip barriers for entrapment. The continuity of the valley-fill trend between reservoirs provides a flow conduit to formation water. As a result, production energy is derived partly from an active water drive.

COMPARISON OF MARINE-BAR AND VALLEY-FILL RESERVOIRS

Marine-bar and valley-fill reservoirs each have distinct sets of characteristics that permit their recognition either from cores or logs.

Certain rock properties indicate the marine-bar or valley-fill origin of reservoirs (Fig. 17). Mineralogic study shows that the marine bars contain some glauconite, whereas the valley fill contains none. Content of clay-size material commonly decreases upward through the marine bar and pebbles are scarce. In the valley fill, clay content increases upward in many wells and pebbles are very common. The sedimentary structures in the marine bar include abundant low-angle cross-stratification and symmetrical, rounded ripples, whereas the valley-fill rocks are characterized by common high-angle cross-stratification and asymmetric, steep ripples. Both the textural and stratification characteristics of these rocks reflect depositional process, and the same characteristics can be used to interpret shallow-marine and fluvial rocks in areas other than western Nebraska. Other useful sedimentary structures are burrows, which are common in the bar-margin facies but scarce in the valley fill, and slump structures, commonly seen in shaly valley fill but absent in the marine bar. The marine bar contains both calcareous and arenaceous foraminifers, spores and pollen, and the brachiopod *Lingula*. Foraminifers and brachiopods are absent in the valley fill, but spore and pollen grains are present.

Striking differences are seen in the geometry of the two types of reservoirs (Fig. 18). The marine bars are 0.5–1.5 mi (0.8–2.4 km) wide, less than 6 mi (9.6 km) long, and about 25 ft (7.6 m)

RESERVOIR GEOMETRY

	MARINE BAR	VALLEY FILL
WIDTH	~0.5–1.5 MILES	~0.3–0.4 MILES
LENGTH	< 6 MILES	> 20 MILES
THICKNESS	25 FEET MAXIMUM	80 FEET MAXIMUM
FORM	SCATTERED ELLIPTICAL LENSES	SINGLE LONG PRISM
LATERAL CHANGE	DEPOSITIONAL, GRADATIONAL	EROSIONAL, ABRUPT
BASAL CONTACT	TRANSITIONAL OR SHARP	EROSIONAL

FIG. 18.—Comparison of reservoir geometry of marine bars and valley fill.

in maximum thickness (Fig. 15). The shape is that of an elliptical lens. The lenses are scattered, generally with no alignment in chainlike patterns. The lateral boundaries are depositional and gradational by interfingering. The basal contact may be either transitional through a small interval or sharp. The valley fill, however, is about 0.3–0.4 mi (0.5–0.6 km) wide, has a maximum thickness of about 80 ft (24 m), and has the form of a single, long prism (Fig. 16). The lateral margins of these sandstone bodies are erosional and abrupt. The basal contact is erosional. The forms of the two reservoir types are thought to be typical of similar reservoirs in other areas. The size or scale of these bodies, however, probably is controlled by local setting; bodies of similar origin can be larger in other areas.

Characteristics of these reservoirs which might be used to guide exploration efforts are shown in Figure 19. In the marine bar, trapping is purely stratigraphic and structure need not be considered in exploration for this type of trap. The barrier rocks are adequate, commonly trapping an oil column greater than 100 ft (30.5 m). Proximity of undiscovered reservoirs perhaps can be detected by gradients in clay content established in nearby wildcat wells. Because of its width, the marine bar offers a reasonably large target for exploration. The location of additional marine-bar reservoirs on the basis of trend projection is uncertain, because these reservoirs generally are not aligned. In exploration for valley-fill reservoirs, the structural setting must be understood because an appropriate combination of structure and stratigraphy is required to form a trap. The barrier facies can be less effective than that of the marine bar. The updip facies in the study area traps an oil column of less than 50 ft (15.3 m). There is no indication of proximity to a valley fill, be-

PRODUCTION CHARACTERISTICS

	MARINE BAR	VALLEY FILL
MAXIMUM EXPECTED IN-PLACE RESERVES	40 MILLION BBL	10 MILLION BBL
ENERGY	SOLUTION GAS	WATER DRIVE
PRIMARY RECOVERY	LOW	HIGH
RESERVOIR HOMOGENEITY	HIGH	LOW

FIG. 20.—Comparison of production characteristics of marine bars and valley fill.

cause the enclosing facies are slightly older and are unrelated genetically to the valley fill. The width of the valley fill described herein is exceedingly small, in places only 1,000 ft (305 m), and provides a small target. Once a valley fill has been located in the study area, the trend projection is good because the valley trends are very linear, although the individual channels which occupy valley floors are generally sinuous.

Production characteristics of the two reservoir types are shown in Figure 20. The maximum in-place reserves in a marine-bar may be as much as 40 million bbl. The reserves of individual valley-fill reservoirs are much smaller and cannot be expected to exceed about 10 million bbl. These reserve estimates are dependent on scale, and hence apply only to the part of the Denver basin discussed. In similar reservoirs elsewhere the maximum reserves might be very different.

Production characteristics which might be more generally applicable to areas outside the Denver basin are those related to energy, primary recovery, and reservoir heterogeneity. In the marine bars, the reservoir energy is derived mainly from solution gas because surrounding shale is impervious, and as a result primary recovery is low. The energy in the valley-fill reservoir is provided by water drawn from a larger area, and higher primary recovery is possible. Within the marine-bar reservoirs the distribution of rock types is relatively simple; a central core of clean sandstone grades outward to shale. The upper part of marine bars has less clay and therefore may be more porous and permeable than the lower part. The relatively high and predictable homogeneity simplifies engineering and secondary recovery techniques. Reservoir homogeneity in the valley fill, however, can be low, because litho-

EXPLORATION GUIDELINES

	MARINE BAR	VALLEY FILL
ENTRAPMENT	STRATIGRAPHIC	STRUCTURE REQUIRED
BARRIER LITHOLOGY	COMMONLY ADEQUATE, OIL COLUMNS >100 FT	LESS ADEQUATE, OIL COLUMNS <50 FT
PROXIMITY	GRADIENTS IN CLAY CONTENT	NO INDICATION
TARGET WIDTH	~0.5–1.5 MILES	~1000–2000 FEET
TREND PROJECTION	UNCERTAIN	GOOD

FIG. 19.—Comparison of exploration guidelines for marine bars and valley fill.

logically the valley fill consists of both sandstone and shale distributed by complex sequences of channel cutting, filling, and abandonment. In contrast to the marine bar, the valley fill generally has more clay in the upper part, hence the best porosity and permeability are commonly near the bottom.

REFERENCES CITED

Boardman, A. C., 1959, Willson Ranch field, Banner County, Nebraska, *in* Symposium on Cretaceous rocks of Colorado and adjacent areas: Rocky Mountain Assoc. Geologists, p. 141–142.

De Witte, L., 1955, A study of electric log interpretation methods in shaly formations: Jour. Petroleum Technology, v. 7, p. 103–110.

Harms, J. C., 1966, Stratigraphic traps in a valley fill, western Nebraska: Am. Assoc. Petroleum Geologists Bull., v. 50, no. 10, p. 2119–2149.

MacKenzie, D. B., and D. M. Poole, 1962, Provenance of Dakota Group sandstones of the Western Interior, *in* Symposium on Early Cretaceous rocks of Wyoming and adjacent areas: Wyoming Geol. Assoc. 17th Ann. Field Conf. Guidebook, p. 62–71.

Petroleum Information Corporation, 1967, 1966 Resume, oil and gas operations in the Rocky Mountain region: sec. c, pt. 2, p. 29–37.

Slack, H. A., and C. Otte, 1960, Electric log interpretation in exploring for stratigraphic traps in shaly sands: Am. Assoc. Petroleum Geologists Bull., v. 44, no. 12, p. 1874–1894.

BELL CREEK FIELD, MONTANA: A RICH STRATIGRAPHIC TRAP[1]

ALEXANDER A. McGREGOR[2] AND CHARLES A. BIGGS[3]

Denver, Colorado 80202 and Billings, Montana 59101

ABSTRACT

The Bell Creek field, one of the most important oil discoveries in the United States in 1967, is in T8 and 9S, 54, and 55E, in Powder River and Carter Counties, southeastern Montana. Daily production, currently 50,000 bbl of oil, should rise to 65,000 bbl, outstripping that of any other Rocky Mountain oil field. Reserves are estimated at more than 200 million bbl.

The field is a Lower Cretaceous stratigraphic trap uncontrolled by structure, along the gently dipping east flank of the Powder River basin. Located near the intersection of northeast-trending littoral marine bars and a northwest-trending delta system (evident from earlier regional stratigraphic studies), the Muddy Sandstone trap was formed in a complex assemblage of local, shallow-water, nearshore environments during a regressive phase between two major advances of Early Cretaceous seas. Very porous and permeable sandstone bodies pinch out eastward, contain organic matter, and are underlain and overlain by organic-rich marine shale, providing an ideal trap for indigenous oil. The sandstone tongues were tilted upward on the east shortly after deposition as a result of the westward subsidence of the basin in which the overlying Mowry Shale was deposited. Later subsidence of the east limb of the Powder River basin further increased the size and efficiency of the trap. At least four individual reservoirs are delineated in the complex facies assemblage by oil-water and gas-oil contacts.

The Bell Creek field is an excellent example of major oil fields remaining to be found in stratigraphic traps in the sparsely drilled Rocky Mountain region. Drilling, rather than other exploration methods, is the most efficient and conclusive test of evolving economically oriented stratigraphic concepts.

INTRODUCTION

The Bell Creek field, with a daily production rate that will soon exceed that of any other Rocky Mountain oil field, was found on June 6, 1967, at a time when most major oil companies were decreasing both exploration and lease holdings in the region.

This paper describes more fully the geologic aspects of Bell Creek field set forth in a preliminary paper (McGregor, 1968), and shows that, despite numerous gaps in available geologic data, the general location of the Bell Creek field could be predicted from earlier regional stratigraphic studies and the application of "imaginative geologic thinking" (Cram, 1966, p. 829). The Bell Creek field may become a classic example of a rich stratigraphic trap.

LOCATION

Bell Creek field is in southeastern Montana on the northeastern flank of the Powder River basin,

[1] Manuscript received, May 28, 1968; accepted, June 8, 1968. Modified from a paper given at the 53rd Annual Meeting of the Association, Oklahoma City, Oklahoma, April 25, 1968.

[2] Chief geologist, Samuel Gary, oil producer.

[3] Partner, Sawtooth Oil Company.

The writers are indebted to Steven S. Oriel and Robert R. Berg for constructive suggestions that improved the manuscript.

and northwest of the Black Hills uplift (Fig. 1). The field is mainly in Powder River County but extends into Carter County, as shown in Figure 2, and is in T8 and 9S, R53, 54, and 55E. The town of Broadus, Montana, is 25 mi northwest of the field and Gillette, Wyoming, is 60 mi south of the field. Other oil fields that produce from about the same part of the section along the east flank of the Powder River basin are shown in Figure 2.

REGIONAL STRATIGRAPHY

Producing sandstone beds at Bell Creek are within the Lower Cretaceous sequence, shown in the diagrammatic cross section extending from northwest Montana to central South Dakota (Fig. 3). Two widespread marine transgressions and an intervening regressive phase are recorded in the sequence.

Seawater spread eastward very early in Cretaceous time and deposited the sand of the Fall River Sandstone on nonmarine Lakota strata, both of the Inyan Kara Group (Bolyard and McGregor, 1966). Farther seaward, dark clay beds were deposited which are now assigned on the east to the Skull Creek Shale, and on the west to the Thermopolis Shale. Even at the maximum extent of the sea, when the Skull Creek was deposited as far east as eastern South Dakota, sand beds of mixed marine and continental origin,

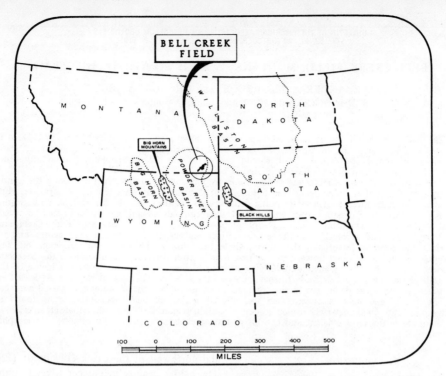

Fig. 1.—Location of Bell Creek field, Montana, Powder River basin, and major structural provinces in western United States.

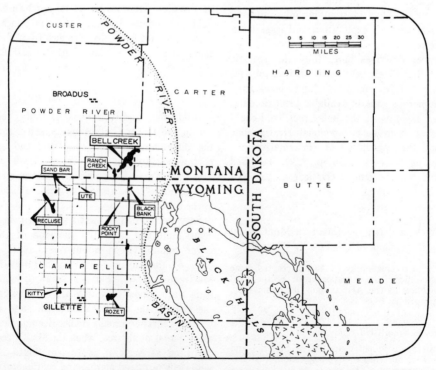

Fig. 2.—Location of Bell Creek field and nearby Muddy and Newcastle oil and gas fields along east side of Powder River basin.

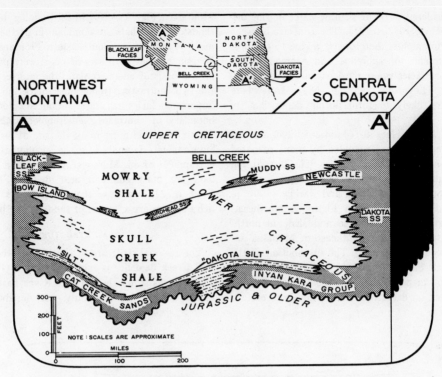

Fig. 3.—Diagrammatic northwest-southeast cross section of Lower Cretaceous rocks from northwest Montana to central South Dakota.

assigned to the Dakota Group, continued to be deposited on the southeast and east margins of the sea.

A regressive phase, critical to the relations at Bell Creek, is recorded in the westward-extending tongue of the Dakota Group (Newcastle Sandstone of the eastern part of the Powder River basin). Discontinuous sandstone bodies in about the same part of the sequence farther west are assigned to the Muddy Sandstone. It is likely that these sandstone bodies are the same age as the Newcastle, but the precise relation has not been established (Cobban and Reeside, 1952). Although the sandstone bodies are discontinuous, and therefore excellent potential traps, they lie within a distinct, literally persistent unit (Eicher, 1960, p. 25; 1962, p. 79). Two other names have been applied to sandstone in about this part of the sequence. The name Birdhead, defined by Thom et al. (1935) as a member of the Thermopolis Shale, has been used by Wulf (1962, Fig. 3, p. 1934) for sandstone beds assigned by others to the Muddy. The name Dynneson was proposed by Wulf (1962, Fig. 3, p. 1396) for a sandstone member at the base of the Mowry Shale, above a regional disconformity. Although data are inadequate to demonstrate the precise relation of the sandstone beds at Bell Creek to those at the type localities of the Newcastle, Muddy, Birdhead, and Dynneson Sandstones, the name Muddy is used in the rest of this paper.

The second marine transgression is recorded in eastward overlap of the Muddy Sandstone by the Mowry Shale. Farther east in eastern South Dakota the Mowry intertongues with the Dakota.

Various depositional environments have been inferred for the sandstones formed during the regressive phase. Units within the Newcastle Sandstone exposed in the Black Hills have been assigned continental, brackish-water, and marine origins, as summarized by Robinson et al. (1964, p. 45) and Wulf (1962, p. 1392, Fig. 16). Wulf concluded that the Newcastle was deposited in an extensive delta distributary environment. Sandstone beds exposed farther west, as in the Big Horn basin, and assigned to the Muddy, generally are considered to be of shallow-water marine origin (Eicher, 1962, p. 91). The contrast between the environments, therefore, is a clue to the possible presence of shoreline and nearshore shallow-water facies where most organic matter accumulates (Levorsen, 1967, p. 649), and where traps are likely to form during deposition (p. 575).

Even better clues to conditions favorable for oil accumulation are provided by maps prepared in regional stratigraphic studies. The distribution of Muddy sandstone beds and equivalents is shown on a map by Haun and Barlow (1962, p. 21) and is reproduced in Figure 4. Northeast-trending belts of sandstone and shale in northeastern Wyoming and southeastern Montana are conspicuous on this map. The distribution of the Newcastle Sandstone and the Skull Creek Shale is shown on maps prepared by Wulf (1962, Figs. 7, 16) based on data from the Dakotas and eastern Montana (Fig. 5). Northwest-trending belts of sandstone and shale are dominant in northeasternmost Wyoming and southeastern Montana. If it is assumed that both sets of maps were prepared on the basis of adequate data, the intersection of the two contrasting trends must mark the site of important facies changes produced along the boundary of contrasting environments (Fig. 6). The intersection must mark the site of nearshore environments favorable for the accumulation and entrapment of oil. Moreover, offshore and barrier bars are most common near deltas (Shepard, 1960, p. 220) because the distributaries of the delta system are a source of sand for the bars.

REGIONAL STRUCTURE

The regional structure of the area in which the Bell Creek field lies is relatively simple (Fig. 7). The east flank of the Powder River basin dips rather gently and uniformly west-northwest at

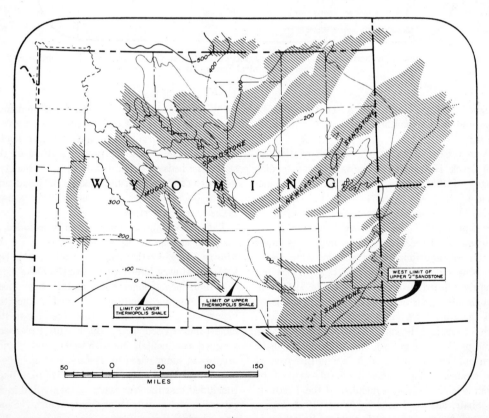

FIG. 4.—Isopach map of Thermopolis Shale showing distribution of Muddy Sandstone and equivalents in Wyoming (from Haun and Barlow, 1962, Fig. 7). CI = 100 ft.

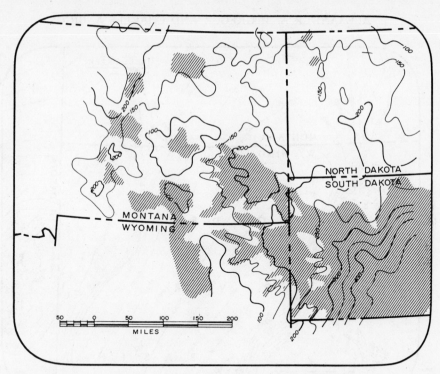

FIG. 5.—Isopach and sandstone-shale ratio map of Skull Creek Shale showing areal distribution of Newcastle Sandstone and equivalents in Dakotas and eastern Montana (from Wulf, 1962, Fig. 7). CI = 50 ft.

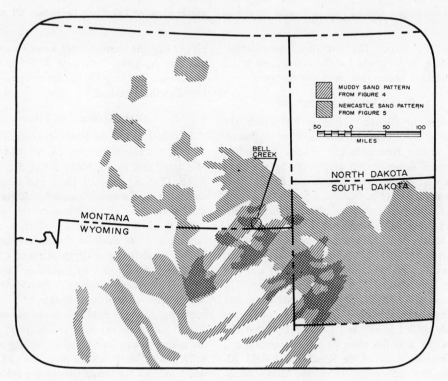

FIG. 6.—Location of Bell Creek field with respect to sandstone distribution shown in Figures 4 and 5.

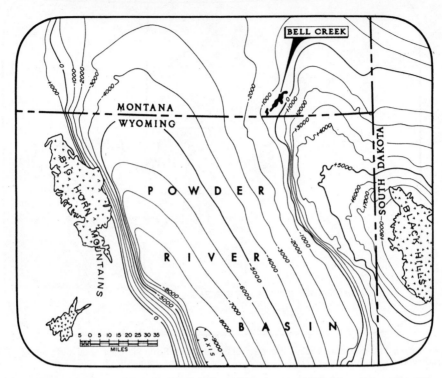

Fig. 7.—Structure contour map of Powder River basin. Datum is top of Fall River Formation. CI = 1,000 ft.

about 100 ft/mi. The structural environment, therefore, is favorable for the entrapment of oil in sandstone bodies that pinch out eastward.

Other Favorable Factors

Before the discovery of oil at Bell Creek, the presence of commercial quantities of hydrocarbons in the Newcastle and Muddy Sandstones had been demonstrated amply in the small oil fields west of the Black Hills, from Rocky Point to Ranch Creek (Fig. 2).

The exploration program that led to the discovery of Bell Creek uncovered additional favorable factors in southeastern Montana (Gary and McGregor, 1968): (1) locally thick, porous, permeable sandstone beds in the Muddy and Fall River Formations in Powder River County; (2) oil stain and a little free oil observed in thinner and less permeable sandstone beds in some previously drilled holes; and (3) the fact that the Bell Creek area lies along both the structural and stratigraphic strike of the Ranch Creek field. Although four wells completed in the Muddy Sandstone at Ranch Creek averaged about 80 bbl of oil per day and showed rapid water encroachment and five dry offset tests were drilled in early development, it is very likely that more oil could be found on the northeast.

Bell Creek Oil Field

The Bell Creek field was discovered June 6, 1967 by the Exeter Drilling Co. No. 33-1 Federal-McCarrell test in the NE¼ NE¼ Sec. 33, T8S, R54E (Fig. 8). Samuel Gary and Consolidated Oil and Gas Company supported Exeter in the drilling of the well. A drill-stem test recovered 2,760 ft of 30° API gravity oil from a porous sandstone 27 ft thick in the upper part of the Muddy Sandstone at a depth of about 4,500 ft.

Shallow depths and easy drilling have led to rapid development of the field. By mid-April 1968, 300 productive wells on 40-acre sites had been completed, and the field was producing more than 50,000 bbl of oil per day. If additional facilities are added to the four pipelines serving the field, daily production should be more than 65,000 bbl, which exceeds the production of any other Rocky Mountain oil field. Proved reserves, including sec-

BELL CREEK FIELD, MONTANA: A RICH STRATIGRAPHIC TRAP

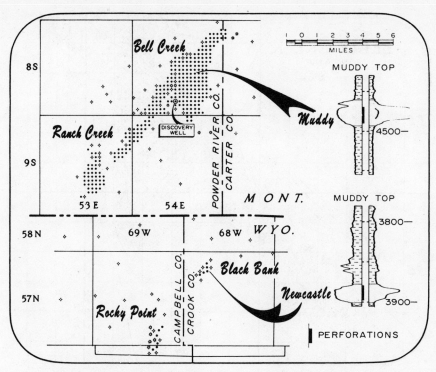

FIG. 8.—Map showing relation of Bell Creek to nearby Ranch Creek, Black Bank, and Rocky Point fields. Electric logs show Bell Creek and Ranch Creek fields produce from upper Muddy Sandstone; Black Bank and Rocky Point fields produce from lower Muddy Sandstone or from Newcastle Sandstone.

ondary recovery, are estimated to be 200 million bbl of oil.

As of April 1968, the Bell Creek field is about 15 mi long and 3.5 mi wide at its widest point (Fig. 8). More than 15,000 acres has been proved productive. Initial oil production rates of wells range from 50 to 1,500 b/d, and the average is 500 b/d of 34° API, sweet, intermediate-base oil. Abnormally low bottom-hole pressures (about 1,190 psi) require artificial lift on all wells.

About 30 wells are being completed each month under current field operations. Active development drilling is concentrated on the east, northeast, and southwest. Only on the northwest, where water is present in the producing sandstone, has the limit of the field been delineated fairly well.

The Bell Creek field has demonstrated amply that it was the most notable discovery made in the United States in 1967. Oil production in Montana has increased by more than 60 percent, contributing greatly to the economy of the state. Moreover, the field is an outstanding example of giant oil fields remaining to be found in the Rocky Mountain region at reasonable cost by exploration-oriented companies. The minimum return on investment estimated for Bell Creek is 20:1, which makes the field as economically attractive as the larger oil fields of the world.

LOCAL STRATIGRAPHY

Rapid development of the Bell Creek field has resulted in the acquisition of a very large volume of geologic data, only part of which has been sifted and analyzed. Thicknesses of the Muddy Sandstone in the field are shown on the isopachous map in Figure 9. Combined thickness of sandstone units in the Muddy ranges from 20 to 30 ft in most of the field, but locally is more than 40 ft and thickens westward. Thickness trends parallel the structural strike and the sandstone units pinch out southeastward, which suggests that the sandstone beds were parts of northeast-trending barrier bars. Sandstone is also absent along northeast-trending belts within the area, possibly representing lagoonal deposits. Narrow transverse or

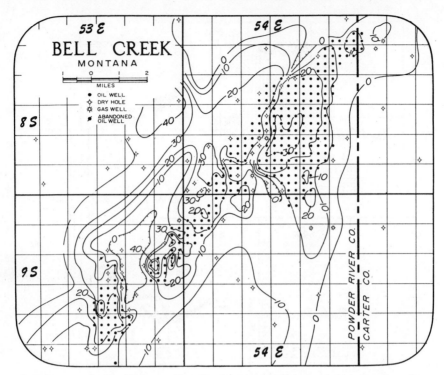

Fig. 9.—Isopach map of gross thickness of Muddy Sandstone in Bell Creek field; based on data available mid-April 1968. CI = 10 ft.

northwest-trending belts of thin sandstone may represent inlets. The southeastward-extending lobes may represent overwash fans into lagoonal areas or they may be part of a delta system.

Properties of the producing sandstone at Bell Creek are well illustrated by the electric log and core analysis shown in Figure 10. The sandstone is 30 ft thick in this well and has porosity greater than 30 percent and a permeability of 10 darcys. Residual oil saturation is about 20 percent, whereas water saturation is markedly varied. However, these oil and water saturation values probably are related more closely to the type and quality of drilling fluid and to the amount of flushing during coring than to properties of the reservoir.

The stratigraphic position of the producing sandstones differs among the nearby fields and within the Bell Creek field itself. The electric logs in Figure 8, for example, show that production in Bell Creek is from sandstone in the upper part of the Muddy Sandstone, whereas at Black Bank it is from the lower Muddy or Newcastle.

Electric logs must be used with caution in stratigraphic and other studies because resistivity is greatly varied among wells within the reservoir or within a single well. Resistivity ranges from 8 ohms to more than 200 ohms within a particular oil reservoir. The resistivity of produced water ranges from 0.25 to 2.2 ohms and is related inversely to the proportion of clay in the sandstone body. Therefore, accurate evaluation of electric logs requires a knowledge of the proportion of clay disseminated through a potential reservoir.

Diagrammatic sections across the Bell Creek field, both from southwest to northeast and from northwest to southeast (Fig. 11), illustrate the lenticular geometry of the main producing sandstone bodies. These bodies may be even more discontinuous than is shown. Drilling indicates at least four different positions for gas caps and two for oil-water contacts. Thus, the field is believed to have at least four separate reservoirs.

Discontinuity of sandstone beds is shown also by the west-to-east electric-log sections in Figures 12 and 13. The sandstone shown in Figure 12 thins eastward and pinches out east of the east-

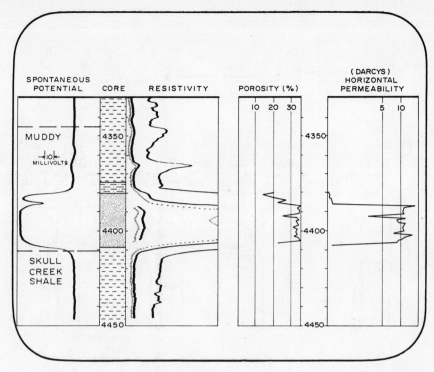

Fig. 10.—Electric-log and core analysis of typical section of productive Muddy Sandstone in Bell Creek field.

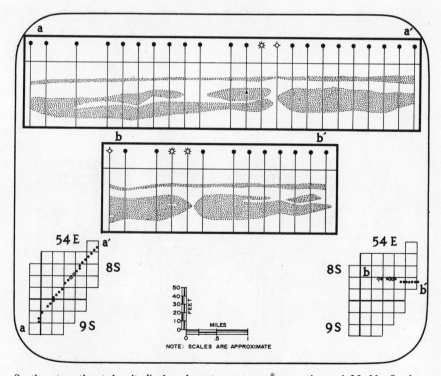

Fig. 11.—Southwest-northeast longitudinal and west-east transverse sections of Muddy Sandstone in Bell Creek field. Illustrates lenticular geometry of main producing sandstone bodies.

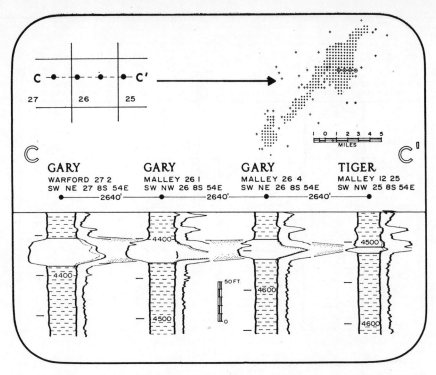

Fig. 12.—West-east cross section of Muddy Sandstone across eastern part of Bell Creek field, showing eastward thinning of reservoir rock.

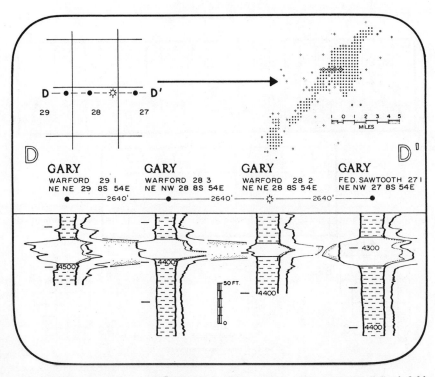

Fig. 13.—West-east cross section of Muddy Sandstone across western part of Bell Creek field showing discontinuity of reservoirs.

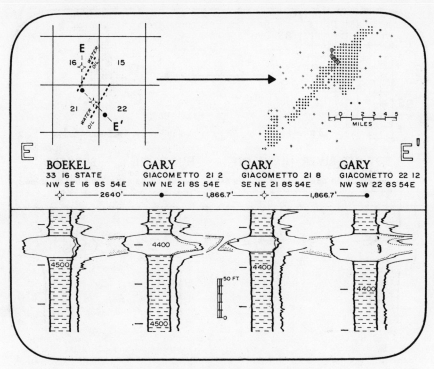

Fig. 14.—North-south cross section of Muddy Sandstone diagonally across two oil-water contacts in western part of Bell Creek field.

ernmost well shown. Figure 13, which is a western continuation of Figure 12, shows two separate reservoirs. The two wells on the west are oil wells with gas-oil ratios of 200:1. The next well east produces gas with a minor amount of condensate, whereas the next well updip on the east is an oil well with a gas-oil ratio of 200:1.

The oil-water contacts along the west side of Bell Creek field are indicated in the electric-log cross section in Figure 14. The northernmost well produced water, but the next well south produced 700 bbl of oil per day. The next well southeast produced water, whereas the easternmost well also produced 700 bbl of oil per day. These data suggest that continuity of the sandstone bodies is disrupted by pinchouts between the wells in the NW¼ NE¼ and in the SE¼ NE¼ of Sec. 21.

Discontinuity of sandstone bodies is indicated also in the longitudinal section shown in Figure 15, extending from Ranch Creek across Bell Creek.

Environments of Deposition

Analysis of data from the Bell Creek field is too incomplete to permit firm conclusions regarding the environments of deposition of producing rocks. However, a few tentative ideas may be of interest.

The available data support the general inference drawn from regional stratigraphic studies that the Bell Creek area marks the site of the intersection between northeast-trending offshore bars and a northwest-trending deltaic distributary system. The possible paleogeography for the area during the time of Muddy Sandstone deposition at Bell Creek is represented diagrammatically in Figure 16. At least four distinct local environments are suggested.

Sandstone in the McCullough No. 2 NW 32 gas well in the NW¼ SE¼ Sec. 34, T57N, R69W, Campbell County, Wyoming (Fig. 16, well No. 1), is illustrated in Figure 17 as a possible example of a meandering-channel facies. The sandstone is white with very fine angular to subangular grains, silty and partly clayey, and includes a trace of finely divided plant fragments, some dark shale laminae, scattered coal fragments, and a few tan siderite veins and pellets. Porosity and

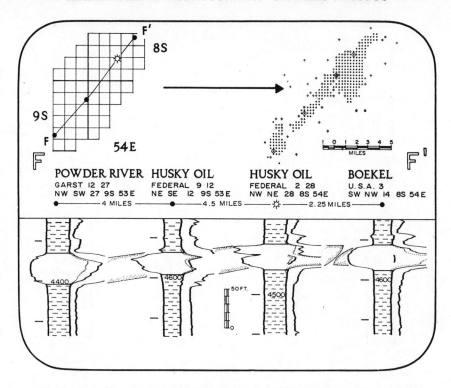

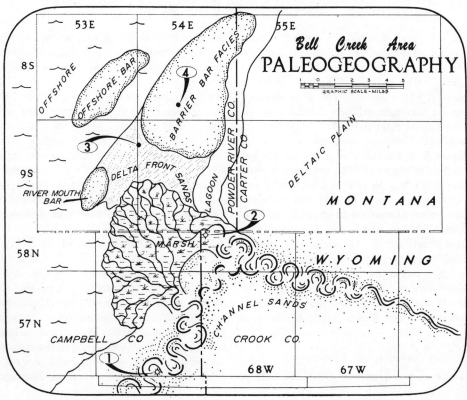

FIG. 15.—Southwest-northeast longitudinal cross section of Muddy Sandstone across Ranch Creek and Bell Creek fields.

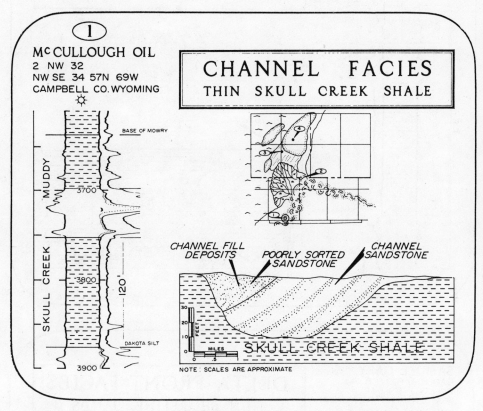

FIG. 17.—Electric log and generalized west-east cross section of meandering-channel deposit. Log is well 1, Figure 16.

permeability values are poor to fair. The grain size increases downward and the contact with underlying claystone is sharp. The Skull Creek Shale is only 120 ft thick here, 50 ft thinner than in nearby areas. Whether this thinning represents actual removal of shale before filling of a channel by Newcastle Sandstone, as shown in Figure 17, or is associated with facies changes, has not been established. Although the Skull Creek Shale is thinner than in other areas, the total interval between the base of the Mowry and the base of the Skull Creek is remarkably uniform.

A possible marsh facies of the Newcastle Sandstone is illustrated in Figure 18, which shows the sandstone in the Exeter Drilling Company No. 19-1 Johnson dry hole in SE¼ SE¼ Sec. 19, T58N, R68W, Crook County, Wyoming (Fig. 16, well No. 2). The sandstone, possibly deposited along a distributary channel near the delta front, includes interlaminae of siltstone, lignite, and other carbonaceous material.

Sandstone in the Samuel Gary Baruch-Foster No. 7-1 oil well in SW¼ NW¼ Sec. 7, T9S, R54E, Powder River County, Montana (Fig.

FIG. 16.—Diagrammatic map of paleogeography of Bell Creek and Rocky Point areas during deposition of the Muddy Sandstone. Four distinct local environments produced deposits illustrated in Figures 17-20.

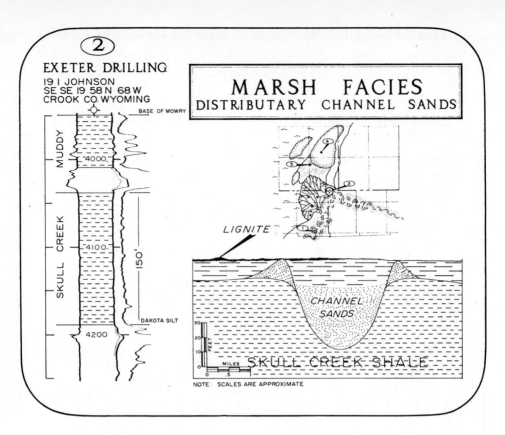

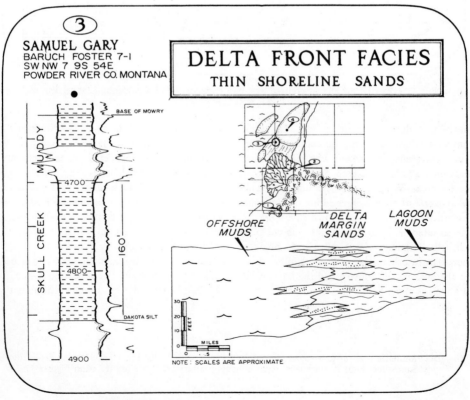

Fig. 18.—Electric log and generalized west-east cross section of distributary-channel deposit in delta-marsh facies. Log is well 2, Figure 16.

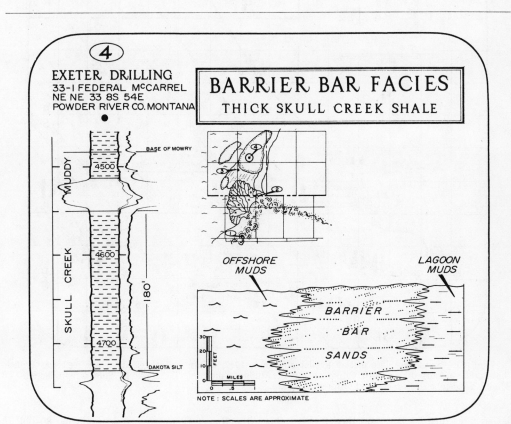

Fig. 20.—Electric log and generalized west east cross section of barrier-bar deposit. Log is well 4, Figure 16.

16, well No. 3), may be a delta-front deposit (Scruton, 1960) as illustrated on Figure 19. The rock consists of fine, subangular, well-sorted grains, includes variable amounts of clay, and is both indistinctly cross-laminated and ripple marked. The sandstone intertongues abruptly with mudstone and apparently consists of reworked sand detritus from the Newcastle.

Sandstone in the Exeter Drilling Company No. 33-1 Federal-McCarrel discovery well, in NE¼ NE¼ Sec. 33, T8S, R54E, Powder River County, Montana (Fig. 16, well No. 4), is illustrated on Figure 20 as a barrier-island or offshore-bar deposit. The sandstone is 30 ft thick, white, contains fine, very well-sorted, subangular grains of both clear and frosted quartz, and contains about 10 percent kaolin which increases upward. Also present in the indistinctly cross-bedded sandstone are glauconite, sparse muscovite, wood, and other carbonaceous fragments. The very porous and permeable sandstone grades downward into the underlying shale. The results of a detailed study of cores from Bell Creek field and an interpretation of the origin of the producing sandstone bod-

Fig. 19.—Electric log and generalized west-east cross section of delta-front deposit. Log is well 3, Figure 16.

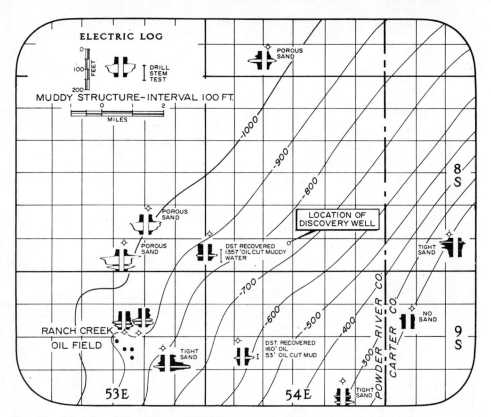

Fig. 21.—Subsurface information available June 1, 1967, before Bell Creek discovery well was drilled. Structural contours drawn with datum at top of Muddy Sandstone. CI = 100 ft. Shows reproductions of Muddy Sandstone electric logs and drillstem test data.

ies are presented by Berg and Davies (1968, this issue of the *Bulletin*).

Local Structure

Early drilling in the Bell Creek exploration program was influenced by reported local structures along the favorable stratigraphic fairway. Anticlines on the north in the Mizpah and Coalwood areas, shown on surface maps by the U.S. Geological Survey (Parker and Andrews, 1939; Bryson, 1952), had been drilled. Other tests were drilled in Custer, Powder River, and Carter Counties on reported seismic highs (Gary and McGregor, 1968). Structures suggested by photogeologic surveys also were drilled. All the holes were dry, but useful stratigraphic information was obtained.

Local information available by June 1, 1967, in addition to the published regional stratigraphic studies, is summarized on Figure 21. Four wells in the Ranch Creek area were producing 80 bbl of oil per day with substantial water. One test had been drilled in the SW¼ SW¼ Sec. 30, T8S, R54E, but this dry hole had a show of oil in the Muddy Sandstone, and a drillstem test recovered 1,357 ft of oil-cut water from a 4-ft porous interval. Tests drilled in T8S, R53E, indicated thicker and more porous sandstone. A drillstem test of a hole drilled in the NW¼ NW¼ Sec. 17, T9S, R54E, yielded 213 ft of free oil and oil-cut mud from a 3-ft porous zone.

The elimination of sites where wells could be drilled partly on the basis of structure led to the drilling of a wholly stratigraphic test. The decision was to drill no more than 3 mi from the hole that yielded free oil, along structural strike but slightly basinward stratigraphically, to test a delta-front or offshore or barrier-bar environment.

Drilling since then, summarized on the structure contour map (Fig. 22), established the absence of

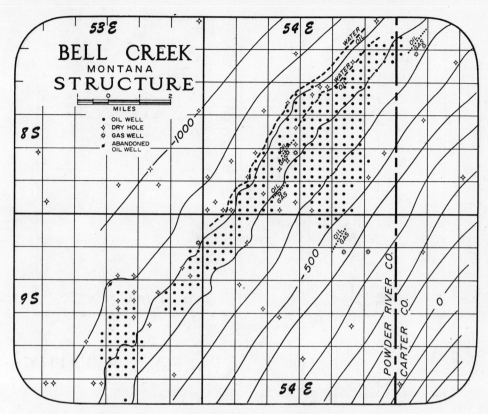

FIG. 22.—Structure-contour map with datum on top of Muddy Sandstone at Bell Creek field. CI = 100 ft. Dip of homocline is remarkably uniform.

local structural control for the field. The field is on a remarkably uniform homocline sloping northwest 100 ft/mi. The local "nosing" in the Ranch Creek area in T9S, R23E, is accentuated by a change in regional strike from just west of north in northern Wyoming to almost due northeast in this area. Also shown on the map are two relatively sharp oil-water contacts and four distinct gas-oil contacts, which suggest the presence of multiple reservoirs.

A map combining present structure with paleogeography at the time of Muddy Sandstone deposition, shown in Figure 23, perhaps best summarizes the controls for the accumulation of oil in the field. Additional data undoubtedly will lead to modifications of the local environments inferred for the area.

CONCLUSIONS

1. The Bell Creek field now produces 50,000 bbl and is capable of producing more than 65,000 bbl of oil per day. Reserves are estimated at 200 million bbl, and Bell Creek soon will produce more than any other Rocky Mountain field.

2. The field is a rich stratigraphic trap, with no prominent local structural control.

3. The stratigraphic trap formed near the strandline during a regressive phase between two major advances of the Early Cretaceous sea.

4. Local environments of deposition are represented by a complex assemblage of marginal-marine and deltaic facies, comparable with those evident in modern sediments (van Andel and Curray, 1960; Miller, 1965).

5. Updip limits of the overall reservoir are formed by lateral facies change from porous and permeable sandstone to lagoonal claystone, siltstone, and silty carbonaceous claystone.

6. The approximate position of the Bell Creek field could have been predicted by delineating the intersection of northeast-trending marine-bar sandstones and northwest-trending delta systems

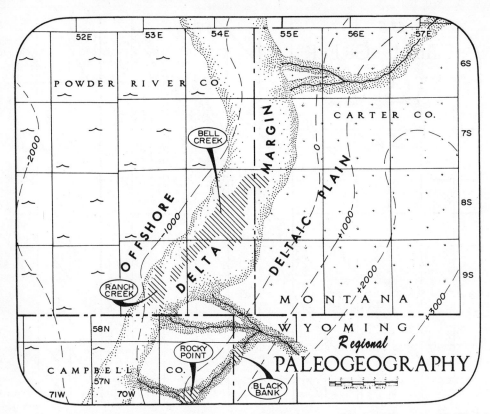

Fig. 23.—Depths of Muddy Sandstone paleoenvironments in Bell Creek and Rocky Point areas. Modified from Figures 16 and 22. Structural datum is top of Muddy Sandstone. CI = 1,000 ft.

that had been established by earlier regional stratigraphic studies.

7. At least four individual reservoirs are suggested by oil-water and gas-oil contacts.

8. The stratigraphic traps formed early. Eastward pinchouts of the sand bodies were enhanced as traps by the depositional basin sinking on the west, as shown by isopachous maps of the overlying Mowry Shale.

9. Oil accumulation in the traps was enhanced further by subsequent development of the Powder River basin during which the sandstone tongues were inclined to still more favorable positions.

10. The sources of hydrocarbons are the local sediments; the subjacent and superjacent marine shales, and the reservoir rock itself are rich in organic material.

11. Regional studies of the Cretaceous System in the western interior indicate that many additional favorable sites comparable with Bell Creek remain to be drilled. Many such sites have been tested by less than one well per township, and it is possible that major oil fields will be found.

12. The estimated return on investment at Bell Creek of 20:1 makes the field as economically attractive as the larger oil fields of the world.

13. Local surface structures in the region are not a clue to the presence of deeper structures, because Tertiary rocks at the surface are separated from underlying Cretaceous strata by an angular unconformity.

14. Apparent seismic highs in the region reflect surface topography rather than subsurface structure.

15. Shallow depths and ease of drilling make drilling the most efficient and unequivocally conclusive exploration method once the favorable geologic environments have been delineated, as at Bell Creek.

References Cited

Berg, R. R., and D. A. Davies, 1968, Origin of Lower Cretaceous Muddy Sandstone at Bell Creek field, Montana: Am. Assoc. Petroleum Geologists Bull., v. 52, no. 10, p. 1888–1898.

Bolyard, W. W., and A. A. McGregor, 1966, Inyan Kara Group, Black Hills area: Am. Assoc. Petroleum Geologists Bull., v. 50, no. 10, p. 2221–2245.

Bryson, R. P., 1952, The Coalwood coal field, Powder River County, Montana: U.S. Geol. Survey Bull. 973-B, p. 23–106.

Cobban, W. A., and J. B. Reeside, 1952, Correlation of Cretaceous formations of western interior of United States: Geol. Soc. America Bull., v. 63, no. 10, p. 1011–1044.

Cram, I. H., 1966, The oldest is the newest: Am. Assoc. Petroleum Geologists Bull., v. 50, no. 5, p. 826–829.

Eicher, D. L., 1960, Stratigraphy and micropaleontology of the Thermopolis Shale: Yale Univ. Peabody Museum of Nat. History Bull. 15, 126 p.

―――― 1962, Biostratigraphy of the Thermopolis, Muddy and Skull Creek Formations, *in* Symposium on Early Cretaceous rocks of Wyoming and adjacent areas: Wyoming Geol. Assoc. 17th Ann. Field Conf. Guidebook, p. 72–93.

Gary, Samuel, and A. A. McGregor, 1968, Exploration philosophy behind the Bell Creek oil field discovery, Powder River County, Montana: Mountain Geologist, v. 5, no. 1, p. 15–21.

Haun, J. D., and J. A. Barlow, Jr., 1962, Lower Cretaceous stratigraphy of Wyoming, *in* Symposium on Early Cretaceous rocks of Wyoming and adjacent areas: Wyoming Geol. Assoc. 17th Ann. Field Conf. Guidebook, p. 15–22.

Levorsen, A. I., 1967, Geology of petroleum: 2d ed., San Francisco, California, W. H. Freeman and Co., 724 p.

McGregor, A. A., 1968, Bell Creek oil field, Powder River and Carter Counties, Montana: Wyoming Geol. Assoc., Earth Science Bull., v. 1, no. 1, p. 29–36.

Miller, D. N., Jr., 1965, Recognition and classification of marginal marine environments, *in* Sedimentation of Late Cretaceous and Tertiary outcrops, Rock Springs uplift: Wyoming Geol. Assoc. 19th Ann, Field Conf. Guidebook, p. 209–218.

Parker, F. S., and D. A. Andrews, 1939 (1940), The Mizpah coal field, Custer County, Montana: U.S. Geol. Survey Bull. 906-C, p. 85–133.

Robinson, C. S., W. J. Mapel, and M. H. Bergendahl, 1964, Stratigraphy and structure of the northern and western flanks of the Black Hills uplift, Wyoming, Montana and South Dakota: U.S. Geol. Survey Prof. Paper 404, 134 p.

Scruton, P. C., 1960, Delta building and the delta c sequence, *in* Recent sediments, northwest Gulf of Mexico: Am. Assoc. Petroleum Geologists, p. 82–102.

Shepard, F. P., 1960, Gulf Coast barriers, *in* Recent sediments, northwest Gulf of Mexico: Am. Assoc. Petroleum Geologists, p. 197–200.

Thom, W. T., Jr., *et al.*, 1935, Geology of Big Horn County and the Crow Indian Reservation, Montana: U.S. Geol. Survey Bull. 856, 200 p.

van Andel, T. H., and J. R. Curray, 1960, Regional aspects of modern sedimentation in northern Gulf of Mexico and similar basins, and paleogeographic significance, *in* Recent sediments, northwest Gulf of Mexico: Am. Assoc. Petroleum Geologists, p. 345–364.

Wulf, G. R., 1962, Lower Cretaceous Albian rocks in northern Great Plains: Am. Assoc. Petroleum Geologists Bull., v. 46, no. 8, p. 1371–1415.

Depositional Systems in Wilcox Group (Eocene) of Texas and Their Relation to Occurrence of Oil and Gas[1]

W. L. FISHER[2] and J. H. McGOWEN[2]
Austin, Texas 78712

Abstract Regional investigation of the lower part of the Wilcox Group in Texas in outcrop and subsurface indicates seven principal depositional systems: (1) Mt. Pleasant fluvial system developed updip and in outcrop north of the Colorado River; (2) Rockdale delta system, present primarily in the subsurface, chiefly between the Guadalupe and Sabine Rivers; (3) Pendleton lagoon-bay system in outcrop and subsurface largely on the southern flank of the Sabine uplift; (4) San Marcos strandplain-bay system in outcrop and subsurface mainly on the San Marcos arch; (5) Cotulla barrier-bar system in the subsurface of South Texas; (6) Indio bay-lagoon system developed updip and in outcrop in South Texas; and (7) South Texas shelf system, an extensive system entirely within the subsurface. The Rockdale delta system consisting of large lobate wedges of mudstone, sandstone, and carbonaceous deposits, is the thickest and most extensive of the lower Wilcox depositional systems. It grades updip to the thinner terrigenous facies of the Mt. Pleasant fluvial system. Deposits of the Rockdale delta system were the source of sediments redistributed by marine processes and deposited in laterally adjacent marine systems. Delineation of depositional systems and, more specifically, delineation of component facies of the several systems permit establishment of several regional oil and gas trends which show the relation of producing fields and distribution of potentially productive trends.

INTRODUCTION

The Wilcox Group (early Eocene) is a thick sequence of predominantly terrigenous clastic sedimentary rocks which is volumetrically a significant part of the large terrigenous fill of the western Gulf Coast province. It is an economically important group of rocks, providing oil, gas, and freshwater reservoirs, as well as deposits of lignite, ceramic clay, and industrial sand.

The study area is approximately 40,000 sq mi in the Texas Gulf coastal plain, extending from the Sabine River on the east to the Rio Grande on the south in a 100- to 300-mi-wide belt including outcrop and subsurface. Subsurface control consisted of approximately 2,500 electric logs, supplemented by lithologic control based on outcrop descriptions, and cuttings and cores of strategically located wells. Fifty stratigraphic dip sections and 15 strike sections were prepared at spacings of 8 to 10 mi to determine three-dimensional geometry and relations of principal lithologic units. Emphasis in outcrop study was on the mapping of principal rock units, interpretation of primary sedimentary structures, and determination of fossil composition.

This paper is part of a continuing, larger regional study of the Wilcox Group in Texas which has been under way at the Bureau of Economic Geology for the past three years. Herein, consideration is restricted to the lower part of the Wilcox Group, including the sequence underlain by marine mudstone and claystone of the Midway Group and normally overlain by marine or lagoonal mudstone of the middle part of the Wilcox Group. The Carrizo Formation and the upper part of the Wilcox Group are not considered.

TERMINOLOGY AND DEPOSITIONAL SYSTEMS

The stratigraphic terminology used by the writers is informal. Reference is made to defined, formal stratigraphic units (chiefly in outcrop) where applicable. The terminology used derives from the concept of depositional systems or complexes, which has been employed recently, for example, in the study of modern

---

[1] Manuscript received and accepted, February 2, 1968. Parts of this paper were presented by W. L. Fisher to the Gulf Coast Association of Geological Societies, San Antonio, Texas, 1967. Present paper is republished with minor modifications from paper by Fisher and McGowen in Gulf Coast Assoc. Geol. Socs. Trans., v. 17, p. 105–125, 1967. This paper was nominated by the Gulf Coast Association of Geological Societies for republication in the AAPG *Bulletin*; publication authorized by the Director, Bureau of Economic Geology, The University of Texas at Austin, and by the Gulf Coast Association of Geological Societies.

[2] Bureau of Economic Geology, The University of Texas at Austin.

The writers thank M. J. Casey, L. F. Brown, Jr., and P. T. Flawn, Bureau of Economic Geology, and Alan J. Scott, Department of Geology, The University of Texas at Austin, for reading the original manuscript and offering many helpful suggestions. Gerald H. Baum, Texas Water Development Board, made available the extensive well-log library of that agency; numerous Gulf Coast operators provided records of specific wells. Peter U. Rodda and J. Stewart Nagle, Bureau of Economic Geology, The University of Texas at Austin, have been helpful in many ways during course of the project. Illustrations were prepared under direction of J. W. Macon, Bureau of Economic Geology, The University of Texas at Austin.

sediments and depositional environments of the South Texas coast (Hayes and Scott, 1964) and in reference to the Frio barrier bar system of South Texas (Boyd and Dyer, 1964). In studies of modern sediments emphasis generally is placed not only on description of specific environments and two-dimensional distribution of sediments and environments, but also on analysis and integration of facies; fabrication of facies tracts and three-dimensional depositional models of known depositional complexes or systems are used. These approaches are significant in the interpretation of ancient sediments, especially in regional analysis. The writers place more emphasis on lithologic and fossil composition, sedimentary structures, and distribution, geometry, and relations of component facies than on regional correlation and persistence of locally established formal stratigraphic units.

The principal unit here employed is the depositional system, recognized by specific criteria and designated by a genetic term, *e.g.*, delta system, fluvial system, barrier-bar system, *etc.* Components of a depositional system are referred to simply as facies, *e.g.*, the delta-front facies or distributary-channel facies of a delta system. Smaller scale components are designated by such terms as units, beds, or deposits. Basically, the concept of depositional systems in ancient sediments, fabricated principally from integration of facies, involves recognition of large-scale natural genetic elements, comparable to modern depositional systems readily apparent from physiographic characteristics.

PRINCIPAL DEPOSITIONAL SYSTEMS

Seven principal depositional systems are recognized in the lower part of the Wilcox Group of Texas (Fig. 1). The dominant depositional element is the large Rockdale delta system, which comprises large lobate wedges of mudstone, sandstone, and carbonaceous deposits, chiefly lignite (Fig. 1). Maximum lower Wilcox sedimentation, with sequences as much as 5,000 ft thick, took place during deposition of this system. It is present principally in the subsurface and grades up the depositional slope and up the structural dip to thinner terrigenous clastic facies of the Mt. Pleasant fluvial system, present in outcrop north of the Colorado River. A large and probably main part of the original fluvial system was deposited updip of the present outcrop, and only a part of the system is preserved in the East Texas basin.

Southwestward transport of sand from the delta system, principally by longshore currents during periods of delta destruction, and the presence of local embayments marginal to the delta system resulted in distinct depositional systems laterally associated with the delta system. These systems include, on the east, the Pendleton lagoon-bay system developed in embayments marginal to the delta system; on the southwest, the San Marcos strandplain-bay system developed proximal to the delta system; and the Cotulla barrier bar and complementary Indio lagoon-bay systems developed more distally from the main area of deltation. Contemporaneous sedimentation in systems marginal and lateral to the delta system was relatively slow, and the sequences generally are less than one-fourth the thickness of the delta system. Gulfward from the Cotulla barrier bar system and partly lateral to the Rockdale delta system is the South Texas shelf system, marked chiefly by mudstone and relatively minor sandstone.

Mt. Pleasant Fluvial System

The Mt. Pleasant fluvial system includes most of the lower part of the Wilcox Group in outcrop north of the Colorado River and in the northern half of the Sabine uplift (Figs. 1, 2). It extends down structural dip to a line running approximately from central Bastrop County through central Lee, northern Burleson, southern Robertson, central Leon, northern Houston, southern Cherokee, and central Nacogdoches Counties. South and downdip from this line the system grades into the Rockdale delta system (Fig. 3); laterally it grades into embayed marine facies developed marginally to the fluvial and delta systems. Three fairly distinct facies are preserved, recognition of which is based on (1) sandstone-body geometry, (2) sandstone-isolith pattern, (3) internal sedimentary structures, (4) vertical sequence in structures and textures, (5) grain size, and (6) relative amounts of channel and overbank deposits (Figs. 1–3). These facies reflect different fluvial environments.

Tributary-Channel Facies

In northernmost Wilcox outcrops, from central Van Zandt County northeast to Texarkana, and in outcrop in the northern part of the Sabine uplift, sandstone deposits are present in definite, separated areas (*e.g.*, Mt. Pleasant, Reilly Springs, Simms deposits); claystone and mudstone beds crop out in intervening areas. Isoliths of sandstone units outline elongate bodies with an overall dendritic tributary pat-

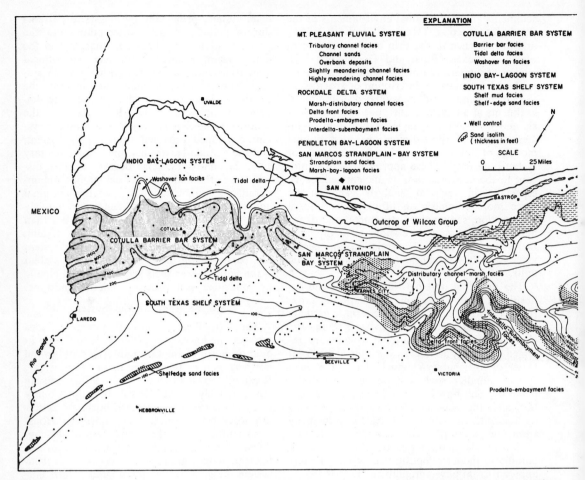

Fig. 1.—Principal depositional systems,

tern (Fig. 2). The fluvial facies defined chiefly by this pattern extends from the outcrop southward in the subsurface to a line from southwestern Smith County to west-central Rusk County. The thickness of the tributary-channel facies ranges from about 300 to 800 ft and commonly increases downdip and toward the axis of the East Texas basin; the facies consists of about 40 percent sandstone or channel deposits and about 60 percent mudstone or overbank deposits. The overbank (or nonchannel) mudstone beds are mostly laminated, light to moderate gray, and contain minor amounts of carbonaceous matter. Individual sandstone units range from 5 to about 80 ft in thickness and are channel deposits with flat tops and convex-downward bases. Bases of channels are characterized by channel lag of granules to fine pebbles, and a few angular clay pebbles. Sandstone in lower parts of the channel deposits contains trough or festoon cross-beds and in upper parts contains wedge sets of cross-beds. Upper parts of channel sandstone are horizontally bedded. Sandstone in this facies is mostly fine grained to locally medium grained; mean sand grain size averages about 0.14 mm, and range in mean is about 0.11 to 0.25 mm; channel sandstone typically becomes finer upward. Channels are cut into underlying and laterally adjacent overbank mudstone layers. Sinuous, elongate, narrow lignite deposits commonly are found laterally adjacent to channel sandstone deposits; they apparently accumulated as peat bogs in abandoned stream courses. Claystone beds within and adjacent to sandstone channel deposits are chiefly kaolinitic; feldspar is mostly weathered. Claystone of overbank deposits, by contrast, is largely nonkaolinitic—mostly illitic, chloritic, and montmorillonitic; feldspar of overbank mudstone deposits gener-

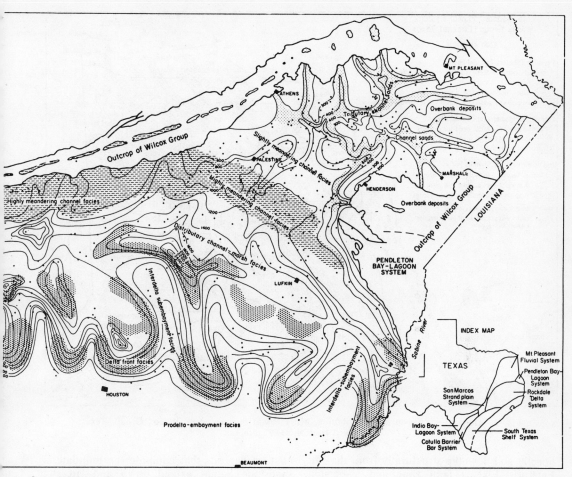

lower part of Wilcox Group (Eocene), Texas.

ally is unweathered. Ceramic clay plants and numerous abandoned lignite mines of East Texas are adjacent to the local Wilcox sandstone deposits because of the close association of kaolinitic clay and lignite to the channel deposits.

Slightly Meandering Channel Facies

Down the depositional slope in the fluvial system a transitional facies is characterized by uniform to slightly dendroid sandstone isolith patterns (Figs. 1, 2), which are less dendritic than those in the tributary-channel facies. This change in pattern probably represents a slightly greater meandering of the original stream system. Channel sandstone deposits are more numerous within this facies than in the tributary-channel facies; they make up about 60 percent of the total facies, and overbank deposits compose 40 percent. The facies ranges in thickness from about 500 to 800 ft; it crops out from central Freestone County, through western Henderson County, into southern Van Zandt County and extends downdip in the subsurface to a line from central Freestone County through central Anderson and central Cherokee Counties, and into northwestern Nacogdoches County. In addition to containing a higher percentage of channel deposits, individual sandstone bodies in this facies are slightly more extensive, and mean grain size is somewhat larger than that of the tributary-channel facies. Primary sedimentary structures of sandstone and relation of channels to clay-mineral distribution and lignite deposits are similar for the two facies.

Highly Meandering Channel Facies

A distinct facies is developed in the Mt. Pleasant fluvial system just downslope deposi-

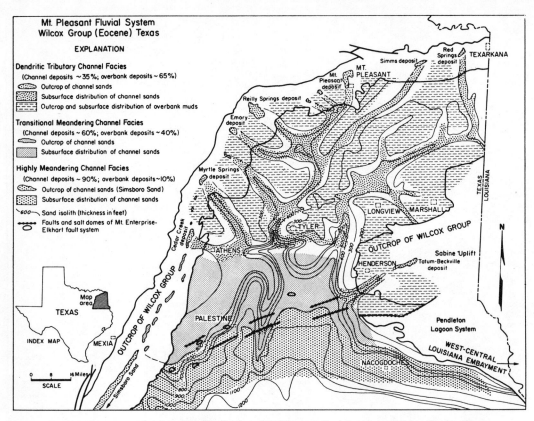

Fig. 2.—Outcrop and subsurface distribution, Mt. Pleasant fluvial system, Wilcox Group, Texas. For well control, see Figure 1.

tionally and downdip structurally from the Mt. Enterprise-Elkhart fault zone and associated salt domes on the southwest (Figs. 1, 2). Within this facies sandstone predominates and overbank mudstone composes less than 10 percent. Sandstone isolith lines generally conform and sandstone in the facies (*e.g.*, Simsboro sandstone outcrop) forms areally persistent units. For such units as these, Pettijohn *et al.* (1965) suggested the term multilateral. The bodies are made up of multiple channels which trend at various angles to the strike of the sandstone isolith lines. Aggregate thickness of sandstone in the facies ranges from 600 to 1,100 ft. Individual sandstone units, comprising numerous superposed and multilateral channel deposits, are as much as 30 mi wide. Areas along the strike between the more extensive sandstone units are made up of greater amounts of overbank mud deposits. Sand grains of the facies have a mean average size about 0.22 mm and are much coarser than sand grains in upslope fluvial facies. Sandstone channel deposits that compose the multilateral sandstone bodies extend down and laterally into underlying and adjacent channel sandstone. Channels are broadly convex downward and are characterized by a basal, locally conglomeratic zone with festoon or trough crossbedding which represents channel-lag deposition. This zone is overlain by a thin unit with tabular cross-beds, in turn overlain by a zone of wedge-set cross-beds (epsilon bedding of Allen, 1963). The tabular and wedge cross-bedded sandstones probably represent lateral deposition as the toe and lower parts of point bars. The horizontally bedded and rippled zones, characteristic of the upper parts of many point-bar deposits, are rarely preserved in this facies, as the channel-lag and lower point-bar deposits are succeeded disconformably by channel deposits of a similar sequence. Within the preserved channel sequence there is only slight, if any, upward fining. The relatively small amount of clay in this facies is very kaolinitic. Lignite deposits, typically associated

Depositional Systems in Wilcox Group (Eocene) of Texas

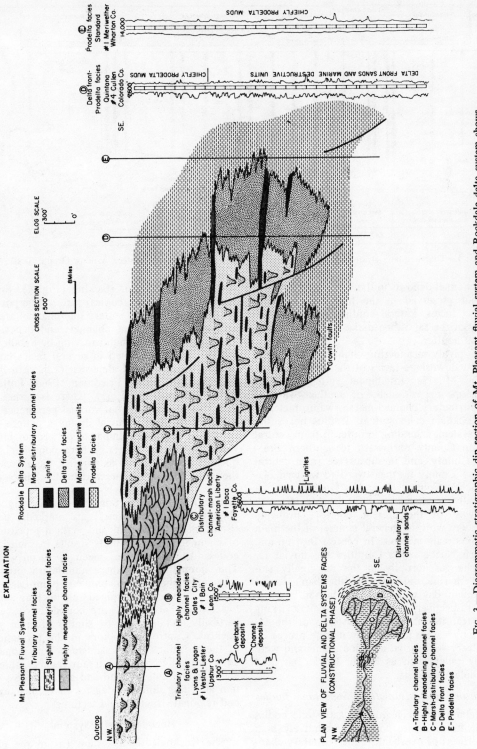

FIG. 3.—Diagrammatic stratigraphic dip section of Mt. Pleasant fluvial system and Rockdale delta system shows relation and character of principal component facies.

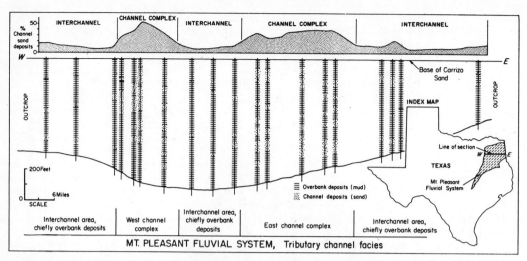

Fig. 4.—Diagrammatic stratigraphic strike section of Mt. Pleasant fluvial system, Wilcox Group, Texas.

with channel deposits in the other fluvial facies, are not preserved in the highly meandering channel facies. Lateral continuity and permeability of the sandstone make this facies a significant aquifer.

The highly meandering channel facies is the farthest downslope facies of the preserved part of the Mt. Pleasant fluvial system, grading downslope depositionally or southeastward to the distributary channel-marsh-swamp facies of the Rockdale delta system. It thus heads the delta system, extending in outcrop from central Bastrop County northward to southern Freestone County, and in subsurface from central Bastrop County northeastward to western Nacogdoches County (Fig. 1). Undoubtedly, it originally extended a considerable distance northwest of the present outcrop, probably coincident with the general courses of the main Wilcox stream systems. In central Bastrop and western Nacogdoches Counties the fluvial facies changes abruptly to the marginally embayed marine mudstone of the San Marcos strandplain-bay and Pendleton lagoon-bay systems, respectively. Within narrow belts marking these transitions, fluvial sandstones of the facies, which were modified by marine processes, have very well-sorted and fine grains and were locally redeposited as beach, barrier-bar, and strandplain deposits.

Multistory and Multilateral Sandstone Bodies

A characteristic feature of sandstone bodies in the fluvial facies containing relatively high proportions of overbank muds (tributary-channel and slightly meandering channel facies) is vertical persistence or stacking (Fig. 4). Successive sandstone bodies are superposed roughly, and complementary interchannel areas are void of significant channel sand deposits. This feature, termed multistorying by Feofilova (1954), has been noted in several facies containing channel sand deposits (*e.g.*, Mueller and Wanless, 1957; Friedman, 1960; Potter, 1963; Brown *et al.*, 1967). There is variation in degree of stacking, but vertical persistence is sufficient to show a fluvial drainage pattern of composite sequences similar to patterns of individual sandstone bodies. Where point-bar sequences are developed as parts of these facies, the sequences are characteristically complete. In contrast, sandstone bodies in fluvial facies containing low proportions of overbank muds (highly meandering facies) are markedly persistent laterally. Lateral and vertical repetition of channel sandstone units results in multilateral sandstone bodies as much as 30 mi wide. The channel units in these multilateral bodies commonly show an incomplete point-bar sequence; the coarse-grained channel-lag and lower point-bar deposits are bounded erosively vertically and laterally by similar channel sequences. Controlling factors in the formation of these distinct fluvial sandstone units apparently are differences in stream regime, differences in amounts of mud and sand in the system affording different degrees of compaction, and regional structure and tectonics.

Rockdale Delta System

The Rockdale delta system (as in the nomenclatural designation of Echols and Malkin,

1948) is the dominant depositional system of the lower Wilcox Group in Texas, comprising areally about 60 percent and volumetrically about 80 percent of the known lower Wilcox (Fig. 1). It was the ultimate depositional site of most sediments transported through the fluvial system and the principal source of sediments redistributed and deposited in associated barrier-bar, strandplain, lagoon-bay, and shelf systems.

The Rockdale delta system is downdip structurally from the fluvial system. The southern limit is marked by a line extending from north-central Karnes County north to central Bastrop County, and the eastern limit is marked by a line from central Nacogdoches County through northern San Augustine and Sabine Counties to the Sabine River. The downdip limit is beyond present well control (Fig. 1). Maximum thicknesses of 3,500–5,000 ft coincide with the central axis of the system which is chiefly along drainage of the modern Brazos River and roughly coincident with structural axis of the Houston embayment. Laterally and along the present structural strike the delta system generally, but not uniformly, thins to about 1,500 ft in the southwestern part of the system and to about 1,000 ft in the eastern part.

Criteria for recognition of this part of the lower Wilcox Group as a delta system include (1) composition and boundary relations of component facies, (2) external geometry of skeletal sand facies, (3) position in relation to laterally associated depositional systems, and (4) contrast in rates of deposition, with thick, rapidly deposited terrigenous clastic sequences (constructional units) alternating with thin, slowly deposited, relatively persistent marine and marsh deposits (destructional units).

Principal Facies

Four regional facies define the Rockdale delta system. Downslope depositionally and downdip structurally from the highly meandering facies of the Mt. Pleasant fluvial system, the lower Wilcox consists of an extensive facies of sandstone units alternating with mudstone, siltstone, and numerous beds of lignite (Fig. 3). Farther downdip this facies changes rather abruptly to thick sandstone, which in turn grades down dip and along structural strike to thick, uniform mudstone sequences. The alternating sandstone and mudstone-lignite sequences are interpreted to be distributary-channel and interdistributary deposits accumulated on a delta plain; the sandstone facies represents delta-front deposits laid down chiefly as distributary-mouth bars; the downdip thick mudstone sequence represents prodelta deposition. Mudstone sequences along strike of the delta-front sandstone facies are interpreted to be interdelta facies.

Distributary channel-marsh-swamp facies.—The facies consisting of alternating sandstone and mudstone-siltstone-lignite units is an areally extensive facies of the Rockdale delta system (Fig. 3). It contrasts markedly with the immediately upslope fluvial facies, chiefly in sandstone-body geometry and relative proportion of channel and nonchannel deposits. The fluvial facies contains cross-bedding structures and textural features indicative of channel-lag and lateral point-bar deposition in highly meandering streams. The result is relatively persistent, uniform, multilateral sandstone units with very few nonchannel deposits preserved. The sandstone changes downdip structurally and downslope depositionally to isolated, elongate, narrow sandstone bodies interspersed in nonchannel mudstone and associated lignite. In these respects it resembles the tributary-channel and slightly meandering facies of the Mt. Pleasant fluvial system. These facies differ significantly, however, in that the channels in the delta system are thicker, are symmetrical in cross section, and show a distributary rather than a tributary pattern in plan. Symmetry in cross section, thickness, and compactional features of underlying mudstone suggest accumulation of the sandstone as bedload deposits in relatively straight, stabilized streams, in which accumulation by downward accretion was facilitated by compactional subsidence of the channel sandstone in a framework of interchannel mudstone. Such features characterize certain modern delta-plain distributary channels and contrast with the lateral deposition in point bars of meandering, very sinuous streams. Mudstone and siltstone, composing about 60 percent of this facies, accumulated as subaerial levee, crevasse-splay, lacustrine, and flood-basin interdistributary deposits.

Two kinds of lignite are present within the distributary channel-marsh facies, both distinct from the elongate lignite deposits of the fluvial system. One lignite, which is tabular in shape but only local in extent, probably represents peat accumulation in small interdistributary depressions similar in origin to interdistributary peat in the Mississippi delta system, as discussed by Fisk (1960) and Frazier (1967). A second type is also tabular in shape but much more extensive; this type developed as regional, landward facies of marine delta destruction. Other distinctions between fluvial and delta-plain lignites are made on the basis of grade and composition. Fluvial lignites are more variable in composition than delta lignites and contain high percentages of woody materials consistent with a swamp origin. Delta lignites are slightly higher in grade and have higher specific gravity, and the relatively low percentages of woody material are more characteristic of marsh growth. Regional variations in Wilcox lignites (Fisher, 1963) reflect differences in depositional origin.

Delta-front facies.—Downdip the distributary channel-marsh-swamp facies changes to a thick sandstone facies in which lignite is absent and mudstone content is low (Fig. 3). The sandstones mostly have fine grains, generally much better sorted than those of fluvial and distributary channel sandstones. Commonly, the sandstone contains thin, multidirectional ripple cross-laminae and other current features. Plant and detrital lignite fragments are present locally. Individual sandstone units may be 200 ft or more thick; in plan, external geometry is lobate or lunate. The units are interpreted to be distributary-mouth bars representing delta-front deposition. The high percentage of sandstone in this facies resulted from deposition of bedload of delta-plain distributary channels debouching into relatively quiet marine water; the muds were carried in suspension farther seaward and accumulated as prodelta deposits. Relative cleanness, good sorting, and multidirectional current features indicate winnowing and reworking of sands in this environment by both channel and open-water currents under shoal conditions in the delta-front environment. Redistribution of these sands and lateral coalescence of the constructive delta-front deposits formed a fairly continuous delta-front sandstone facies. The relative amount of redistribution and lateral coalescence within the delta-front facies in the different Rockdale deltas resulted in delta lobes ranging from elongate to rounded.

Prodelta facies.—The farthest downdip sections of lower Wilcox penetrated by wells generally consist of relatively uniform, thick sequences of dark mudstone (Fig. 3). They contain moderately large amounts of very finely comminuted and disseminated organic matter, laminations marked commonly by slight textural and color variations, and a marine fauna exceedingly low in number and species diversity. This mudstone facies interfingers with, and grades updip to, the delta-front sandstone facies and represents deposition from suspension seaward from the delta front and river mouth as prodelta facies. Thickness and downdip extent of the facies are difficult to determine, primarily because few wells have penetrated significant thickness of the lower Wilcox in the downdip areas. However, the few deep wells indicate that the facies is probably the thickest in the delta system; approximately 4,000–5,000 ft has been recorded in some deep wells.

Interdelta, submembayment facies.—Sequences in areas between principal delta lobes are similar in some respects to the prodelta facies (Fig. 1). They consist principally of mudstone, much of which is laminated, but are thinner than the prodelta facies and in addition contain several thin, burrowed mudstone units or thin units of fossiliferous siltstone and very fine-grained sandstone with small amounts of glauconite. These thin units accumulated more slowly than the laminated mudstone, in areas not directly affected by deltation. The interdelta facies grades updip structurally to marsh facies of the delta plain and laterally along strike to delta-front sandstone facies.

Delta-Destruction Units

Delineation of a thick terrigenous clastic sequence such as the Rockdale delta system into component and individual deltas is a time-consuming task. Basically, it involves recognition and correlation of thin marine units formed in the process of marine destruction of a constructive delta mass. During periods of great sediment influx, the tract of delta facies (distributary channel-marsh-swamp, delta front, prodelta) is established and, during progradation and basinal and compactional subsidence, a thick terrigenous sequence accumulates. With cessation of sediment influx due either to diminishment of the source area or, more commonly, to upstream avulsion resulting in abandonment of the previous drainage system and establishment of a new network, compaction of the water-saturated sediments of the constructed delta mass exceeds the rate of deposition. The delta lobe, or at least marginal parts of it, is subjected to marine encroachment involving reworking of the constructive delta surface and underlying delta sediments. Rate of sedimentation during destruction is very low and as a consequence units are relatively thin. In the Rockdale delta system, delta-destruction units consist of very well-sorted, fine-grained, slightly glauconitic, commonly bioturbated sandstone, shell detritus, and mudstone either burrowed or containing a marine fauna. These units were deposited chiefly on the lower parts of the abandoned delta plain. Landward destruction of the plain involved extensive marsh growth, coincident with compactional subsidence, which resulted in areally extensive lignite deposits. This lignite contrasts with the thinner, local lignite bodies that formed as interdistributary peat. The delta-destruction units apparently represent deposition in environments comparable with delta-destruction environments in the modern Mississippi system. The marine sandstone originated under conditions similar to those of the Chandeleur Islands, and the burrowed or fossiliferous mudstone originated as open-bay deposits similar to those of Breton Sound.

Recognition of delta-destruction units and their facies relations in delta systems should be followed by studies of their areal extent by careful correlation of closely spaced wells in a large area. By this means the delta system can be subdivided and smaller delta units mapped. Determination of sand content of such subdivisions permits isolith mapping of skeletal elements of the delta (distributary-channel and delta-front facies) and, with mapping of component facies, allows delineation of specific delta lobes. Most delta-destructive units are restricted areally to a specific, underlying, constructive delta lobe. Because delta construction is a relatively rapid process and delta destruction a slow process, destruction of one delta lobe may persist during the construction, subsidence, and subsequent initiation of destruction of an adjacent lobe; thus destructional units of two or more lobes may be partly contemporaneous. Recognition of widespread destructional units permits larger scale, though only approximate, subdivision of delta systems.

Individual Deltas in Rockdale Delta System

The writers have mapped and delineated 16 principal deltas within the lower Wilcox Rockdale delta system (Fig. 5). These have the component facies and facies pattern heretofore described. Striking, and significant in another context, are the vertical stacking of individual deltas and the persistence of interdelta subembayments within the delta system. Further, the Rockdale deltas show a marked coincidence with modern coastal-plain streams (Fig. 5).

Maps of several delta-destruction units have made possible the delineation of three main phases of deltation in the Rockdale system, in addition to the outlining of principal individual deltas. The initial phase of deltation resulted in three relatively small deltas situated roughly along drainage of the modern Colorado, Brazos, and Trinity Rivers (Fig. 5). They range in area from 200 to 350 sq mi, and thicknesses are on the order of 300 to 500 ft. A second or intermediate period of delta building resulted in the largest, thickest, and most prograded deltas of the system. The larger deltas of this phase, as much as 2,000 sq mi in area, were constructed along the axes of the modern Guadalupe, Colorado, and Brazos Rivers; the smaller deltas, 500–700 sq mi in area, developed along the axes of the modern Sabine and Neches Rivers. The Brazos and Colorado deltas (Fig. 5) are the thickest, with maximum thickness of 1,500 ft, exclusive of the prodelta facies. A terminal phase involved delta construction in all areas of intermediate-phase deltation, as well as construction of a large delta situated primarily in Angelina, Houston, and Trinity Counties (Angelina delta). These terminal deltas are comparable in size and thickness with the underlying, intermediate-phase deltas, but do not extend as far downdip structurally.

Although individual deltas within the Rockdale system are varied in size and sediment volume, a more significant variation is in the shape of the delta lobes (Fig. 6). One type, shown by the intermediate-phase delta constructed in the area of the Trinity River, the four intermediate and terminal deltas constructed in the area of the Neches and Sabine Rivers, and, to some extent, the three initial-phase deltas, has the elongate, narrow shape of the modern birdfoot delta and certain initial deltas of older complexes delineated by Frazier (1967) in the Mississippi delta system. A second type, shown by the other deltas of the Rockdale system, has a generally lobate or rounded shape similar to that of premodern, abandoned deltas of the Mississippi system. If a delta line is drawn in the manner in which bayline maps commonly are made (Lowman, 1949), the elongate, narrow birdfoot deltas of the Trinity, Neches, and Sabine Rivers extend considerably beyond the line; thus, progradation into deeper waters, as in the modern deep-water birdfoot delta of the Mississippi system (Fig. 6), is suggested. These deltas are associated with a sequence relatively high in mudstones, a feature that strengthens the analogy with the sand-poor, deep-water birdfoot delta of the Mississippi. The initial-phase deltas similarly represent progradation over un-

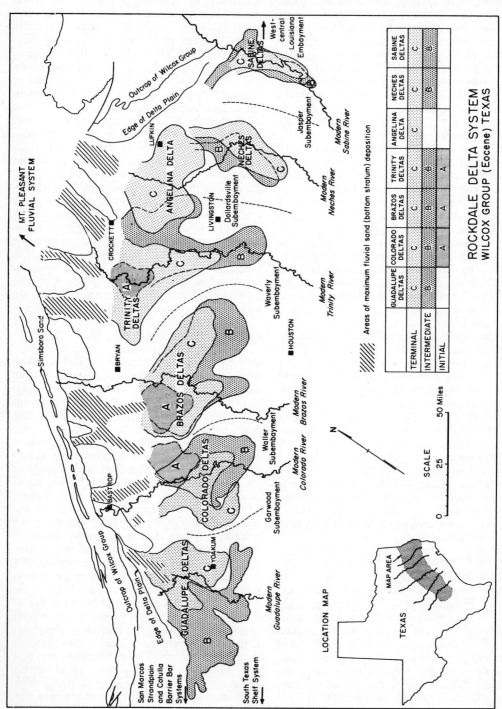

FIG. 5.—Distribution of principal deltas, Rockdale delta system, Wilcox Group, Texas; lobal outline drawn at distal margin of delta front facies. For well control, See Figure 1.

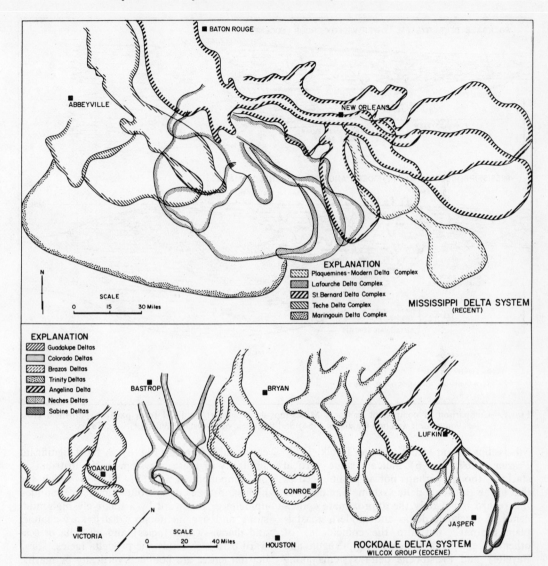

Fig. 6.—Comparison of size, distribution, and arrangement of principal delta lobes of the Holocene Mississippi delta system, southeastern Louisiana (modified from Frazier, 1967), and Rockdale delta system, Wilcox Group (Eocene, Texas).

derlying thick mudstone. By contrast, the lobate or rounded deltas of the Rockdale system consist of relatively large proportions of sandstone, a characteristic that suggests construction as shoal-water deltas analogous to the sand-rich, lobate, shoal-water deltas of the modern Rhone (Kruit, 1955) and premodern Mississippi (Fisk, 1955, 1961; Frazier, 1967).

Vertical stacking or approximate superposition of individual deltas within the system, persistence of interdelta subembayments, and coincidence of Rockdale deltas and modern drainage have been mentioned. The controlling factor or factors, like those of the multistoried channel sandstone bodies in upslope fluvial facies of the Mt. Pleasant system, are not fully understood. Such features apparently are common in fluvial systems containing relatively high percentages of mud and may be common in delta systems, although few ancient delta systems have been mapped and studied. Maps of delta systems in the Pennsylvanian of the Eastern Interior basin (Wanless et al., 1963) and the modern and premodern deltas of the

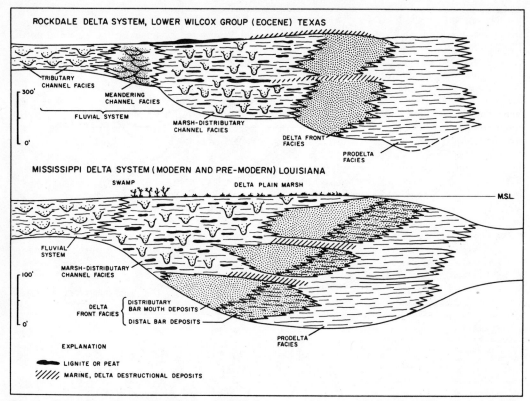

Fig. 7.—Comparison of component facies of Mississippi delta system (modified from partly hypothetical cross section of Coleman and Gagliano, 1964) and Rockdale delta system, Wilcox Group, Texas.

Mississippi system (Kolb and van Lopik, 1966; Frazier, 1967; Fig. 6) indicate some vertical stacking, though perhaps not as pronounced as that in the Rockdale delta system. Significant in this regard is not only the approximate coincidence of Rockdale deltas and modern coastal-plain drainage, but also the coincidence of other Eocene delta systems (*e.g.*, Yegua and Fayette) and Pleistocene delta systems along these axes. Certainly, compactional subsidence is a possible factor, although the more important control probably is tectonic. Structural control is indicated by the presence of growth-fault zones coincident with the updip limits of the thicker facies of the Rockdale delta system (Fig. 3).

Comparison with Mississippi Delta System

The Eocene Rockdale and the Holocene Mississippi delta systems, both built into the Gulf Coast basin, are similar in shape, occurrence, position of shoal-water lobate deltas and deep-water birdfoot deltas, and vertical stacking of individual deltas. The two systems are comparable in area (Fig. 6). A most significant similarity of the two systems is the distribution of component facies (Fig. 7), in which the high-sandstone delta-front facies is bounded up-delta or landward by a relatively high-mudstone and organic facies (distributary-channel and marsh-swamp facies) and off-delta or seaward by a high-mudstone prodelta facies. These detrital lenses are bounded vertically by marine and landward-equivalent destructional units. The significance of bounding facies in delta sequences has been emphasized by Coleman and Gagliano (1964); delineation of component facies, together with determination of external geometry of the skeletal elements (distributary-channel and delta-front sandstone bodies) and delineation of delta-destruction units, are judged to be the best regional criteria for recognition of ancient deltas or delta systems.

Pendleton Lagoon-Bay System

Marginal to the Rockdale delta system and along the southern part of the Sabine uplift (chiefly Shelby and northern parts of San Augustine and Sabine Counties), the Wilcox Group is predominantly a mudstone facies. This facies includes parts or all of several formally defined stratigraphic units (Pendleton, Marthaville, Hall Summit, Lime Hill, Converse, Cow Bayou, Dolet Hills, and Naborton Formations). It is present in outcrop and in subsurface. The facies, designated informally as the Pendleton lagoon-bay system (Fig. 1), is approximately 1,000 ft thick in Texas. It is comparable in thickness with fluvial facies along the strike, but thicker than updip fluvial facies and thinner than downdip delta facies.

The Pendleton system consists chiefly of laminated to locally bioturbated mudstone, and local lenses and beds of very fine- to fine-grained, sparingly glauconitic, massive to cross-bedded sandstone and fine-grained, broken, massive to cross-bedded, lignitic sandstone. Certain of the mudstones are gypsiferous; clay-ironstone concretions are common. Lignite is locally common, but relatively impure and in thin, discontinuous beds. The system contains a few fairly discrete units with marine fossils. Faunal composition is chiefly molluscan, with a few foraminifers and ostracods; fossils are present mostly in the sparingly glauconitic units. Species diversity is moderate to low, at least for fossiliferous units, but not so low as to suggest very abnormal salinity. Mollusks comprise largely a bank and mud fauna, characetized by several thick-shelled forms. The fauna of the Pendleton system has been described by Barry and LeBlanc (1942).

The system is underlain by marine clay of the Midway Group (Porters Creek Formation) and overlain by sandstone of the Carrizo Formation. A large part of the facies is contemporaneous with delta facies (Sabine deltas of the Rockdale system) on the west and south, although the upper part is contemporaneous with other Wilcox depositional systems not considered here. The Pendleton system grades west and north into the various fluvial facies of the Mt. Pleasant system; it extends east into Louisiana for an undetermined distance.

The lithologic and biologic composition of the Pendleton system and its situation marginal to the Rockdale delta system suggest that deposition took place in low-energy lagoons, lakes, bays, and mudflats flanking the area of maximum deposition in the delta system, much like the marginal environments of the Mississippi delta system. Deposition was chiefly marine and circulation ranged from restricted to open; the system opened seaward primarily on the southeast, and there was greater brackish-water and freshwater influence on the west and north, where the system grades into fluvial facies of the Mt. Pleasant system.

San Marcos Strandplain-Bay System

In outcrop south of the Colorado River, the highly meandering channel sandstone facies of the Mt. Pleasant system changes abruptly to a facies primarily of mudstone which locally contains marine fossils (Fig. 1). This facies persists in outcrop through southern Bastrop, Caldwell, Guadalupe, and northern Wilson Counties and extends a short distance downdip into the subsurface. It includes the Hilbig facies of Claypool (1933). The faunal and lithologic composition of the facies is very similar to that of the Pendleton system. A similar origin is suggested, with deposition in embayments and on mudflats along the opposite margins of downslope fluvial facies of the Mt. Pleasant system and delta facies of the Rockdale system. In outcrop, mud deposits of this facies are continuous below the downslope facies of the Mt. Pleasant system (highly meandering channel facies or Simsboro sandstone) about as far north as midway between the Brazos and Trinity Rivers, where the mudstone grades into fluvial and delta facies of earliest Rockdale deltation (initial phase, Fig. 5). The northern extent of this mudstone facies includes at least parts of units formally designated the Hooper and Seguin Formations. Southwestward, displacement of the facies is contemporaneous with large-scale deltation in the lower Wilcox Rockdale system (intermediate and terminal phases, Fig. 5).

Downdip and marginal to the shoal-water deltas of the southwestern part of the Rockdale system (Guadalupe deltas), mudstone grades into a facies consisting of interbedded sandstone and mudstone. The sandstone units generally have fine, well-sorted grains, are calcareous and sparingly glauconitic, and commonly contain *Ophiomorpha* burrows, a few detrital shell fragments, and small armored mud balls. They are present as elongate bodies parallel with the depositional strike and as tabular or sheet sandstone characterized by low-angle cross-bedding. Individual sandstone and mudstone units range up to 50 ft in thickness. Mudstone units are similar to those of the updip mudstone facies, but are commonly pyritic and slightly gypsiferous. Carbonaceous materials are present locally as thin lignitic stringers. This interbedded sandstone and mudstone facies is approximately 1,200 ft thick, and is present chiefly in southwestern Gonzales, Wilson, southwestern Karnes, and eastern Atascosa Counties. Composition of the facies and its position marginal to shoal-water deltas of the Rockdale system suggest an origin at least mechanically analogous to that of the Holocene chenier system of southwestern Louisiana. The mud accumulated in low-energy environments ranging from mudflat to open bay, contemporaneous with periods of active deltation. During marine destruction of the adjacent sand-rich, shoal-water deltas, sands were redistributed laterally by prevailingly southwesterly longshore currents; elongate and sheet sands were deposited in a strandplain marginal to the delta system. Transgression and regression of relatively thin marine units in this facies thus were related to periods of active deltation and delta destruction. Subsurface sandstone units such as the Poth and Bartosch are typical of this facies. In reference to their origin and area of principal occurrence, the updip mudstone and downdip sandstone-mudstone facies along the southwestern margin of the Rockdale delta system are designated the San Marcos strandplain-bay system.

Cotulla Barrier-Bar and Indio Lagoon-Bay Systems

Southwestward, sheet and elongate sandstone bodies of the San Marcos system grade into facies of elongate sandstone bodies arranged parallel with the depositional strike. The sandstone bodies generally have fine, very well sorted grains and are locally glauconitic; internal structures are not known because of lack of cores. Thickness of individual sandstone units ranges as much as 100 ft; aggregate sandstone thickness is 400–1,000 ft for the facies. Sandstone isolith lines show an extensive elongate sandstone unit oriented parallel with the depositional strike. This sandstone belt trends northeast-southwest in western Atascosa, southern Frio, northwestern LaSalle, southeastern Dimmit, and northwestern Webb Counties (Fig. 8). Updip from the sandstone belt, in southern Bexar, northern Atascosa, northern Frio, Zavala, western Dimmit, and northwestern Webb Counties and extending to the outcrop, is a predominantly mudstone facies, including most of the Indio Formation (Fig. 8). The principal lithologic component of this facies is mud, although locally sand is predominant. Mudstone is mostly gray to buff, calcareous to locally gypsiferous, and pyritic, and some is glauconitic or dolomitic. Structurally, it ranges from distinctly laminated to extensively burrowed. Marine fossils are chiefly thick-shelled mollusks; a few units contain large amounts of a very few species of foraminifers, mostly textulariids and globigerinids. Concentric mounds with internal structures, probably of algal origin, commonly are associated with the mudstone. Lignite beds of this facies are thin, discontinuous, and commonly impure, in contrast with the lignite of the fluvial and delta systems. Sandstone in this facies is of varied texture, but most has fine, moderately well-sorted grains; it is developed locally in relation to distinct depositional features, including tidal deltas, washover fans, and small-scale delta and fluvial units.

The lower Wilcox sandstone and mudstone facies of South Texas are judged to have originated as extensive barrier-bar and complementary lagoon-bay systems and are designated here, respectively, the Cotulla barrier-bar system and the Indio lagoon-bay system. The sand was transported laterally from shoal-water deltas of the Rockdale system on the northeast by prevailing southwesterly longshore currents. In the San Marcos strandplain system, situated intermediate to the delta and barrier-bar systems, the adjacent delta system served not only as a source of sand but also, through periods of active and inactive deltation, as a control for the distribution of the sheet sand. Accumulation of sand in the barrier-bar system was too distant from the delta system for the latter to have had direct control on sandstone-body geometry. The principal distinction in sand accumulation between these two systems is that sheet or tabular units formed in the San Marcos and elongate bodies formed in the Cotulla. Similar patterns of sand dispersal are present in modern and premodern marginal-marine systems southwest of the Mississippi delta system. Accumulation of sand in relatively narrow elongate belts in the barrier-bar system, as opposed to sheet-sand accumulation in the strandplain, facilitated development of extensive, landward mud facies—the Indio lagoon-bay system. Development of a distinctive lagoon-bay system is thus a complement of the barrier-bar system.

Other features substantiate interpretation of the South Texas, lower Wilcox facies as barrier-bar and lagoon-bay systems. First, sandstone-isolith patterns depicting the lagoonal side of the barrier-bar system are generally more irregular than those of the seaward side; further, the transition from sandstone to mud-

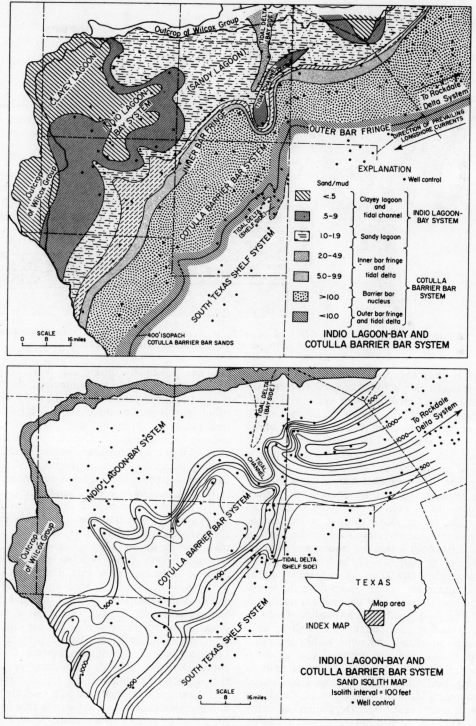

FIG. 8.—Sand-mud ratio map and sand isolith map of Cotulla barrier-bar system and Indio lagoon-bay system, Wilcox Group, Texas.

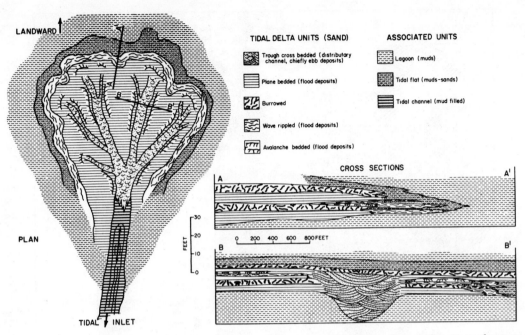

Fig. 9.—Principal rock units, lagoon-side tidal delta, Indio lagoon-bay system, Wilcox Group, southeastern Medina (outcrop) and northeastern Frio (subsurface) Counties, Texas.

stone is more abrupt on the lagoonal than on the seaward side (Fig. 8). Such features are typical of modern barrier-bar systems. Second, in southeastern Frio and northeastern LaSalle Counties, sandstone isolith lines of the Cotulla barrier-bar system are indented sharply on the lagoon side; a less pronounced indentation is present on the seaward side. Position and configuration of these indentations suggest the former presence of a tidal channel, a feature common to modern barrier bars. Flood and ebb tidal surges move through a channel or inlet in the bar, accreting sand in the form of tidal deltas both within the lagoon and onto the foreshore or shelf side of the bar. Presence of an anomalous thickness of sandstone in the Indio lagoon-bay system updip from the terminus of the inferred tidal channel (in northeastern Frio and southeastern Medina Counties and extending into outcrop), as well as sedimentary structures and units within the sandstone bodies, indicate accumulation as a lagoonal tidal delta (Fig. 8). Principal units of the lagoonal tidal deltas developed in outcrop of southern Medina County include flood and ebb deposits (Fig. 9). Deposition occurred both within tidal distributary channels and on interchannel flats. Channel deposits are trough or festoon cross-bedded sandstone, with axes of cross-bed sets trending generally south; deposition was principally during ebb flow. Toward the barrier bar, and in the subsurface, the tidal channel becomes relatively straight and nondistributive; this part of the channel is filled with mudstone and apparently was filled after the channel became inactive. Deposits of the interchannel flats are chiefly plane-bedded sandstone grading marginally through megarippled sandstone to tidal-flat and lagoonal mudstone; locally, thin, avalanche-bedded sandstone is associated with the plane-bedded deposits. Bedding, structures, and directional features of the interchannel deposits indicate that deposition took place mainly during flood surge. Plane-bedded sandstone shows graded bedding with a vertical increase in the amount of clay. The upper units are extensively burrowed, apparently as the result of destruction while parts of the tidal delta became emergent. Lobate extension of sandstone on the foreshore side of the bar and just south of the inferred tidal channel (east-central LaSalle County) probably represents a delta constructed by ebb-tide transport of sand through the tidal channel. On the assumption that these features originated as a tidal channel and complementary tidal deltas, the position of the deltas in respect to the channel, as well as the shape and configuration of the tidal channel, substantiate the inference of a prevailing southwestward transport of sediment by longshore currents. Lobate lagoonward extensions of sandstone in northeastern and north-central Dimmit County (Fig. 8), without associated isolith indentations indicative of tidal channels, possibly represent deposition as washover fans, with sand carried into the lagoon by storm surges. This feature is common in modern barrier bars (Andrews, 1966).

South Texas Shelf System

Downdip and seaward of the Cotulla barrier-bar system, a predominantly mudstone system characterizes the lower Wilcox of South Texas. This system extends across about 3,000 sq mi in southwestern Karnes, northern Goliad, northern Bee, north Live Oak, McMullen, northwestern Duval, Webb, Zapata, and western Starr Counties (Fig. 1). It is relatively uniform in thickness through much of its extent, ranging from about 1,000 to 1,500 ft, but thickens northeastward toward the Rockdale delta system and southward along the Rio Grande. Principal lithologic component is mudstone; aggregate thickness of sandstone in the system is generally less than 200 ft (Fig. 1). Individual sandstone units are varied in number, but commonly 8–10 are present. Thickness ranges generally from 10 to a maximum of 100 ft; average thickness of sandstone units is about 30 ft. Mudstones are of varied composition, but generally are light or medium gray to olive or buff, and laminated to extensively burrowed; marine fossils and glauconite pellets are common. Mudstones are commonly calcareous, though some are pyritic or carbonaceous. The sandstone is very fine to fine grained, well sorted to muddy, slightly glauconitic, and commonly contains detrital shell fragments. Faunal diversity is moderate; composition is chiefly molluscan, but thick-shelled forms are not predominant as in other lower Wilcox marine facies. Foraminifers are numerous and a few echinoderms are present.

The comparative uniformity of the system, the predominantly mudstone composition, marine fauna, and position relative to other depositional systems suggest that deposition took place on an extensive, shallow-water continental shelf; hence the system is designated the South Texas shelf system. Part of the Falls City Formation, as proposed and described by Hargis (1962), is included in this system. Vertical variations in abundance and diversity of marine fossils, amount of glauconite, extent of burrowing, and presence of sandstone units probably are related to variations in rate of deposition coincident with filling and foundering of the shelf; these cycles presumably relate to phases of delta construction and destruction in the Rockdale delta system along strike on the northeast.

Origin of the relatively thin sandstone units in the South Texas shelf system is not clear. Their external geometry is not well known, but present control suggests that there are both laterally persistent sheet sandstones, more numerous updip toward the Cotulla barrier-bar system, and elongate, NE-SW-trending sandstone bodies. The source of sand in the South Texas shelf system, like that of updip sandstone systems, probably was the Rockdale delta system on the northeast, but mode of transport and dispersal is unknown. Relatively sharp indentations in sandstone isolith lines of the southwestern lobes of the Guadalupe deltas of the Rockdale delta system (Fig. 1) suggest that submarine channeling of the deltas may have occurred during periods of delta destruction, with possible dispersal of sand across the shelf by southward-flowing turbidity currents. (Submarine channeling in the Wilcox as reported by Hoyt in 1959 is associated with an upper Wilcox system.) A more probable explanation of the origin of these sandstone units, is accumulation as barrier bars during periods of initial marine destruction of gulfward-projecting delta lobes and subsequent to shelf filling coincident with principal delta construction. The origin thus is similar to that of the Cotulla barrier-bar system, but there is only minor accumulation downdip and principal accumulation updip in the Cotulla system. Sheet sandstone of the shelf system may be related to transgression associated with periods of continued delta destruction; shelf-edge marine reworking of foundered barrier-bar sands during this period was likely. Extensive mud deposition in this system presumably was contemporaneous with periods of active deltation in the Rockdale delta system, although mudstones of the two systems are compositionally distinct. The amount of mud deposition on the shelf was small in comparison with the amount of contemporaneous prodelta mud deposition. Differences in composition and amount of mud in the two systems probably are related to marked differences in rates of accumulation.

SAND-DISPERSAL PATTERNS

The dispersal pattern of coarser fraction sediments (generally sand) in a terrigenous clastic system provides significant clues to the origin of specific facies and also provides a basis for the integration of various facies and systems.

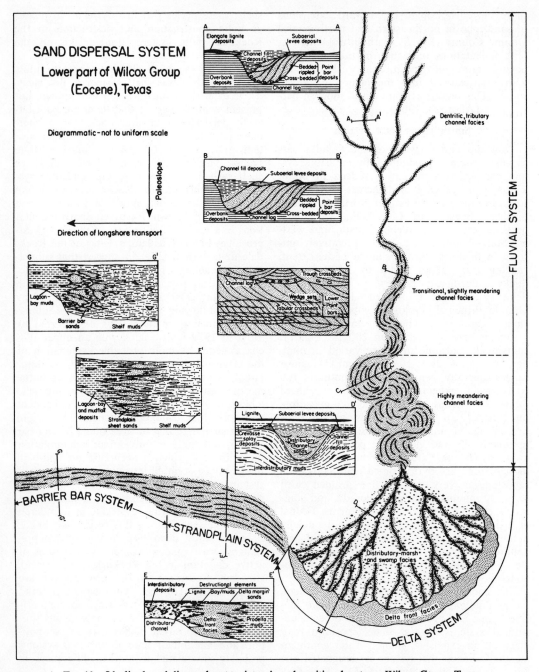

Fig. 10.—Idealized sand-dispersal system in various depositional systems, Wilcox Group, Texas.

Transport, dispersal, and deposition of the coarser fraction sediments normally involve higher energy elements, and thus the geometry of sandstone bodies effectively outlines skeletal units of the system. Within the various depositional systems of the lower Wilcox, several modes of sand transport and deposition occurred; each resulted in sandstone units of distinct geometry, structure, and texture (Fig. 10). The pattern involves dispersal through several fluvial facies, in which only a relatively small part of the sand was permanently stored or preserved, and through separate delta facies in which most deposition involved permanent storage, especially in the deep-water, elongate deltas. Marine redistribution of a relatively small part of the sand deposited in the delta-front facies, chiefly in shoal-water deltas, resulted in accumulation of sand in marginal strandplain, barrier-bar, and shelf systems (Fig. 10). Sandstone bodies of the fluvial system and of distributary channels of the delta system have regional trends normal to the depositional strike; sandstone bodies of the delta-front facies and those of the shelf system roughly parallel the depositional strike; those of the barrier-bar and strandplain systems closely parallel the depositional strike.

The highest upslope element preserved in the lower Wilcox sand-dispersal pattern, apparently only a minor part of the original fluvial system, consists of elongate channel sandstone with dendritic tributary distribution; ratio of nonchannel or overbank mudstone to channel sandstone is high. Downslope depositionally, the ratio of nonchannel mudstone to sandstone decreases, and individual sandstone bodies increase in width as a result of deposition by increasingly meandering streams. A zone of highly meandering stream deposition, as judged from the dominance of a point-bar sequence, resulted in a fluvial facies composed chiefly of extensive, multilateral sandstone bodies with very little associated nonchannel mudstone. This fluvial facies is present downslope and as the preserved part of the main fluvial system. Upslope this facies is bounded by a fault system (present Mt. Enterprise-Elkhart system) or threshold (Tyler basin threshold) which influenced deposition in other Cenozoic and Mesozoic units. It suggests the possible presence of a fall line, or at least a local gradient change that resulted in increased sand deposition. Once such a zone was established, weak bank conditions of streams traversing it would contribute to a high degree of meandering or lateral shifting. A similar high-sandstone fluvial facies in sand-dispersal patterns of Upper Pennsylvanian units in the Eastern Interior basin is reported by Potter (1963).

The next downslope element in the sand-dispersal pattern consists of elongate sandstone bodies in a distributive network; these are associated with relatively large amounts of nonchannel mudstone (Fig. 10). Channel sand accumulation was chiefly by downward accretion through compaction, as indicated by the nonerosive bases of the distributary channels and compactional features of underlying mudstone.

Principal deposition of sand in the delta system occurred in the delta-front facies, built up of distributary-mouth bars. Where delta-front sands form elongate, narrow sand facies (barfinger sands of Fisk, 1961) in relatively high-mud frameworks and under deeper water conditions (birdfoot deltas), sand storage is generally permanent. This is apparently true of the elongate delta fronts characteristic in the eastern part of the Rockdale system where they are not associated with marginal barrier-bar and strandplain sands; interdelta subembayment facies contain very few sandstone bodies. Where the delta front is constructed as a lobate to rounded facies of sand-rich, shoal-water deltas (typical of deltas in the southwestern part of the Rockdale system), some of the sand may be redistributed laterally by marine destruction of the deltas and transported by longshore currents. In such situations two main types of sandstone bodies accumulated, though both have a common source. Adjacent to the delta system, accumulation generally resulted in relatively thin, areally persistent sheet sands which made up a strandplain system. Down the longshore current and more distal from the delta system, sand accumulated in more narrowly restricted, elongate barrier bars with the complementary development of a distinct landward lagoon-bay system.

A final pattern of sand dispersal, not shown on Figure 10 but possibly a part of lower Wilcox depositional systems, is transport and deposition by turbidity currents. Sharp indentations in sandstone isolith lines of certain of the shoal-water deltas in the Rockdale system (Guadalupe and Colorado deltas) may reflect mud-filled, submarine canyons that were cut and subsequently filled during specific stages of delta construction and destruction. Such a feature has been documented in a higher Wilcox system (Hoyt, 1959), and submarine channeling is well known from several modern delta systems. Comparable features in lower Wilcox systems are probable. Under such conditions delta sands eroded during channeling would be carried down depositional slope by turbidity currents and deposited as fans at the first major flattening of the slope. Mode of dispersal would be similar to that outlined by Walker (1966, Fig. 18) for near-source submarine fans associated with the Pennine delta in the upper Carboniferous of northern England. Verification of such fans in the lower Wilcox awaits the drilling of more deep downdip wells.

DEPOSITIONAL SYSTEMS AND BASIN TECTONICS

The writers have used the concept of depositional systems not only as an approach to interpreting certain rock units and facies of the Wilcox Group, but also as a basis for establishing criteria for recognizing large-scale, regionally developed depositional units. Many sedimentary basins, and notably the Gulf Coast basin, contain very thick and large masses of terrigenous clastic rocks. Application of diagnostic criteria for recognition of specific depositional environments and the utilization of detailed depositional models are facilitated first by outlining major depositional systems involved in the basin fill.

The nature of a sedimentary basin, especially its structural or tectonic attitude, is critical to the concept of depositional systems because of its effect on (1) volume of sediment or magnitude of sequences involved, and (2) variation in distribution of component facies. First, in a relatively unstable basin where subsidence is generally a response to sediment load (as is judged to be the case of the Gulf Coast basin during Wilcox deposition), sites of deposition of particular facies tend to be maintained. Thus a barrier-bar system, for example, can be constructed of several individual barrier bars, the main features of which are similar to those of the larger system. Essential difference between the two is size. In magnitude, the Cotulla barrier bar system here recognized in the lower Wilcox of South Texas is closely comparable to other barrier-bar systems recognized elsewhere in the Gulf Coast basin—Frio barrier-bar system (Neogene) of South Texas (Boyd and Dyer, 1964) and Terryville barrier bar (Jurassic) of southern Arkansas and northern Louisiana (Thomas and Mann, 1966).

Second, the tectonic nature of the basin markedly influences the distribution of component facies within a depositional system, although composition and character of component facies of a given depositional system may be very similar in different structural basins. For example, delta systems developed in relatively stable basins, such as the Pennsylvanian of North Texas (Eastern Interior basin) and the western interior United States, commonly show fairly extensive progradation. Sections within such systems contain a vertical succession of component facies with prodelta mudstone overlain by delta-front sandstone, in turn succeeded by delta-plain and fluvial facies. By contrast, delta systems formed in rapidly subsiding, unstable basins, such as the Gulf Coast basin, commonly do not contain a pronounced vertical succession of facies, but rather have pronounced lateral facies that generally persist vertically. Within any particular part of the delta system, therefore, a vertical sequence generally shows only a single component facies of the system.

OIL AND GAS OCCURRENCE

A well-established generalization in Gulf Coast petroleum geology is that occurrence of oil and gas is controlled regionally by depositional facies, and local entrapment is by structures coincident with depositional facies that contain optimum reservoirs. The environmental control of oil and gas occurrence in terrigenous clastic rocks has been emphasized particularly by Rainwater (1963). Distribution of depositional facies in the context of the regional system of which they are a part explains control of known accumulations and provides a basis for outlining regional, potentially productive trends. In the lower Wilcox of the Texas Gulf Coast, five or six main oil- and gas-producing trends (Fig. 11) are defined in relation to depositional systems established herein: (1) delta trend (chiefly delta-front constructional and marine destructional sandstone), (2) strand-plain trend, (3) barrier-bar trend (including tidal deltas), (4) shelf trend, (5) shelf-edge trend, and (6) a possible submarine-fan trend not yet defined.

Oil and gas accumulation in the lower Wilcox from north-central Karnes County northeastward to the Sabine River is controlled closely by distribution of specific deltas of the Rockdale delta system. Principal reservoirs are delta-front sandstone and closely associated marine-reworked sandstone from delta destruction; known fields are largely coincident with a relatively narrow belt determined by the more distal fringes of delta-front sandstone of specific delta lobes (Fig. 11). Interfingering of delta-front sandstone with adjacent marine prodelta mudstone provides the multipay reservoirs typical of many of the large delta-front fields (*e.g.,* Falls City, Sheridan, Columbus, Lake Creek, New Ulm, and Quicksand Creek). Marine clay deposited over delta-front sand during stages of marine destruction of delta lobes locally provides an effective seal for sandstone reservoirs. Numerous large-scale structures along the structural trend of lower Wilcox pro-

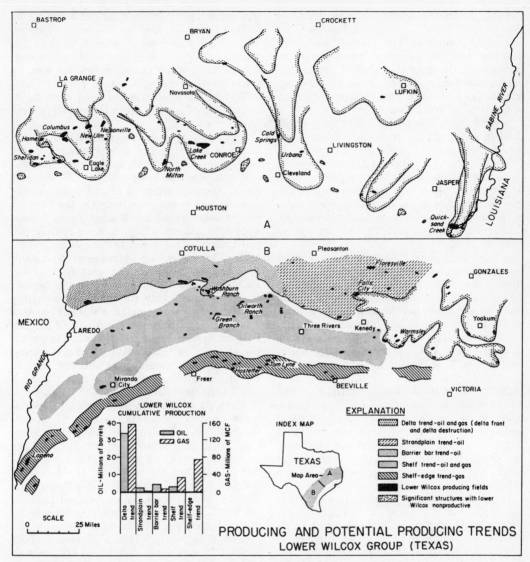

Fig. 11.—Productive and potentially productive oil and gas trends, lower part of Wilcox Group, Texas. For well control, see Figure 1.

duction within the delta system are notably nonproductive because of their location in sandstone-poor interdelta subembayment facies (Fig. 11). Growth faults and diapiric structures are associated with zones of thick deposition, such as the delta front; structures associated with these features are locally important traps.

Marginal to the delta trend is the standplain trend, controlling oil accumulation chiefly in the lower Wilcox of northwestern Gonzales, Wilson, northwestern Karnes, and eastern Atascosa Counties. Areally extensive sheet sandstone interbedded with dominantly marine mudstone forms a wide productive trend, which includes the farthest updip lower Wilcox production.

Sandstone units on the seaward side of the Cotulla barrier-bar system constitute the barrier-bar trend which extends in a narrow zone from northern McMullen County southwestward across LaSalle County through western Webb County. Most of the established production from this trend is in east-central LaSalle

County, associated mainly with a shelf-side tidal delta of the Cotulla system. Provided that entrapping structures are present, additional discoveries in the barrier-bar trend are likely; significantly, the trend is at shallow depth. Production has been chiefly oil; however, some significant gas wells have been completed recently in this trend in Webb County.

Several lower Wilcox fields have been developed in the sandstone reservoirs of the South Texas shelf system. With the exceptions of Green Branch and Dilworth Ranch, however, fields are small. Individual sandstone units within this extensive system are sporadic in distribution and not easily definable; many sandstone units are muddy with low permeability. Production has been primarily gas with small amounts of oil.

A final, main lower Wilcox trend is referred to here as the shelf-edge trend (Fig. 11). Sandstone bodies of this trend are part of the South Texas shelf system, though their origin is not fully understood. They form a fairly well-defined, narrow zone extending from the southwestern margin of the Rockdale delta system through central Bee, central Live Oak, northwestern Duval, southeastern Webb, eastern Zapata, and western Starr Counties. Productive sandstone units presumably originated as barrier bars similar to elongate sandstone bodies updip on the remaining part of the shelf system and in the Cotulla system. Their characteristics and downdip occurrence suggest that shelf-edge reworking and cleaning may have been a result of the higher energies associated with such a zone.

If the presence of submarine-fan deposits is indicated by deep drilling, an additional lower Wilcox trend may be established. Most likely area of such sandstone development would be downdip of the major shoal-water deltas of the Rockdale system, chiefly in Goliad, Victoria, Jackson, and Wharton Counties. Identification of such a trend will require extensive deep drilling.

Comparison of Lower Wilcox and Northwest Gulf of Mexico Depositional Systems

Comparison of the lower Wilcox with the Holocene of northwest Gulf of Mexico shows many similarities both in the kinds and distribution of depositional systems and in the component facies of these systems (Fig. 12). This is expected, because both are developed in the same structural basin. Prominent in the northwest Gulf is the extensive Mississippi delta system made up of five distinct delta complexes (Fig. 6) and at least 16 subdeltas (Frazier, 1967), with total area of about 20,000 sq mi. The lower Wilcox Rockdale delta system consists of seven delta complexes and at least 16 subdeltas (Fig. 5) with a total area of about 24,000 sq mi. Similarities in component facies and shapes of specific deltas in the two systems have been noted.

The Mississippi delta system is fed by the large, interior-draining fluvial system of the Mississippi River. Much of the analogous fluvial system of the lower Wilcox (Mt. Pleasant fluvial system) extended inland from present outcrop and is not preserved. On the basis of comparable size of the Rockdale and Mississippi delta systems, a large, interior-draining fluvial system is presumed to have been present during lower Wilcox deposition. Maximum thickness of the preserved part of the Mt. Pleasant fluvial system is between the modern Colorado and Trinity Rivers, and presumably the larger streams entered the Gulf in that area. The principally preserved part of the Mt. Pleasant system in northeast Texas was probably a minor or secondary part of the system, with streams comparable to modern streams east of the Mississippi River such as the Pearl and Tangipahoa Rivers.

Marginal to the area of maximum development of the gulfward-extending Mississippi River and delta systems are relatively low-energy basins or depressions in which freshwater to marine deposition occurs (Lake Pontchartrain, Lake Borgne, and part of Mississippi Sound on the east, Vermillion Bay, West Cote Blanche Bay, White Lake, and Grand Lake on the west). Lower Wilcox analogues include the predominantly marine mudstone systems of the Pendleton lagoon-bay system developed along the east flank of the Rockdale and Mt. Pleasant systems and the updip part of the San Marcos system on the southwest or opposite side.

A net southwestward longshore transport of sediments from both the modern Mississippi and the Wilcox Rockdale delta systems formed comparable depositional systems along strike of these systems. The southwestern Louisiana chenier system, adjacent to the Mississippi delta system, is at least mechanically analogous to the San Marcos strandplain system, which developed just down the longshore drift of the Rockdale delta system. More distally along the strike of the Holocene Mississippi delta system is a barrier-bar system, of which the modern Bolivar-Galveston barrier bar is a part. This system is considered analogous to the bar-

Depositional Systems in Wilcox Group (Eocene) of Texas

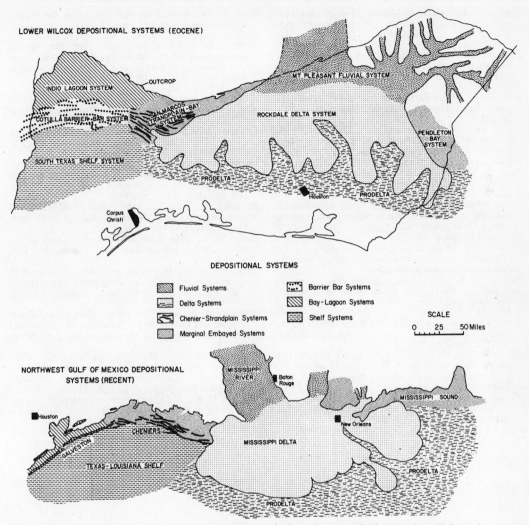

Fig. 12.—Comparison of lower Wilcox (Eocene) and northwestern Gulf of Mexico (Holocene) depositional systems.

riers making up the Cotulla barrier-bar system of the South Texas lower Wilcox. Landward complement of the modern barrier bars is a bay-lagoon system, represented at present by Galveston Bay. Deposits of this system are the modern analogue of the Indio lagoon-bay system. The extensive Louisiana-Texas continental shelf, situated gulfward of the barrier bar and chenier system and partly marginal to the Mississippi delta system, is the modern counterpart of the lower Wilcox South Texas shelf system.

References Cited

Allen, J. R. L., 1963, The classification of cross-stratified units, with notes on their origin: Sedimentology, v. 2, p. 93–114.

Andrews, P. B., 1966, Facies and genesis of a hurricane washover fan, St. Joseph Island, central Texas coast: Unpub. Ph. D. dissert., Texas Univ. at Austin, 238 p.

Barry, J. O., and R. J. LeBlanc, 1942, Lower Eocene faunal units of Louisiana: Louisiana Geol. Survey Bull. 23, 208 p.

Boyd, D. R., and B. F. Dyer, 1964, Frio barrier bar system of South Texas: Gulf Coast Assoc. Geol. Socs. Trans., v. 14, p. 309–322.

Brown, L. F., et al., 1967, Role of compaction in development of geometry of superposed elongate sandstone bodies (abs.): Am. Assoc. Petroleum Geologists Bull., v. 51, p. 455–456.

Claypool, C. B., 1933, The Wilcox of central Texas: Abs. of Ph. D. thesis, Illinois Univ., 12 p.

Coleman, J. M., and S. M. Gagliano, 1964, Cyclic sedimentation in the Mississippi River deltaic plain: Gulf Coast Assoc. Geol. Socs. Trans., v. 14, p. 67–80.

Echols, D. J., and D. S. Malkin, 1948, Wilcox (Eocene) stratigraphy, a key to production: Am. Assoc. Petroleum Geologists Bull., v. 32, p. 11-33.

Feofilova, A. P., 1954, The place of alluvium in sedimentation cycles of various orders and time of its formation: Akad. Nauk SSSR Geol. Inst. Trudy, v. 151, Coal ser., no. 5, p. 241-272.

Fisher, W. L., 1963, Lignites of the Texas Gulf coastal plain: Texas Univ. Bur. Econ. Geology Rept. Inv. 50, 164 p.

―――― and J. H. McGowen, 1967, Depositional systems in the Wilcox Group of Texas and their relationship to occurrence of oil and gas: Gulf Coast Assoc. Geol. Socs. Trans., v. 17, p. 105-125.

Fisk, H. N., 1955, Sand facies of recent Mississippi delta deposits: 4th World Petroleum Cong. Proc., Sec. 1-C, p. 377-398.

―――― 1960, Recent Mississippi River sedimentation and peat accumulation: 4th Internat. Cong. Carboniferous Stratigraphy and Geology, Heerlen 1958, Compte Rendu, p. 187-199.

―――― 1961, Bar-finger sands of the Mississippi delta, in Geometry of sandstone bodies: Am. Assoc. Petroleum Geologists, p. 29-52.

Frazier, D. E., 1967, Recent deltaic deposits of the Mississippi River: their development and chronology: Gulf Coast Assoc. Geol. Socs. Trans., v. 17, p. 287-311.

Friedman, S. A., 1960, Channel-fill sandstones in the middle Pennsylvanian rocks of Indiana: Indiana Geol. Survey Rept. Prog. 23, 59 p.

Hargis, R. N., 1962, Stratigraphy of the Carrizo-Wilcox of a portion of South Texas and its relationship to production: Gulf Coast Assoc. Geol. Socs. Trans., v. 12, p. 9-25.

Hayes, M. O., and A. J. Scott, 1964, Environmental complexes, South Texas coast: Gulf Coast Assoc. Geol. Socs. Trans., v. 14, p. 237-240.

Hoyt, W. V., 1959, Erosional channel in the middle Wilcox near Yoakum, Lavaca County, Texas: Gulf Coast Assoc. Geol. Socs. Trans., v. 9, p. 41-50.

Kolb, C. R., and J. R. van Lopik, 1966, Depositional environments of the Mississippi River deltaic plain —southeastern Louisiana, in Deltas in their geologic framework: Houston Geol. Soc., p. 17-61.

Kruit, C., 1955, Sediments of the Rhone delta; I, Grain size and microfauna: Koninkl. Nederlandsch Geol. Mijnbouwk Gen. Verh., Geol. Ser. v. 15, p. 397-499.

Lowman, S. W., 1949, Sedimentary facies in the Gulf Coast: Am. Assoc. Petroleum Geologists Bull., v. 33, p. 1939-1997.

Mueller, J. C., and H. R. Wanless, 1957, Differential compaction of Pennsylvanian sediments in relation to sand-shale ratios, Jefferson County, Illinois: Jour. Sed. Petrology, v. 27, p. 80-88.

Pettijohn, F. J., et al., 1965, Geology of sand and sandstone: Indiana Geological Survey and Dept. Geology, Indiana Univ. Conf., 205 p.

Potter, P. E., 1963, Late Paleozoic sandstones of the Illinois basin: Illinois Geol. Survey Rept. Inv. 217, 92 p.

Rainwater, E. H., 1963, The environmental control of oil and gas occurrence in terrigenous clastic rocks: Gulf Coast Assoc. Geol. Socs. Trans., v. 13, p. 79-94.

Thomas, W. A., and C. J. Mann, 1966, Late Jurassic depositional environments, Louisiana and Arkansas: Am. Assoc. Petroleum Geologists Bull., v. 50, p. 170-182.

Walker, R. G., 1966, Shale grit and Grindslow shales: transition from turbidite to shallow water sediments in the upper Carboniferous of northern England: Jour. Sed. Petrology, v. 36, p. 90-114.

Wanless, H. R., et al., 1963, Mapping sedimentary environments of Pennsylvanian cycles: Geol. Soc. America Bull., v. 74, p. 437-486.

Recognition of Barrier Environments[1]

DAVID K. DAVIES,[2] FRANK G. ETHRIDGE,[3] and ROBERT R. BERG[4]
Columbia, Missouri 65201, Carbondale, Illinois 62901,
and College Station, Texas 77843

Abstract The vertical succession of sedimentary structures and textures at Galveston Barrier Island, Texas, is identical with vertical successions in two ancient barrier complexes, one in the Lower Cretaceous of Montana and the other in the Lower Jurassic of England. Within both Holocene and ancient examples, there is a gradation upward from (1) irregular interlaminations of siltstone and claystone at the base, through (2) burrowed and generally structureless sandstone, to (3) low-angle and microtrough cross-laminated sandstone, terminating in two of the examples in (4) structureless and rooted sandstone. This sequence represents deposition in (1) lower shoreface, (2) middle shoreface, (3) upper shoreface-beach, and (4) eolian environments, respectively.

Analyses of quartz size and content of the Holocene and ancient barriers yield textural and compositional parameters that are environmentally sensitive. Plots of these parameters demonstrate that each of the environments may be distinguished on the basis of thin-section analyses. Consequently, full diameter cores, which show sedimentary structures, may not be necessary for precise environmental interpretation in the subsurface. Indeed, thin sections of sidewall cores may yield significant and reliable environmental interpretations in barrier sandstones.

Textural and sedimentary structural similarities between Galveston Island and the ancient examples permit a general model of barrier sedimentation to be developed.

INTRODUCTION

One of the most impressive developments in sedimentology during the past decade has been the establishment of criteria for the recognition of depositional environments. Sandstone-body morphology, sedimentary structures, and vertical grain-size changes are parameters that have been used with considerable success in diverse environments (Allen, 1962, 1965; Exum and Harms, 1968; Potter, 1967; Shelton, 1967; Visher, 1965). The burgeoning of environmental studies has made it apparent that correct identification of depositional environments is requisite to any successful search for stratigraphic traps. Recent discoveries in the Muddy Sandstone of the Powder River basin have emphasized the importance of environmental analysis in stratigraphic exploration (McGregor and Biggs, 1968, 1969).

This paper focuses on one of the potentially most productive sand bodies—the barrier. Because of deposition in relatively shallow, and often agitated waters, barriers tend to be composed of mature sediments that are highly porous and permeable (Berg and Davies, 1968). Flanking lagoonal and offshore sediments are characterized by finer grained, organic-rich deposits that act not only as natural traps but also as potential source beds for petroleum.

Although barriers are not uncommon in the geologic record, relatively few have been studied in sufficient detail to allow recognition of the individual environmental facies of which they are composed. Differentiation of shoreface, beach, dune, and washover subenvironments is rarely undertaken, and these subenvironments generally are grouped together as bar or barrier (Sabins, 1963; Weidie, 1968). The

---

[1] Manuscript received, July 27, 1970; accepted, October 7, 1970.
Part of this paper was presented at the Eastern Montana Symposium of the Montana Geological Society, October 22, 1969 (Davies and Berg, 1969) and pertinent sections are reproduced here with the permission of the Montana Geological Society.

[2] Department of Geology, University of Missouri.

[3] Department of Geology, Southern Illinois University.

[4] Department of Geology, Texas A&M University.

The cores and samples on which this study is based were provided by many individuals and companies, and we are appreciative of their cooperation. Among those whose help has been particularly important are A. A. McGregor, C. A. Biggs, W. H. Broman, R. J. LeBlanc, Sr., L. D. Meckel, M. A. McBrayer, J. M. Parker, R. J. Stanton, and M. A. Warner. Shell Development Company kindly provided samples and cores from Galveston Island. Chevron Oil Company, Kirby Petroleum, and Samuel Gary provided cores, thin sections, and information concerning the Cretaceous of the Powder River basin.

Acknowledgment is made to the donors of the Petroleum Research Fund, administered by the American Chemical Society, for partial support of this research. Other organizations which kindly provided financial support include the Department of Geology, Texas A&M University, and the Natural Environment Research Council of Great Britain.

William R. Bryant and Michael R. Facundus reviewed the manuscript.

© 1971. The American Association of Petroleum Geologists. All rights reserved.

greatest drawback to environmental investigations in ancient sequences is the scarcity of favorably located outcrops or cores, which prevents detailed study of sedimentary structures and textures.

The study of coastal constructional features such as barriers is fraught with terminological difficulties. Coastal geomorphologists studying Holocene sediments prefer to restrict the term "bar" to constructional features that are submerged at high tide, and to use the terms "barrier islands," "barrier beaches," and "spits" for those which are emergent (Hoyt, 1967; Price, 1951, 1968; Shepard, 1952). The use of such precise terminology demands a knowledge of the detailed configuration of the coastline, and whether the "bar" was emergent or submerged. However, in the geologic record it is generally difficult to decide how a particular constructional feature was related and oriented with respect to the coastline. Penecontemporaneous erosion may have resulted in the destruction of emerged parts of such features, leaving little indication as to the emergent or submerged nature of the "bar."

These difficulties are illustrated by some examples of coastal features. One example, from the Jurassic of England, was at no time parallel with the coastline. In the other example from the Cretaceous of Montana, there is no indication of the detailed configuration of the coastline, nor is there evidence to suggest that it was either attached to the shore (spit) or separated laterally by tidal channels (barrier island). However, both examples were emergent during their history, and both represent coastal constructional features and hence are called "barriers" in this paper.

It is the purpose of this paper to demonstrate that barrier environments can be recognized readily in both Holocene and ancient strata. The remarkable similarity of internal structures and textures in the examples discussed indicates that correct environmental interpretations can be made, even with limited outcrops and cores. In fact, in certain instances it appears possible to recognize and segregate specific environments through thin section analysis alone. Three examples, chosen to illustrate the characteristic sedimentary features of barrier complexes, include one example from the Holocene (Galveston Island), one from the Cretaceous of Montana, and one from the Jurassic of England. The similarity of ancient and modern sequences enables a general model of barrier sedimentation to be developed.

METHODS

Four full diameter cores, representative of an almost complete sand sequence, were available from Galveston Island. Small gravity cores from the adjacent lagoon, together with surface exposures of dunes enabled the full barrier sequence to be studied and sampled. Full diameter cores from the entire Muddy Sandstone interval were available from seven wells through the Bell Creek (Lower Cretaceous) barrier. All cores were cut vertically to show details of bedding and structure. The Jurassic strata of England were studied in outcrop.

Because these sandstones are friable, selected samples were impregnated with epoxy resin and thin sectioned. In each section, the longest axis of 50 randomly selected quartz grains was measured, and the mean of these measurements is used as an expression of grain size for each sample. The selection of a single mineral for measurement, such as quartz, eliminates grain-size variation that results from a change in mineral composition. Such a standardized procedure is critical if the objective is to compare several sets of samples to define a size difference for environmental interpretations. It would be meaningless to measure the axes of several different minerals and to use the mean of such measurements as an expression of grain size. If this were done, differences between samples would reflect differences in mineral composition. For example, the mean size for a pure quartz sandstone would be significantly increased by the addition of mica, a mineral with an exaggerated long axis (Griffiths, 1967, p. 64-66). Therefore, the term "grain size," as used herein, refers to the mean size of the longest axis of quartz measured in thin section.

The composition of each sample was evaluated by point counting 100 grains per thin section. This technique yields an unbiased estimate of the volumetric abundance of each of the mineral species present (Chayes, 1956).

GALVESTON ISLAND—A RECENT BARRIER

As a result of the considerable detail published by Bernard et al. (1962), Galveston Island is one of the best documented of barrier complexes (Fig. 1): The internal sequence of textures and sedimentary structures of this Holocene barrier serves as a useful standard with which the internal features of ancient barrier sequences may be compared. Because of the limited circulation of Bernard et al.'s paper, their conclusions are not widely known and are summarized here.

552 David K. Davies, Frank G. Ethridge, and Robert R. Berg

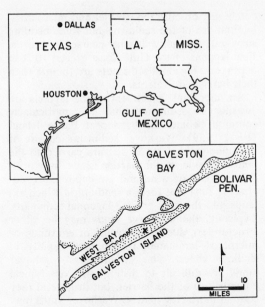

FIG. 1.—Index map of northwest Gulf of Mexico showing location of Galveston Island (X indicates location of core R–3963, illustrated in Fig. 3).

Galveston Island barrier sands are lenticular, with a maximum thickness of 50 ft. They are flanked both landward and seaward by silts and clays representing typical lagoonal and offshore deposits respectively. Sections across the island, together with radiocarbon dating, have revealed an offlap sequence that resulted from the seaward accretion of the barrier during stationary sea level. The vertical sequence of sedimentary features from bottom to top of the island is characteristic, and allows ready subdivision into four distinct units: (1) lower shoreface, (2) middle shoreface, (3) beach-upper shoreface, and (4) eolian (Fig. 2).

The lower shoreface, or shoreface-toe, sediments are deposited seaward of the break in offshore slope, within the depth interval of 30–40 ft. These sediments consist of interbedded, burrowed, and churned (bioturbated), very fine-grained sand, silt, and silty clay which may reach a thickness of 6 ft.

The overlying middle shoreface sediments were deposited shoreward of lower shoreface sediments. They consist of very fine-grained sand which is so extensively bioturbated that sedimentary structures are only rarely preserved. Some cross-laminated, shelly-sand layers are present sparingly as are interlaminae of silty clay. In general, therefore, middle shoreface sands are structureless and bioturbated.

They range in thickness from 10 to 34 ft, and are deposited in from 5 to 30 ft of water.

Beach and upper shoreface sediments gradationally succeed and lie shoreward of middle shoreface deposits. They consist of fine to very fine-grained, well-laminated sands 3–10 ft thick. The most characteristic sedimentary structure is planar, low-angle, cross-lamination with localized developments of microcross-lamination. Burrowing is scarce, and shells may be locally abundant.

Although Bernard et al. (1962) distinguished between beach and upper shoreface at Galveston, the two environments are difficult to distinguish solely on the basis of sedimentary structures and textures. At Galveston Island the distinction between upper shoreface and beach is based on a knowledge of sea level, the upper shoreface environment being restricted to the shallow water seaward of the beach, but this distinction is not easily made in cores. As a result, the beach and upper shoreface environments are combined in Figure 2.

Eolian sediments that cap the Galveston barrier are generally cross-laminated (festoon and planar) and parallel-laminated. These structures are progressively destroyed with increasing age as a result of plant and animal action, as well as by weathering and the movement of groundwater. Thus, older eolian deposits tend to be structureless or massive, but with definite traces of plant rooting and thin soil zones. Eolian sediments are fine- to very fine-grained sands, generally 2–8 ft thick.

The characteristic vertical sequence of structures is well developed in all cores from Galveston Island. A petrographic analysis of the vertical variation in grain size was undertaken. The results for one core (R-3963), illustrated in Figure 3, show that the vertical change in sedimentary structures is matched by a typical gradational change in grain size from the lower

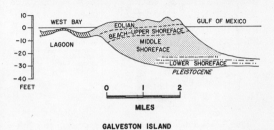

FIG. 2.—Diagrammatic cross section of Galveston Island, Texas, showing principal environments of barrier island complex. Modified from Bernard et al. (1962, Fig. 12, p. 204).

shoreface (very fine sand) to the eolian environment (fine sand). The increase in grain size from the beach to eolian environment at Galveston Island is apparently anomalous. A vertical decrease of grain size (based on sieve analyses data) from beach to eolian has been reported from other barrier systems such as Mustang and Padre Islands, Texas (Dickinson and Hunter, 1970; Mason and Folk, 1958), along the Atlantic Coast of America (Giles and Pilkey, 1965), and New South Wales, Australia (Gibbons, 1967).

Landward from Galveston barrier island are lagoonal sediments. They consist of interbedded, burrowed, and churned clay and silt with some fine-grained sand. These shallow water deposits are commonly succeeded by rooted clay and silt of bordering marshes. Both animal burrows and roots generally are vertical and commonly obscure the parallel and microcross-laminations of the sediment, but some bedding survives, especially in the coarser interbeds. Few lagoons of the Gulf Coast exceed 10 ft in depth, and lagoonal sediments are thinner than their barrier equivalents.

An important constituent of the lagoons are "washover" sediments that are particularly common behind the barrier at Padre Island (Fisk, 1959). Washover sediments consist of very fine-grained sands that are carried to the landward side of the barrier island by storm-driven water or wind and are deposited as coalescing fans that form a sand apron which extends into the finer-grained lagoonal sediments. Typically, the washover sands may be either structureless, due to burrowing or churning, or microcross-laminated, if deposited rapidly in shallow water. The structureless washover sands are difficult to distinguish from typical eolian sands of the barrier, but in general they contain more clay matrix. Washover sands may grade vertically into true marsh sediments which are structureless, extensively rooted silts and silty clays.

CRETACEOUS BARRIER

One of the greatest drawbacks to detailed environmental investigations in ancient sandstone bodies is the lack of sufficient outcrops or cores. Only through good sample coverage may ancient and modern barrier complexes be compared. One ancient barrier complex that has been drilled and cored extensively is in the Lower Cretaceous of Montana, and part of this complex forms Bell Creek oil field (Fig. 4).

At Bell Creek field the Lower Cretaceous Muddy Sandstone forms a prolific stratigraphic trap. High porosity and permeability values combine to make it an excellent oil reservoir. The sandstone body is parallel with regional strike, and is comprised of quartz-rich sandstone with an average thickness of 22 ft. Two distinct lenses form the reservoir, and sandstone thicknesses decrease northwest and southeast (Fig. 5).

Bell Creek field has been interpreted as having originated as a barrier bar complex, flanked on the northwest by marine siltstone and shale, and on the southeast by typical lagoon deposits (Berg and Davies, 1968; McGregor and Biggs, 1968). Details of stratigraphic relations and reservoir characteristics have been published (Berg and Davies, 1968; McGregor and Biggs, 1968, 1969; Haun and Barlow, 1962).

The succession of sedimentary structures and

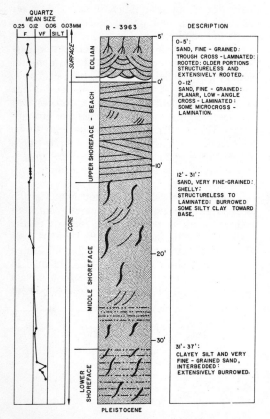

FIG. 3.—Sedimentary structures, textures, and lithology of typical barrier island sediments in core R-3963 from Galveston Island. Eolian sediments were not cored; eolian section was measured close to original core site in order to complete vertical sequence of environments.

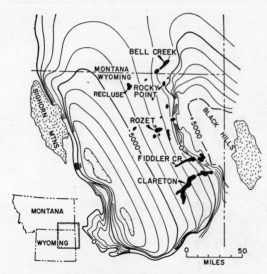

FIG. 4.—Index map of Powder River basin, Wyoming and Montana, showing location of some principal oil fields in Muddy Sandstone, including Bell Creek field. Structural datum is on top of Dakota sandstone. C. I. = 1,000 ft.

textures in the Muddy Sandstone at Bell Creek field is similar to the succession outlined for Galveston barrier island (Fig. 6). As at Galveston, four distinct units may be recognized readily in a complete vertical sequence, and these are, in ascending order, (1) lower shoreface, (2) middle shoreface, (3) beach-upper shoreface, and (4) eolian. Boundaries between each of these units are gradational, and in no place is the sequence disrupted by the intercalation of nonadjacent units, such as 3 on 1 or 4 on 2. If units are absent, they are generally the eolian and beach-upper shoreface units.

The basal unit of the Muddy barrier sequence is composed of 2–3 ft of interlaminated siltstone and claystone. Laminations are irregular, but may be wavy to regular; microcross-lamination is present in the form of discrete "pods" (X on lowest photograph, Fig. 6). Silt-filled burrows up to ¼-in. in diameter parallel the bedding. Fine-grained rocks of this unit have at least 50 percent matrix and are interpreted as lower shoreface sediments deposited seaward of the barrier.

The basal unit grades upward into very fine-grained sandstone through an interval of less than 1 ft. Overlying sandstones are structureless to sparsely mottled, and in the core illustrated, reach 5 ft in thickness (Fig. 6). The structureless nature of the sandstone is believed to reflect intense burrowing activity which has homogenized the sediment. This conclusion is supported by the presence of indistinct mottling and a few discontinuous or broken laminae. Distinct burrows are absent, probably because the sandstone is well sorted and lacks strong grain-size contrasts. The structureless, very fine-grained sandstone originated in the middle shoreface zone of the barrier sequence, seaward of the beach but landward of the finer offshore siltstones.

Structureless sediments of the middle shoreface unit grade upward into well-laminated, fine-grained sandstone about 12 ft thick (Fig. 6). Laminations in this unit are planar, low-angle, and individual sets may be 1 ft thick. Thin, structureless zones, 2 in. thick, are interbedded with the planar cross-laminated sets. Microcross-lamination is present sporadically and represents small, asymmetric ripples that probably developed under local conditions of reduced energy. Burrowing is sparingly present, and rooting is absent. These sediments probably originated on a beach or in adjacent shallow water, landward of their middle shoreface equivalents. They are thus equated with sediments from the beach-upper shoreface unit at Galveston Island.

The Muddy Sandstone may be capped with 2–3 ft of structureless, rooted, very fine-grained sandstone. The transition from the well-laminated to overlying structureless sandstone unit is gradational, and is marked by an increase in

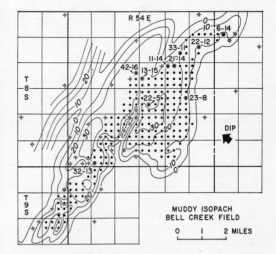

FIG. 5.—Isopach map of Muddy Sandstone at Bell Creek field showing gross thickness of Muddy. C. I. = 10 ft. Core samples were obtained from numbered wells. Modified from McGregor and Biggs (1968, Fig. 9).

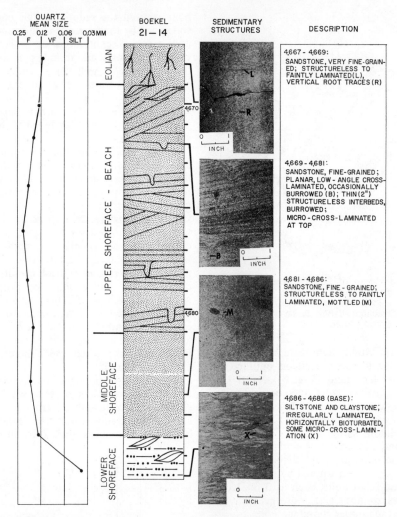

Fig. 6.—Sedimentary structures, textures, and lithology of typical barrier-bar reservoir sediments in well 6–14, Bell Creek field, Montana.

matrix content and a decrease in grain size. The structureless sands of the uppermost unit contain some subtle evidence of lamination or an irregular pseudolamination. This pseudolamination is the result of diffused accumulation of organic material not coincident with original structure. Such concentrations of organic material in eolian sands of Galveston Island are the result of the downward movement of surface water. The organic material is deposited at the water table and forms indistinct pseudolaminae (Bernard et al., 1962, p. 200).

Within the barrier sandstones there is an upward increase in grain size from lower shoreface to upper shoreface-beach, and then a decrease in grain size to the eolian. The decrease in grain size upward from beach to eolian is in contrast to the upward increase on Galveston Island, but is identical with the change observed in other Holocene barrier systems of the United States and Australia (Dickinson and Hunter, 1970; Gibbons, 1967; Giles and Pilkey, 1965; Mason and Folk, 1958).

The barrier sandstones are flanked on the southeastern (updip) edges by interbedded fine- to very fine-grained sandstone, siltstone, and shale. These sediments have been interpreted as lagoonal deposits and as washover fans landward of the barrier bar, and form an effective trap to petroleum migration.

The vertical succession of sedimentary structures in the trap sediments updip from Bell Creek field is not as distinctive as in the barrier sequence. Variability is the rule, and siltstone and claystone alternate irregularly with sandstone in any vertical sequence. This reflects variability of sediment supply behind the barrier, for the coarser sediments are derived from the erratic activity of storm-driven winds and waters. The finer siltstone and claystone represent more persistent depositional conditions within the lagoon.

In a core through typical trap sediments (Fig. 7), three distinct units may be recognized within the Muddy interval. These are, from bottom to top, (1) lagoon, (2) washover, and (3) marsh.

The basal unit is composed of 6 ft of siltstone, claystone, and sandstone. The siltstone and claystone may be parallel-laminated or bioturbated (Fig. 8C). Sandstone interbeds are common, have erosive bases characterized by scour and fill structures, and are microcrosslaminated (Fig. 8D). Vertical burrows are numerous, particularly toward the base (Fig. 8D). The finer grained sediments represent true lagoonal deposits, whereas sandstone interbeds reflect the periodic influx of coarser material from the adjacent barrier.

Laminated and burrowed siltstone, claystone, and sandstone of the lagoon unit, grade upward into 4 ft of structureless silty, clayey, very fine-grained sandstone (Fig. 8A, B). Distinct burrows are observed in this unit (Fig. 8B), but are restricted to the lower part. The structureless nature of this unit may reflect the activity of burrowing and churning organisms. The high silt and clay content of these rocks indicates a dominantly low-energy environment of deposition, such as might occur in washover fans in the lee of barrier systems.

The structureless, calcitic and sideritic sandstone beds of the washover unit grade upward into structureless silty and bentonitic claystone as much as 3 ft thick (Fig. 7). Root structures are common, and they are marked by accumulations of organic material along thin, irregular tubes of random attitude. Sedimentary structures have been destroyed by the intense rooting.

The vertical sequence from lagoon through washover to marsh represents the gradual filling of the lagoon adjacent to the barrier. Because of the shallow-water conditions that characterized the lagoon, the Muddy interval updip from the barrier is reduced in thickness from about 20 ft at the margin of the barrier sandstones to 10 ft or less on the northeast edge of the field. Washover sandstone is also thin, and it is oil productive only where it is continuous with thicker barrier sandstones. Other washover lenses produce only gas, probably because of reduced permeability to oil.

Barrier-bar reservoir and lagoonal trap sediments at Bell Creek field therefore may be recognized readily through a study of sedimentary structures and textures. These sedimentary characteristics are similar to those of modern barrier sequences of the Texas Gulf Coast.

JURASSIC BARRIER

A thick barrier complex in Lower Jurassic (Toarcian) strata of southern England has been described in detail by Davies (1969). This barrier is considered to have originated as a land-tied spit which was locally emergent and cut by prominent tidal channels. Because of the refined biostratigraphic framework developed for the Jurassic rocks of this area, it is possible to map the geographic distribution of environments for a series of isochronous units. Figure 9 is a paleogeographic reconstruction of southern England for one of these isochronous units (*Dumortieria moorei* subzone). These quartzose siltstones and sandstones were deposited in four distinct environments of deposition each of which may be equated with modern equivalents at Galveston Island. These environments include (1) the bar (upper shoreface-beach), (2) fore-bar (lower and middle shoreface), (3) back-bar (washover), and (4) tidal channel (Figs. 9, 10). Although the spatial relations and sediments characteristic of these environments are akin to those at Galveston Island, the more general terms "bar," "fore-bar," and "back-bar" are preferred for each environment because this was not a barrier system in the strict sense, being flanked on north and south by open marine waters.

1. Rocks of the bar environment are very fine-grained sandstones, characterized by well-preserved sedimentary structures. These structures are generally small scale, and consist of a low angle cross-lamination, or microtrough cross-lamination (Figs. 10, 11). Paleocurrent analyses indicate a bimodality of flow in this environment, such as is characteristic of modern barriers (Klein, 1967).

2. Fore-bar siltstones are excessively bioturbated, and sedimentary structures are rarely preserved intact (Fig. 10). These bioturbated siltstones are blue, calcareous, and medium

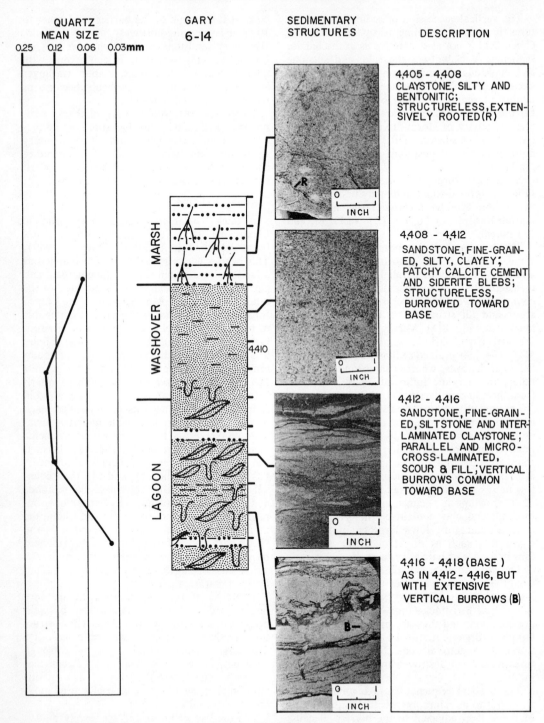

Fig. 7.—Sedimentary structures, textures, and lithology of lagoon sediments in well 6-14, Bell Creek field, Montana.

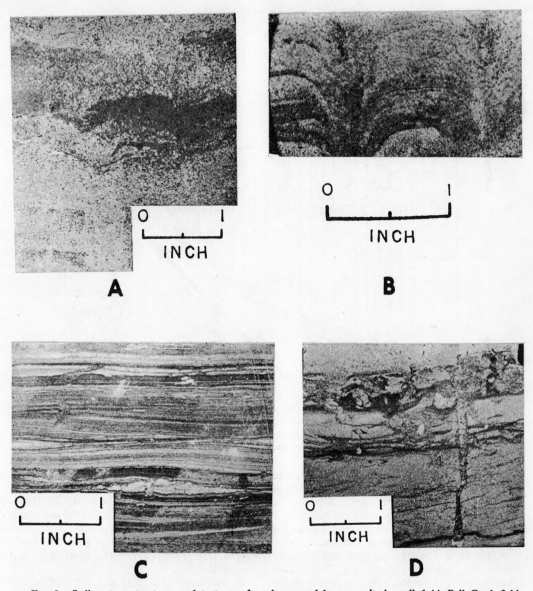

Fig. 8.—Sedimentary structures and textures of washover and lagoon units in well 6–14, Bell Creek field. **A.** Burrowed and churned washover sandstones (4,411 ft.). **B.** Prominent burrow in microcross-laminated washover sandstone (4,412 ft). **C.** Microcross-laminated, parallel laminated, and burrowed lagoonal siltstones and claystones (4,414 ft). **D.** Microcross-laminated lagoonal sandstones and siltstones with prominent vertical burrow (4,416 ft).

grained at the base of any vertical sequence, and become yellow, calcareous, and coarse-grained upward. The underlying, marine, blue siltstones, sometimes referred to as the "Upper Lias Clays," are considered to represent the lower shoreface environment; succeeding yellow siltstone beds represent the middle shoreface environment.

3. Rocks of the back-bar environment are almost indistinguishable from those of the forebar. Because they originated from transport either over the bar or through tidal channels,

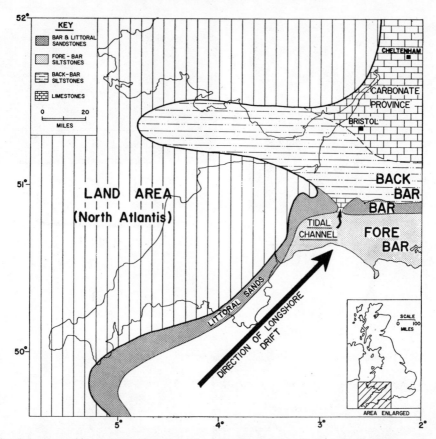

Fig. 9.—Paleogeographic reconstruction of southern England during Early Jurassic (*Dumortieria moorei* subzone). Recent evidence suggests that Wales was not emergent during Jurassic (Blundell *et al.*, 1970).

they are analogous to washover systems in the Holocene. Inasmuch as this Jurassic example originated as a spit, these washover deposits are not lagoonal. Rather they were deposited in a relatively open marine environment, but in the lee of the main constructional feature.

4. Rocks of the tidal channel environment are coarse, fragmental biosparites. Flow directions in one of these channels indicates a northeasterly direction of flow. These rocks are characterized by a vertical decrease in grain size and a vertical decrease in the scale of sedimentary structures from large scale festoon crossbedding near the base to microtrough crosslamination toward the top of the channel sequence (Davies, 1969, Fig. 11).

This Jurassic spit and its associated barrier islands migrated southward with time—a result of the constant supply of sediment to its southern side, and a failure of subsidence to keep pace with sedimentation. The rate of migration is estimated at 1 mi every 80,000 years. Lateral migration resulted in the preservation of vertical sequences commencing with lower shoreface sediments, succeeded by middle shoreface, then upper shoreface-beach sediments, capped with washover sediments (Fig. 10). The whole barrier system is underlain and overlain by marine sediments.

Sedimentary structures and textures of this Lower Jurassic example can therefore be equated with those developed for barriers in the Cretaceous of Montana and along the present Gulf Coast.

PETROGRAPHY

In each of the systems discussed, environmental interpretations have been based on sedimentary structures and textures displayed either in numerous outcrops across a large area or in full diameter cores. A third parameter, however, that of rock composition, is also

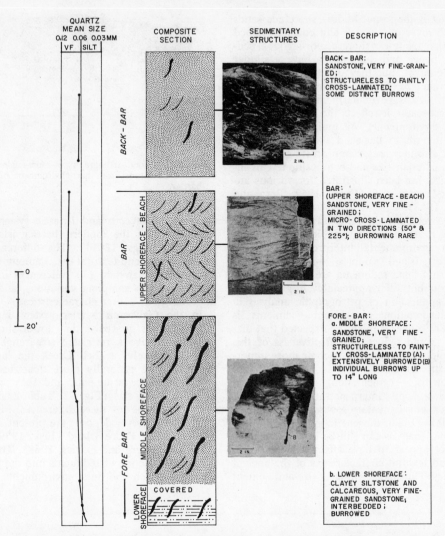

Fig. 10.—Sedimentary structures, textures, and lithology of sediments from Lower Jurassic of southern England. Owing to limited vertical exposures, section in figure is composite of three more complete exposures in *Dumortieria moorei* subzone.

available in support of environmental interpretations. Once petrographic criteria have been established in any area, an environmental interpretation may be based on even a small sample without the use of sedimentary structures or knowledge of complete textural sequence. Thus, even a limited number of sidewall cores or cuttings might be used in subsurface interpretations when full diameter cores are not available.

The value of petrographic analysis is demonstrated by the Galveston Island samples. These samples were collected from an area in which the environment of deposition was well established, and subenvironments were recognized largely on the basis of textural sequence and sedimentary structures (Bernard et al., 1962). Fifty samples were selected from cores representing the five principal environments of deposition, and standard thin sections were prepared. The compositional analysis yields data which are segregated according to environments (Fig. 12). As shown by the plot, quartz size and content vary according to environment. Lagoonal sands are less quartzose than lower shoreface sands, but in general the range of quartz is the same. Middle shoreface sands are both more quartz-rich and coarser grained than either the lower shoreface or lagoonal sediments. Beach and dune sands have a similar quartz content, but they differ appreciably in mean grain size (Fig. 12, Table 1).

Grain size is particularly important in this example because it reflects the energy of depositional environments. As grain size increases within the silt to fine-sand size range, quartz content also increases, demonstrating the interrelation of grain size and composition. This interrelation has been illustrated for various ancient sediments also (Ferm, 1962; Berg and Davies, 1968).

These relations show the advantage of considering more than one variable, such as grain size, in environmental interpretations. In the Galveston Island samples, upper shoreface-beach, and dune sediments are almost equally quartzose, but differ appreciably in grain size.

The application of petrographic analysis in the interpretation of ancient sediments is equally effective. It might be expected that diagenesis would reduce the effectiveness of this technique. Changes which become more important with increasing age include secondary overgrowths, cementation, rock-fragment breakdown, and interpenetration of grains. However, in sedimentary rocks which have suffered all but major diagenetic modification, petrographic analyses can still be used. Diagenesis does not appear to have affected gross petrologic and textural characteristics of the ancient examples. Thus plots of quartz size and content

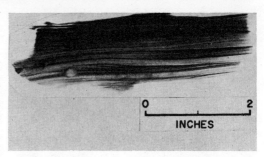

Fig. 11.—X-ray radiograph of low-angle cross-laminated sandstone from bar environment of Jurassic barrier.

for both Cretaceous and Jurassic systems are comparable to the modern barrier island at Galveston (Figs. 12–14). The differences in compositional and textural detail among the examples are related to (1) differences in local conditions, (2) sampling variations, and most important, (3) source characteristics.

In the Galveston barrier system lagoonal sands are less quartzose than lower shoreface sands and have a restricted size range (Fig. 12). Conversely, at Bell Creek the lagoonal sediments are generally more quartzose than lower shoreface sediments, and show a wide range of grain size (Fig. 13, Table 2). There are two reasons for these differences. First, no extensive washover deposits are present at Galveston, unlike Padre Island (Fisk, 1959) and Bell Creek (Berg and Davies, 1968). The continual introduction of sediments into the lagoon from the beach and dune environments would

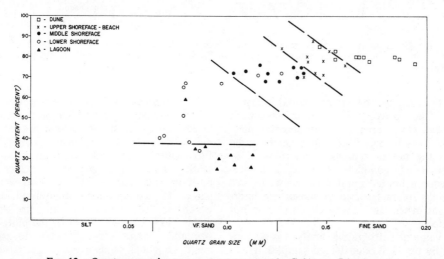

Fig. 12.—Quartz mean size versus quartz content in Galveston Island sands.

Table 1. Quartz Size and Content of Depositional Environments in Galveston Barrier Complex[1]

Depositional Environment	Quartz Mean Size (mm)	Quartz Content (%)
Lagoon	0.096	31.70
Lower shoreface	0.087	54.60
Middle shoreface	0.124	72.15
Upper shoreface-beach	0.144	78.00
Eolian	0.169	80.10

[1] All values are means calculated from 10 samples in each environment.

Table 2. Quartz Size and Content of Depositional Environments in Cretaceous Barrier at Bell Creek Field[1]

Depositional Environment	Quartz Mean Size (mm)	Quartz Content (%)
Lagoon	0.115	62.67
Lower shoreface	0.041	42.00
Middle shoreface	0.144	76.00
Upper shoreface-beach	0.181	91.67
Eolian	0.171	92.67

[1] All values are means.

result in a wide range of grain sizes available for resedimentation in the lagoon, as in the Cretaceous example. Second, samples available from the Cretaceous barrier system were taken near the junction of the lagoon and barrier island, where some hope exists for oil or gas production. Thus, most of the Cretaceous samples are not from the center of the lagoon, but biased in favor of the washover fans on the landward side of the barrier island.

A further important difference in the examples is a result of source characteristics, and this is illustrated by a comparison of analyses from Galveston Island and the Cretaceous and the Jurassic sandstones (Figs. 12–14). In both the Holocene and Cretaceous examples, a wide range of quartz grain sizes was available for deposition. As a result, the quartz size ranges from silt to fine sand. In the Jurassic example, however, the size of quartz available was limited to the coarse silt and very fine sand fraction. Consequently, although beach sediments are only 0.025 mm coarser than lower shoreface sediments, they contain 43 percent more quartz (Table 3).

In spite of differences in detail, these results indicate that petrographic analysis can play an important supportive role in the delineation of environments in ancient barrier complexes. The technique should not supplant regular sedimentary structural and textural studies of cores and outcrops, but should assist through quantifying environments on the basis of composition and texture. Perhaps the most important limitation of the technique is that the size and composition of locally available sediments must be established.

MODEL FOR BARRIER SEDIMENTATION

By use of the sedimentary characteristics of Galveston Island together with evidence from ancient examples and the literature, a model for dominantly regressive barrier sedimentation may be proposed (Fig. 15). This model has 11 characteristics.

1. For any given time interval, the sands will be lenticular and thin landward and seaward.

2. The sand body will be elongate parallel with depositional strike.

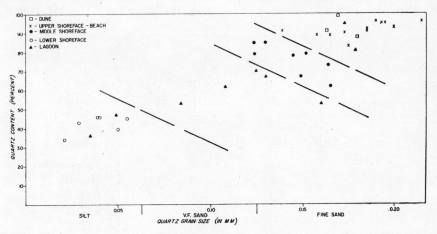

FIG. 13—Quartz mean size versus quartz content in Lower Cretaceous barrier bar at Bell Creek field, Montana.

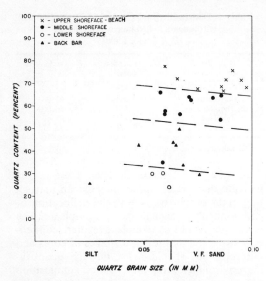

FIG. 14.—Quartz mean size versus quartz content in Lower Jurassic barrier, southern England.

Table 3. Quartz Size and Content of Depositional Environments in Jurassic Barrier[1]

Depositional Environment	Quartz Mean Size (mm)	Quartz Content (%)
Back bar	0.059	38.57
Lower shoreface	0.058	28.00
Middle shoreface	0.070	58.00
Upper shoreface-beach	0.083	71.22

[1] All values are means.

3. The barrier sands may be underlain by either marine or nonmarine sediments, or both. In the barriers discussed by Hoyt (1967), the sands are, rather surprisingly, underlain by nonmarine sediments. However, some barriers, particularly those originating as spits, may be underlain solely by marine sediments—such as the Jurassic example discussed in this paper, and a hypothetical example discussed by Hoyt (1967, p. 1129).

4. The base of the barrier sands should be nonerosive. However, if some barrier systems originate as beaches abutting a coastal plain (as suggested by Hoyt, 1967) localized basal erosion should be evident. Ferm and Hobday (personal commun.) have noted barriers with evidence of localized basal erosion in the Pennsylvanian of the Appalachian region.

5. The barrier sands should show an internal sequence of sedimentary structures similar to that of Galveston Island, representing upward transition from the lower shoreface environment to successively shallower marine environments. An interruption of sediment supply combined with continued subsidence could terminate bar development in any one of these environments. However, a reduction of sediment supply or an increase in relative rate of subsidence could initiate a reversal of sequence, and thus middle shoreface sediment could be overlain by lower shoreface sediment. Bars of this nature are not uncommon, particularly under delta-destructional conditions. Berg and Davies (1970) have termed such bars "transgressive." However, the general model developed herein presupposes dominant regression.

6. Tidal channels may be present and interrupt the typical vertical sequence of sedimentary structures and textures. Such channels will present a vertical succession of structures and textures akin to fluvial channels (Davies, 1969; Klein, 1965). Lateral migration of the channels may obliterate much of the typical middle

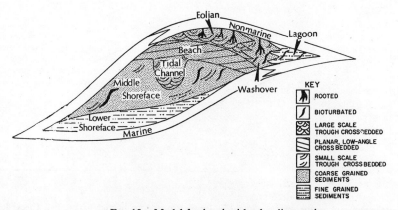

FIG. 15.—Model for barrier island sedimentation.

shoreface sediment and cause extensive areas of the barrier to be characterized by fluvial-type sequence (Hoyt and Henry, 1967).

7. In a typical regressive situation, grain size should increase upward from lower shoreface to beach, providing no source changes or sequence interruptions have occurred.

8. True barrier sands will be flanked on their landward side by lagoonal sediments. The absence of such sediments would indicate that the particular constructional feature was not a barrier in the strict sense. However, the presence or absence of lagoonal sediments does not appear to affect the vertical sequence of sedimentary structures and textures (compare the vertical sequences in the Cretaceous and Jurassic systems discussed herein).

9. The thickness of the sands will depend upon the complex interrelation of sediment supply and rate of subsidence. Galveston Island sands are 50 ft thick; Jurassic sandstones in southern England locally are 150 ft thick.

10. The barrier sands will be capped by either marine or nonmarine sediments, depending upon the tectonic situation and sediment supply at the close of barrier formation.

11. The model is independent of composition and should apply to both carbonate and noncarbonate sediments. The only prerequisites are that the material is granular and that it is not of uniform grain size.

If this model has catholic application, then it should demonstrate its efficacy in sediments of contrasting age and location. The three examples discussed demonstrate that, at least in these instances, the model is sound.

Conclusions

The vertical and lateral successions of sedimentary structures and textures at Galveston Island have resulted in the delineation of several distinct but interrelated environments in this Holocene barrier island. These environments may be recognized readily in ancient sediments, notably in examples from the Cretaceous of Montana and Jurassic of England. Recognition is based on sedimentary structures and vertical grain-size change. The similarities between the Holocene and ancient examples have facilitated development of a model of barrier sedimentation which presupposes a locally regressive situation during barrier growth (but not necessarily during initiation).

Petrographic analyses of quartz mean size and quartz content of several barriers yield textural and compositional parameters which may be as environmentally sensitive as sedimentary structures. Graphic plots of quartz size versus quartz content demonstrate that each of the six barrier bar environments may be distinguished on the basis of thin section analyses. Consequently, full diameter cores, taken to show details of sedimentary structures, may not always be requisite for subsurface environmental interpretation. In fact, thin sections of sidewall cores under specific conditions may yield significant and reliable environmental interpretations in ancient barrier sandstones.

References Cited

Allen, J. R. L., 1962, Petrology, origin, and deposition of the highest lower Old Red Sandstone of Shropshire, England: Jour. Sed. Petrology, v. 32, p. 657–697.

——— 1965, A review of the origin and characteristics of recent alluvial sediments: Sedimentology, v. 5, p. 91–191.

Berg, R. R., and D. K. Davies, 1968, Origin of Lower Cretaceous Muddy Sandstone at Bell Creek field, Montana: Am. Assoc. Petroleum Geologists Bull., v. 52, p. 1888–1898.

——— and ———, 1970, Depositional environments of Muddy reservoir sandstones (Lower Cretaceous) in Powder River basin, Montana and Wyoming (abs.): Am. Assoc. Petroleum Geologists Bull., v. 54, p. 835.

Bernard, H. A., R. J. LeBlanc, and C. F. Major, 1962, Recent and Pleistocene geology of southeast Texas, field excursion no. 3, in Geology of the Gulf Coast and central Texas and guidebook of excursions: Geol. Soc. America Ann. Mtg. Guidebook, p. 175–224.

Blundell, D. J., F. J. Davey, and L. J. Graves, 1970, Surveys over the south Irish Sea and Nymphe Bank: Geol. Soc. London Circ. 161, p. 2.

Chayes, Felix, 1956, Petrographic modal analysis—an elementary statistical appraisal: New York, John Wiley and Sons, 113 p.

Davies, D. K., 1969, Shelf sedimentation: an example from the Jurassic of Britain: Jour. Sed. Petrology, v. 39, p. 1344–1370.

——— and R. R. Berg, 1969, Sedimentary characteristics of Muddy barrier-bar reservoir and lagoonal trap at Bell Creek field: Montana Geol. Soc., 20th Ann. Conf., Eastern Montana Symposium, p. 97–105.

Dickinson, K. A., and R. E. Hunter, 1970, Grain-size distributions and the depositional history of northern Padre Island, Texas (abs.), in Programs with abstracts for 1970: Geol. Soc. America, v. 2, no. 4, p. 280.

Exum, F. A., and J. C. Harms, 1968, Comparison of marine-bar with valley-fill stratigraphic traps, western Nebraska: Am. Assoc. Petroleum Geologists Bull., v. 52, p. 1851–1868.

Ferm, J. C., 1962, Petrology of some Pennsylvanian sedimentary rocks: Jour. Sed. Petrology, v. 32, p. 104–123.

Fisk, H. N., 1959, Padre Island and the Laguna Madre flats, coastal south Texas, in R. J. Russell, chm., Second Coastal Geography Conference: Louisiana State Univ. Coastal Studies Inst., p. 103–151.

Gibbons, G. S., 1967, Shell content in quartzose beach and dune sands, Dee Why, New South Wales: Jour. Sed. Petrology, v. 37, p. 869–878.

Giles, R. T., and O. H. Pilkey, 1965, Atlantic beach and dune sediments of the southern United States: Jour. Sed. Petrology, v. 35, p. 900–910.

Griffiths, J. C., 1967, Scientific method in analysis of sediments: New York and London, McGraw-Hill, 508 p.

Haun, J. D., and J. A. Barlow, Jr., 1962, Lower Cretaceous stratigraphy of Wyoming, in Symposium on early Cretaceous rocks of Wyoming and adjacent areas: Wyoming Geol. Assoc., 17th Ann. Field Conf. Guidebook, p. 15–22.

Hoyt, J. H., 1967, Barrier island formation: Geol. Soc. America Bull., v. 78, p. 1125–1136.

——— and V. J. Henry, Jr., 1967, Influence of island migration on barrier-island sedimentation: Geol. Soc. America Bull., v. 78, p. 77–86.

Klein, G. deV., 1965, Dynamic significance of primary structures in the Middle Jurassic Great Oolite Series, southern England, p. 173–192, in G. V. Middleton, ed., Primary sedimentary structures and their hydrodynamic interpretation: Soc. Econ. Paleontologists and Mineralogists, Spec. Pub. 12, 265 p.

——— 1967, Paleocurrent analysis in relation to modern marine sediment dispersal patterns: Am. Assoc. Petroleum Geologists Bull., v. 51, p. 366–382.

Mason, C. C., and R. L. Folk, 1958, Differentiation of beach, dune, and eolian flat environments by size analysis, Mustang Island, Texas: Jour. Sed. Petrology, v. 27, p. 211–226.

McGregor, A. A., and C. A. Biggs, 1968, Bell Creek field, Montana: a rich stratigraphic trap: Am. Assoc. Petroleum Geologists Bull., v. 52, p. 1869–1887.

——— and ——— 1969, Physical relationships of the Muddy Sandstone in the northeast Powder River basin: Montana Geological Soc., 20th Ann. Conf., Eastern Montana Symposium, p. 107–119.

Potter, P. E., 1967, Sand bodies and sedimentary environments—a review: Am. Assoc. Petroleum Geologists Bull., v. 51, p. 337–365.

Price, W. A., 1951, Barrier island, not "offshore bar": Science, v. 113, p. 487–488.

——— 1968, Bars, in R. W. Fairbridge, ed., Encyclopedia of geomorphology: New York, Reinhold Book Corp., p. 55–58.

Sabins, F. F., Jr., 1963, Anatomy of stratigraphic trap, Bisti field, New Mexico: Am. Assoc. Petroleum Geologists Bull., v. 47, p. 193–228.

Shelton, J. W., 1967, Stratigraphic models and general criteria for recognition of alluvial, barrier-bar, and turbidity-current sand deposits: Am. Assoc. Petroleum Geologists Bull., v. 51, p. 2441–2461.

Shepard, F. P., 1952, Revised nomenclature for depositional coastal features: Am. Assoc. Petroleum Geologists Bull., v. 36, p. 1902–1912.

Visher, G. S., 1965, Use of vertical profile in environmental reconstruction: Am. Assoc. Petroleum Geologists Bull., v. 49, p. 41–61.

Weidie, A. E., 1968, Bar and barrier island sand: Gulf Coast Assoc. Geol. Socs. Trans., v. 18, p. 405–415.

Genetic Units in Delta Prospecting[1]

DANIEL A. BUSCH[2]
Tulsa, Oklahoma 74103

Abstract Deltas generally are formed at river mouths during stillstands of sea level under conditions of cyclic transgression or regression. Consequently, they are rarely isolated phenomena, but form in multiples in a predictable fashion. Reservoir facies consist of continuous and discontinuous, bifurcating channel sandstones and deltafront sheet sandstones. The channel sandstones generally thicken downward at the expense of the underlying prodelta clays and may replace selected parts of the deltafront sheet sandstones.

The lithologic components of a deltaic complex are interrelated and are referred to collectively as one type of Genetic Increment of Strata (GIS). The GIS is a vertical sequence of strata in which each lithologic component is related genetically to all the others. It is defined at the top by a time-lithologic marker bed (such as a thin limestone or bentonite) and at the base by either a time-lithologic marker bed, an unconformity, or a facies change from marine to nonmarine beds. It generally consists of the total of all marginal marine sediments deposited during one stillstand stage of a shoreline, or it may be a wedge of sediments deposited during a series of cyclic subsidences or emergences. An isopach map of a GIS clearly shows the bifurcating trends of the individual distributaries and the shape of the delta, regardless of the variable lithology of the channel fills.

A Genetic Sequence of Strata (GSS) consists of two or more contiguous GIS's and, when isopached, clearly defines the shelf, hinge line, and less stable part of a depositional basin. An isopach map of the McAlester Formation of the Arkoma basin is a good example of a GSS. The oil-productive Booch sandstone is a good example of a deltaic complex occurring within a GIS of this GSS. The upper Tonkawa, Endicott, and Red Fork sandstones of the Anadarko basin are identified as deltaic accumulations within different GIS's.

A hypothetical model serves as a basis for establishing the criteria for (1) recognizing successive stillstand positions of a shoreline, (2) predicting paleodrainage courses, (3) predicting positions of a series of deltaic reservoirs, (4) locating isolated channel sandstone reservoirs, and (5) tracing related beach sandstone reservoirs.

INTRODUCTION

Within the past 15–20 years petroleum explorationists have come to recognize the economic significance of deltas. With this recognition has come a realization of the variabilities of deltas, both external and internal, as well as

[1] Manuscript received, July 14, 1970; accepted, January 15, 1971. Presented before the Association at Calgary, June 24, 1970.
[2] Consulting geologist.

© 1971. The American Association of Petroleum Geologists. All rights reserved.

the constructive and destructive energy agents that are responsible for their development. For decades little more was known about ancient deltas than that published by Barrell (1912) on the classic Catskill-Chemung delta of the Appalachian geosyncline and the much smaller deltas of glacial Lake Bonneville described by Gilbert (1885). In more recent years several significant papers concerning ancient deltas have appeared in the literature. Busch (1953, 1959) described the Pennsylvanian Booch delta of the Arkoma basin; Nanz, Jr. (1954), the Oligocene Seeligson pool of the Gulf Coast; Weimer (1961), the Late Cretaceous Rawlins delta; Halbouty and Barber (1961), an Oligocene delta of the Gulf Coast; Rainwater (1963), a Gulf Coast Miocene delta; and Swann (1964), a Late Mississippian Michigan River delta. No two of these deltas are alike, hence it is apparent that a knowledge of their modern analogues is an essential prerequisite to their recognition and delineation in the subsurface.

Modern deltas have certain distinct differences, too, which are related to such variables as "(1) sediment input in terms of amount, caliber, rate, variation in rate, and suspended/bedload ratio, and these in turn on extent and nature of drainage area and climate; (2) nature of the discharging and reservoir water bodies particularly in regard to relative water densities; (3) reservoir energy (waves, currents, tides), both degree and kind, especially in relation to amount of sediment input; (4) depth of water into which delta is prograding; (5) nature of substrata of reservoir body in regard to compactional subsidence and storage of prograding sands; and (6) structural nature of the depositional basin" (Scott and Fisher, 1969).

The depositional environments and sediments of the delta of the Mississippi River have been extensively investigated. As a result, the Mississippi River deltaic plain undoubtedly is the most completely understood modern depositional environment anywhere in the world. Many of the sedimentary processes in operation in this area today have worldwide application in understanding both modern and ancient deltas. Ironically, however, the modern birdfoot delta of the Mississippi River is not typical of the majority of modern and ancient deltas. This Holocene delta is the result of a combination of environmental variables that exist in only a few places in the world today.

The major emphasis of this discussion is the use of conventional geologic tools in the identification and delineation of deltas in subsurface exploration for hydrocarbons. The tools readily available to all geologists are samples, cores, and more than 30 kinds of mechanical logs. These tools, when used in conjunction with a background knowledge of modern deltas and principles of stratigraphic analysis, are adequate for delta exploration.

GENETIC RELATIONS OF DELTAIC FACIES

Deltas generally are formed at river mouths during stillstands of sea level under conditions of either cyclic transgression or regression. Consequently, they are rarely isolated phenomena but, rather, form in multiples in a predictable fashion. Reservoir facies consist of continuous and discontinuous, bifurcating channel sandstones and delta-front sheet sandstones. The channel sandstones generally thicken downward at the expense of prodelta clays and locally replace selected parts of the delta front sheet sandstones.

Genetic Increment of Strata (GIS)

All marine deltas, with the exception of deep-water submarine fans, are deposited in the marginal-marine and directly adjacent subaerial environments. All the lithologic components of a delta are related genetically to each other and can be referred broadly to two basic genetic depositional units. The smaller of these is the Genetic Increment of Strata (GIS), which is defined as a vertical sequence of strata in which each lithologic component is related genetically to all the others. It is defined at the top by a time-lithologic marker bed (such as a thin limestone or bentonite) and at the base by a time-lithologic marker bed, an unconformity, or a facies change from marine to nonmarine beds. It generally consists of the total of all marginal marine sediments deposited during one stillstand stage of a shoreline, or it may be a wedge of sediments deposited during a series of cyclic subsidences or emergences (Fig. 1). An isopach map of a GIS involving a deltaic system clearly shows the bifurcating trends of the individual distributaries and the shape of the delta, regardless of the variable lithology of the channel fills.

Genetic Sequence of Strata (GSS)

A Genetic Sequence of Strata consists of two or more contiguous GIS's (Fig. 2) and represents more or less continuous sedimentation; the stratigraphic interval must be devoid of angular unconformities, but disconformities of limited extent may be present. An isopach map of a GSS serves to define clearly the shelf, hinge line, and the less stable part of a depositional basin, as shown in Figures 2 and 4. The principal purpose of this type of map is the reconstruction of the paleodepositional configuration of a basin. Most sedimentary basins are so large that only a part may be of interest. This type of isopach map can be constructed with widely scattered control points and is, there-

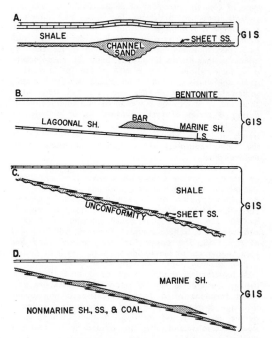

FIG. 1.—Diagrammatic illustrations of Genetic Increments of Strata (GIS). A. Channel and sheet sandstones with overlying shale and limestone; upper limit of GIS is defined by time-lithologic marker bed, base by unconformity. B. Sandstone bar bordered by lagoonal shale on one side and by marine shale on other; upper and lower limits of GIS are defined by time-lithologic marker beds (bentonite and limestone). C. Sheet sandstone, resulting from cyclic subsidence, overlain by basinward-diverging marine shale; upper limit of GIS is defined by time-lithologic marker bed, base by unconformity. D. Sheet sandstone showing two areas of locally thicker sandstone at former stillstand positions of shoreline; upper boundary is defined by time-lithologic marker bed, base by facies change from continental to marine beds.

fore, of considerable value in preliminary studies and evaluations of stratigraphic sections of parts of a depositional basin early in an exploration program. The GIS, on the other hand, is generally most applicable for detailed studies of individual sandstones.

In defining the GIS and GSS, the lithologic-time marker bed is critical because it represents contemporaneity of deposition. The most commonly used marker beds are a thin bed of limestone or bentonite. Coal, black shale, and siltstone also may serve as marker beds where it can be ascertained that they do not cross time lines. Of the five, bentonite is considered to be most reliable. The thickness of a shale sequence between two marker beds generally is fairly uniform within a local area, but normally diverges systematically in a basinward direction on a regional scale. Under starved conditions, which are somewhat exceptional, it could converge basinward. Abrupt changes in such intervals generally are due to facies changes, unconformities (at the base), or growth faulting, and if any of these occur, they must be clearly recognizable before using the GIS and GSS concepts. In essence, the GIS is a particular case of a limited interval isopach applicable to individual reservoir sandstone analysis, whereas the GSS is a case of gross interval isopach for regional stratigraphic analysis.

Subsurface Examples of Delta Analysis

Booch Delta

In Early Pennsylvanian time a large delta complex was deposited in the eastern Oklahoma part of the Arkoma basin. The bulk of this Booch delta covers an area of approximately 2,000 sq mi. It was deposited largely under shoal conditions in a shelf environment. In analyzing the sandstones making up this

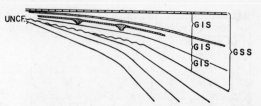

Fig. 2.—Relation of Genetic Sequence of Strata (GSS) to Genetic Increments of Strata (GIS).

delta, it is necessary first to consider the Genetic Sequence of Strata of which they are component parts. To do this, a series of stratigraphic profiles, such as that illustrated in Figure 3, was constructed across each of the tiers of townships shown in Figure 4. These profiles then were tied together by a series of intersecting north-south stratigraphic profiles. All wells not used in the construction of the correlation grid then were correlated with the nearest wells of the grid. In so doing, the upper and lower limits of the McAlester Formation (Genetic Sequence of Strata) were determined as a prelude to constructing an isopach map of this formation.

Figure 3 is an electric log cross section of the McAlester Formation in the Greater Seminole District of Oklahoma, illustrating the rate of thickening of this unit across the northwest shelf area of the Arkoma basin. This wedge of sediments is predominantly shale, interrupted by four members of the Booch sandstone. The middle member of the Booch sandstone increases abruptly in thickness to as much as 250 ft. In so doing, it merges vertically with the lower middle member of the Booch and even with the lower member of the Booch. The principal direction of thickening is downward, at the expense of the shale which encases the lower middle and lower Booch.

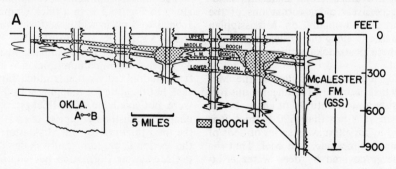

Fig. 3.—Correlation of Booch sandstone members of McAlester Formation (Pennsylvanian) of eastern Oklahoma part of Arkoma basin.

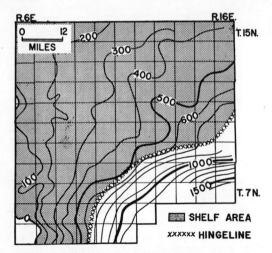

Fig. 4.—Isopach map of McAlester Formation (GSS), Greater Seminole district, eastern Oklahoma. Hinge line separates shelf environment on northwest from more rapidly subsiding basin area on southeast.

The McAlester is overlain conformably by the Savannah Formation; the shales of these two formations are lithologically indistinguishable. The contact between them has been placed arbitrarily at an easily recognized reentrant "angle" of the short normal resistivity curve of the electric logs. This "angle" is present on more than 90 percent of the 650 electric logs available for this study and occurs about 30 ft below the base of the zone known as the "Brown lime." This reentrant "angle" is the reference datum of the stratigraphic profile in Figure 3. The base of the McAlester Formation is conformable with the Atoka shale in the more basinward part of the Arkoma basin, but unconformable on much of the shelf part of this basin. It overlies unconformably successively older units on the west, in the direction of the Hunton arch. For example, from southeast to northwest it overlies Atoka shale and sandstone, on successively older formations of the Morrow Series, and finally on Mississippian shale and limestone along the western margin of the basin where post-Atoka uplift and erosion were greatest.

Electric log and sample studies of the McAlester show that basinward (southeast) this formation consists almost entirely of shale, but in the area where it is less than 600 ft thick it is characterized by an alternating series of beds of shale, sandstone, limestone, and coal. That the shelf was never covered by deep water is attested clearly by the presence of coal beds.

The isopach map of the McAlester (Fig. 4) was drawn to (1) reconstruct the shape of the northwestern part of the Arkoma depositional basin, (2) determine the trend and position of the "line of flexure" (hinge line) that existed during McAlester deposition, and (3) ascertain the principal direction of transport of McAlester sand and mud.

The McAlester Formation thickens abruptly southeast of the hinge line (Fig. 4) in the more rapidly subsiding part of the basin. In the shelf environment, northwest of the hinge line, the rate of southeast divergence is much more gradual. The greatest accumulation of sediments within the mapped area is more than 1,600 ft in T7N, R16E. The strike of the isopach contour lines changes progressively from northeast-southwest to a little west of north. This change indicates that the southeastern part of the mapped area was progressively more downwarped during McAlester deposition than the area north and northwest.

Figure 5 is a composite isopach map of the Booch sandstone units of the McAlester Formation. In the northeastern part of the area (T13–15N, R14–16E) the upper Booch sandstone is the principal unit present and is a series of delta distributary channel fills. More darkly shaded areas in the north-central part of the mapped area represent vertical merging of all four Booch sandstones. In the region west of R12E, where the Booch is less than 60 ft thick, the middle Booch is the principal sandstone present, all other members being shaly to calcareous siltstone or poorly developed sandstone. The latter (all units except the middle Booch) can be correlated only as mappable "zones" rather than as lithologic units. The lower middle and lower Booch sandstones are not present except where merged vertically with the middle Booch in the channel areas.

The most prominent development of sandstone in Figure 5 is in a thick channel, ranging from 3 to 5 mi in width, and trending southeast from T15N, R11E, to T11N, R14E. The mapped length of this channel sandstone is over 35 mi, and the maximum thickness exceeds 240 ft. It undoubtedly extends much farther southeast, but the data for mapping it in this direction were not available. The thickest parts of this channel sandstone comprise over 50 percent of the total thickness of the McAlester; the rest of the section is predominantly shale. The shale of the McAlester Formation has compacted sufficiently from the weight of the overburden, that

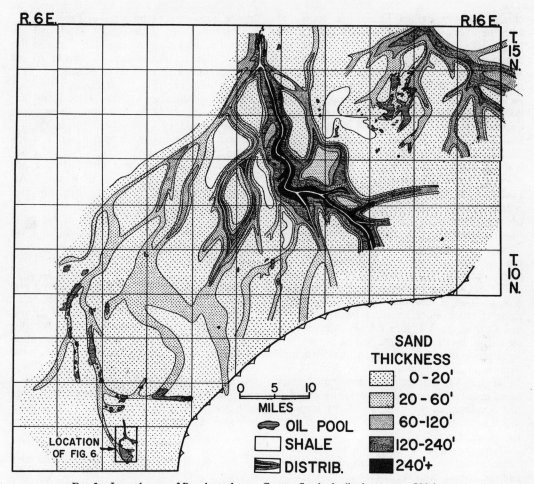

FIG. 5.—Isopach map of Booch sandstone, Greater Seminole district, eastern Oklahoma.

the location and trend of this major sandstone can be determined readily from the isopach map of the McAlester Formation (Fig. 4). As sandstone is relatively noncompactible, the northwest convexity of the 300-, 400-, and 500-ft contours is due primarily to the presence of the prominent distributary channel sandstone. This major channel sandstone and the numerous smaller channels of sandstone make up a distributary network of a large delta which had its apex on the north. It is not likely that all the distributaries shown in Figure 5 were formed simultaneously, for this illustration represents a composite of all stages of distributary development during McAlester deposition. In fact, some of the smaller distributaries close to the largest channel sandstone probably represent sand accumulation of the crevasse splay type.

The isopach map of the Booch delta, shown in Figure 5, is based on several thousand electric and drillers' logs. Such a map serves as little more than a geologic framework from which more detailed studies can be made. For example, Figure 6 is a more detailed analysis of the Booch sandstone in T6N, R8E. Production is from the middle member. The Booch here is overlain by a coal bed which consistently has a characteristic "kick" on the long normal resistivity curve of the electric logs. The structure contours of Figure 6A describe a sinuous, arcuate high over the pool proper (on the south), and a narrow, bifurcating structural ridge on the north. The maximum relief over the pool (south half) is approximately 75 ft. In the northern part of this figure the relief exceeds 120 ft. When the trends of the structural "highs" (heavy dashed lines) are compared with the axes of maximum sandstone thickness (Fig. 6B), it is apparent that the structure is due in part to variations in sandstone thickness. From logs of wells outside of, and marginal to, the pool, it is known that the Booch is lenslike and grades to shale in all directions from the pool. The base of the previously mentioned coal bed serves as the means of identifying the Booch outside the productive area. Figure 6C is an isopotential map of the Booch in which a contour interval of 25 bbl/24 hours was employed. The wells were drilled prior to the advent of hydro-frac, making the estimated daily potentials of the wells more meaningful. Trends of maximum initial potential (heavy dashed lines) coincide extremely well with trends of maximum sandstone thickness (Fig. 6B). This is to be expected inasmuch as the initial potential is partly a function of the relative permeability and amount of sandstone at any one borehole position. The deltaic aspect of this sandstone is much more apparent from Figure 6B and C than from Figure 6A. Postdepositional tectonism locally has modified the area; the most pronounced effect being a local structural depression in contiguous parts of Secs. 16 and 21 (Fig. 6A). Isopach and isopotential maps assist in carrying out a systematic step-out drilling development program. From the isopachs, drillsite locations most likely to extend axial trends are selected.

There are significant petrographic differences between the distributary channel sandstones and the interdistributary sheet sandstones, especially in the southwestern half of the delta shown in Figure 5. Where channel sandstones exceed 20 ft in thickness, they generally have larger geometric mean grain diameters than those of the interchannel areas, the averages being 99 and 86 μ, respectively. The channel sandstones also are better sorted than those of the interchannel. In addition to these differences, the sandstones of the interchannel areas have an average clay content of 20 percent, compared with 15 percent in the channels. The higher percentage of clay and smaller grain size of the sandstone in the interchannel areas might be responsible for the relative absence of oil in these areas.

The Booch sandstones are very fine grained, and of the graywacke type; that is, they contain an abundance of fragments of preexisting rocks, and a relatively large clay matrix. The grain-size distribution and standard deviation are illustrated in Figure 7. Petrographic data for the Booch are summarized in Table 1.

Most oil pools of the Booch sandstone are not in the thick channel sandstones, but in the thinner channel sandstones. A cluster of oil pools of irregular outline is present in T13, 14N, R14E (Fig. 5), about midway between two main distributary systems. If sufficient log data were available in this area, these pools probably could be shown to occur in thinner distributary channel sandstones close to muds of the backswamp environment. Several elongate pools are in channel distributary sandstones in the southwestern part of Figure 5. All of these bear little relation to structure and probably are the result of permeability seals surrounding more porous and permeable parts of channel sandstones. Several significant oil pools (not shown on Fig. 5) occupy very selective parts of the thicker channel sandstones (middle of Figure 5). In this area the channel sandstones are medium to coarse textured and contain less interstitial clay. Such pools occur where the axes of west-plunging structural noses intersect depositional axes of major distributary channel sandstones. In this situation oil-water contacts are readily determined on the north and south sides of the pools, and abrupt loss of permeability occurs on the east and west margins where the channel sandstones merge with the silty, very fine-grained interdistributary sheet sandstones.

Tonkawa Delta

The uppermost member of the Tonkawa (upper Missourian Series of the Pennsylvanian) is an excellent example of a deltaic gas-bearing sandstone reservoir. It is present in Beaver County in the Oklahoma Panhandle part of the Anadarko basin. This delta is described and illustrated in a recent paper by Khaiwka (1968), in which he points out that the upper member of the Tonkawa sandstone is overlain by the thin (3-ft), persistent "Haskell" limestone. Jordan (1957) noted this thin stratigraphic unit as a marker bed; it is a convenient datum of reference in a stratigraphic analysis of the directly underlying upper member of the Tonkawa. Figure 8 is an electric and microlog illustration of the three sandstone members of the Tonkawa showing their relation to the "Haskell" limestone. The downward thickening of the upper member of the Tonkawa sandstone, relative to the "Haskell" limestone reference datum, is illustrated in the lower half of Figure 8. This single stratigraphic profile is typical of many which have

Genetic Units in Delta Prospecting 1143

A.

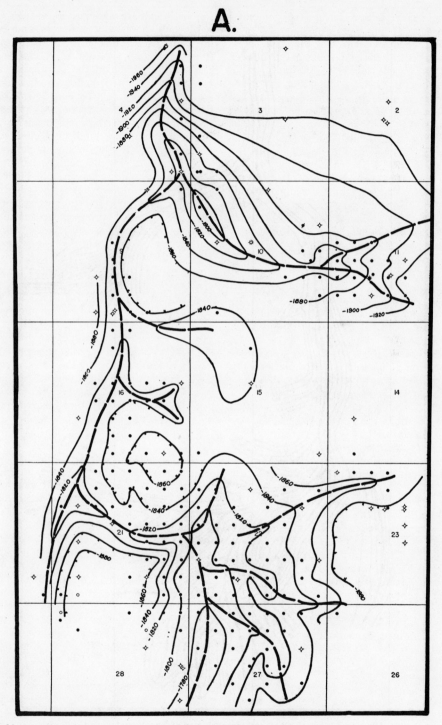

Fig. 6A.—Hawkins pool, central part of T6N, R8E, Hughes County, Oklahoma. Structure, top of Booch sandstone; C.I. = 20 ft. Figure 6B is isopach map of Booch sandstone; C.I. = 5 ft. Figure 6C is isopotential map of Booch sandstone; C.I. = 25 bbl/24 hours.

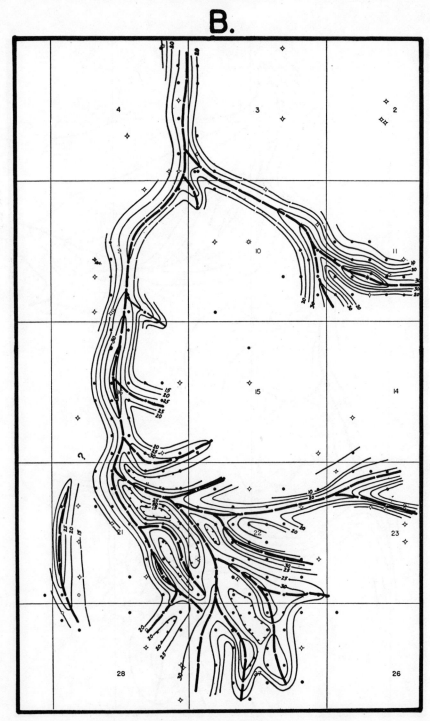

Fig. 6B.—Hawkins pool, Oklahoma. Isopach map of Booch sandstone, C.I. = 5 ft.

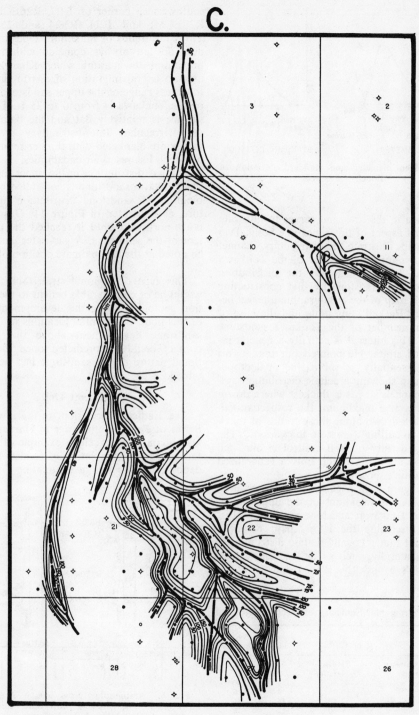

Fig. 6C.—Hawkins pool, Oklahoma. Isopotential map of Booch sandstone. C.I. = 25 bbl/24 hours.

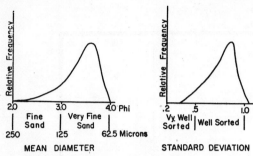

FIG. 7.—Mean diameter and standard deviation of Booch sandstone.

been drawn in the area and serves to illustrate the channel aspect of this sandstone body. Two isopach maps illustrate the distributary channel network of the upper member of the Tonkawa Sandstone. Khaiwka showed the distribution pattern of this sandstone by first constructing an isopach map of the stratigraphic interval between the "Haskell" limestone and the base of the upper member of the Tonkawa sandstone (Fig. 9). This interval is actually a genetic increment of strata. He pointed out that, "This map is essentially a submarine, paleotopographic map showing a deltaic distributary system of channels . . . it is thickest where downcutting was at a maximum; the respective valley axes are drawn along these trends of maximum down-cutting (greatest thickness)." The maximum density of well control is one well per section and, as a result, only a generalized interpretation is possible.

With Figure 9 as a guide, an interpretation of the thickness variations of the upper member of the Tonkawa sandstone was made (Fig. 10). In discussing the latter map, Khaiwka pointed out that, "Locally deltaic distributaries migrated laterally, such as in the 'four-corner area' of T's 3 and 4N, R's 24 and 25 ECM;

Table 1. Petrographic Summary of Booch Sandstones

	Range	Average (μ)	No. of Samples
Mean grain diameter	71–91μ	83	28
Stand. deviation	0.34–1.01ϕ	0.76	28
Clastic quartz	45–90%	54	36
Carbonate	0–33%	4	36
Silt-size clay material	2–40%	16	36
Rock fragments	9–50%	22	36
Secondary silica	9–50%	22	36
Other minerals	0–3 %	2	36

southwestern corner of T4N, R26ECM; T3N, R27ECM; and T2N, R's24 and 25 ECM." Thus, the trends of maximum sandstone thickness do not always coincide with trends of maximum downcutting which Khaiwka attributed to lateral migration of distributaries. The thickness range of the upper sandstone member of the Tonkawa is from 0 to 35 ft. The upper surface is relatively flat and the basal surface quite irregular. Interdistributary areas commonly are blanketed with thinner sandstone, all of which has reservoir capabilities.

The gas distribution in the upper member of the Tonkawa sandstone is controlled by a combination of sandstone distribution and structure, as illustrated in Figure 11. Gas migrated north-northwest until it reached the updip extent of the sandstone. A gas-water contact may be noted at the approximate position of −3,270 ft.

This type of regional stratigraphic analysis would be of considerable benefit to the exploration geologist during the developmental stages of this pool. Preferential locations with thickest sandstone development above the gas-water contact could be predicted once the depositional nature of the sandstone had been ascertained.

Endicott Delta

The Endicott formation, in the lower Virgil Series of Pennsylvanian age in Harper County, Oklahoma, is an excellent example of a subsurface deltaic accumulation of sandstone. It was deposited in a series of bifurcating distributary

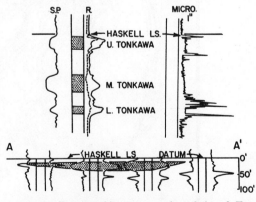

FIG. 8.—Stratigraphic cross section A-A' of Tonkawa sandstone, showing downward thickening at expense of underlying shale. Reference datum is top of "Haskell" limestone. See Figure 9 for location of cross section (modified after Khaiwka, 1968).

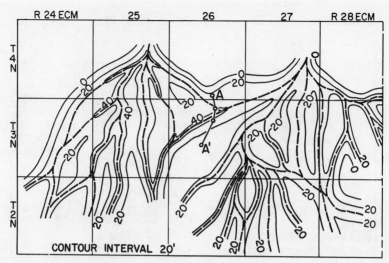

Fig. 9.—Isopach map of GIS between top of "Haskell" limestone and base of upper Tonkawa sandstone, Beaver County, Oklahoma. This map shows two laterally coalescing distributary systems (modified after Khaiwka, 1968).

channels eroded into the north flank of the Anadarko basin. The Endicott sandstone overlies unconformably either the Toronto limestone or the unnamed shale directly underlying the Toronto, as illustrated in the two profiles of Figure 12. These cross sections clearly show that the principal direction of thickening of the Endicott is downward. As the Oread Limestone serves as reference datum for these profiles, the Lovell sandstone, "Haskell" limestone, and Tonkawa sandstone, respectively, are bowed downward directly under the wells exhibiting maximum Endicott channel fill. This is attributed primarily to differential compaction of the shale section separating the Toronto limestone from the Lovell sandstone. Differential compaction of this shale was concurrent with the deposition of the Heebner shale (above the Endicott sandstone) as indicated by a tendency for the Heebner to thin slightly over areas of thick Endicott sandstone and, conversely, to be thicker in areas of thinner Endicott. This is the

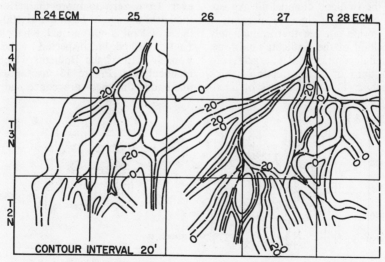

Fig. 10.—Isopach map of upper Tonkawa sandstone, Beaver County, Oklahoma (modified after Khaiwka, 1968).

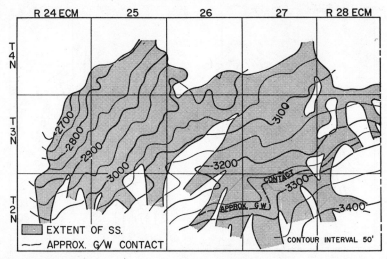

Fig. 11.—Structure map of top of upper Tonkawa sandstone, Beaver County, Oklahoma (modified after Khaiwka, 1968).

result of compensatory deposition in areas of slightly deeper water. The single apparent exception to this principle, in the easternmost well of profile A-A' of Figure 12, is explained by the fact that a remnant of the Toronto limestone is preserved below the Endicott in this well and in all probability would not compact any more than the Endicott itself.

The Oread Limestone is an excellent marker bed in this area and serves as the upper limit of a GIS that extends down to the unconformity at the base of the Endicott channel fill. An isopach map of this GIS (Fig. 13) presents a "cast" of the erosion surface that immediately preceded deposition of the Endicott sandstone. It was this irregular surface that largely controlled the positions of linear trends of thick Endicott sandstone. The axial trends of maximum downcutting in pre-Endicott time are indicated by bifurcating dashed lines. The composite picture is that of part of a deltaic complex consisting of two laterally merging distributary systems which fan out southward.

Khaiwka (1968) pointed out that "Capps (1959), and later Winter (1963), depicted the Endicott as a blanket-type sandstone." This is true if one considers only the generally smooth, rather extensive upper surface of the Endicott. The thickness of this formation ranges, however, from zero to more than 180 ft, and a thickness change of as much as 100 ft in a distance of 1 mi is not unusual. Khaiwka prepared (not published, but inspected by the writer) an isopach map of the Endicott which is almost identical with Figure 13 insofar as trends of maximum thickness are concerned. He stated

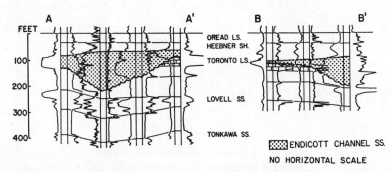

Fig. 12.—Stratigraphic profiles showing channel nature of Endicott sandstone, northwestern Oklahoma. Note that Toronto limestone was eroded away in positions of maximum sandstone thickness. See Figure 13 for locations of profiles (modified after Khaiwka, 1968).

"This similarity indicates that there was little or no migration of the deltaic distributaries. The thalwegs of the individual distributaries, with their bifurcating pattern, almost coincide on the two isopachous maps." He believes that, "The Endicott delta probably was deposited adjacent to a coastal plain of relatively high relief because sand bodies form major parts of its framework."

Limited sample and thin section studies indicate that the Endicott is a gray, fine-grained, poorly sorted, calcareous micaceous sandstone. Khaiwka (1968) has noted 85–90 percent quartz. Locally, the muscovite and chlorite compose up to 10 percent of the bulk volume. In other areas as much as 10 percent chert and 5 percent feldspar (mostly plagioclase) have been estimated from thin-section studies. The porosity ranges from fair to good, the maximum value observed being 18 percent. The sandstone is laminated with light- to dark-gray carbonaceous shale and is overlain by dark-gray and black pyritic shale with coal streaks at the base.

The structure of the Endicott is that of a south-dipping homocline devoid of any significant structural noses or closures. Although the Endicott within the area of Figure 13 has good reservoir characteristics, it apparently is devoid of the necessary trapping mechanism, such as updip permeability seal or structural closure. The only known Endicott production occurs from several wells in the Oklahoma and Texas Panhandle parts of the Anadarko basin.

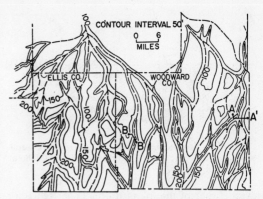

FIG. 13.—Isopach map of GIS between top of Oread Limestone and base of Endicott sandstone, northwestern Oklahoma. This is "cast" of eroded surface upon which Endicott sandstone was deposited in area of two laterally coalescing deltaic distributary systems (modified after Khaiwka, 1968).

Red Fork Delta

The nature of the Red Fork Sandstone of Noble County, Oklahoma, has been debated by geologists for many years. It has been variously interpreted as offshore bar, isolated channel sandstone, and delta. Figure 14 is a structure map of the base of the Oswego limestone, which is a readily mappable unit about 175 ft above the Red Fork. The general structure is that of an irregular monoclinal slope toward the west. A south-southwest-plunging syncline interrupts this slope in the west and is related to the conspicuous Tonkawa fault in the north-

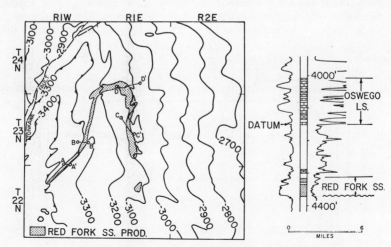

FIG. 14.—Structure of base of Oswego limestone, Noble County, Oklahoma, showing location of four stratigraphic profiles (modified after Scott, 1970).

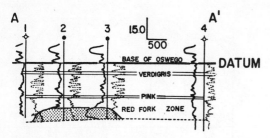

FIG. 15.—Sandbar interpretation of Red Fork Sandstone, South Ceres pool, Noble County, Oklahoma (after Neal, 1951).

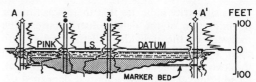

FIG. 16.—Channel interpretation of Red Fork Sandstone, South Ceres pool, Noble County, Oklahoma. See Figure 14 for location of cross section (modified after Scott, 1970).

west corner of the mapped area. The Red Fork Sandstone production describes a horseshoe-shaped area, indicated with stippling on Figure 14. The western and northern parts of this horseshoe-shaped area produce oil, whereas the eastern limb is primarily gas-productive. All production is from a very lenticular Red Fork sandstone. The influence of differential compaction (on structural configuration) of the shales laterally adjacent to the Red Fork is apparent along the west limb of the horseshoe. It is here that the −3,300-ft structural contour line of the base of the Oswego exhibits a pronounced southwestward nosing where it crosses the pool.

The locations of several stratigraphic cross sections are designated A-A', B-B', C-C', and D-D' on Figure 14. Figure 15 is a published stratigraphic profile "hung" on the base of the Oswego limestone (Neal, 1949, 1951). By this interpretation the Red Fork Sandstone increases in thickness at the expense of the shale which directly overlies it and, on this basis, has been interpreted as an offshore bar. Figure 16 (Scott, 1970) is a different interpretation of the same four electric logs used in Figure 15. The marker bed, near the base of the Red Fork Sandstone, is the key to a correct interpretation of this lenticular sandstone. It is present in wells No. 1 and 4, and eroded away in wells No. 2 and 3 where the Red Fork has its maximum thickness. By this interpretation the Red Fork is clearly a channel-type of reservoir sandstone.

Figure 17 is stratigraphic cross section B-B' (see Fig. 14 for location) and shows an amount of downcutting similar to that of Figure 16. More than half the channel fill is silty shale. The marker bed is clearly evident in all but the No. 3 well, where it has been eroded away.

A much thicker development of channel sandstone is present along stratigraphic profile C-C' (Fig. 18), where the maximum thickness is about 50 ft. The effects of differential compaction can be observed in the interval between the marker bed and the Pink limestone reference datum. This stratigraphic interval is thinnest in wells No. 1 and 5, where it consists primarily of compacted shale. In the other three wells noncompactible sandstone predominates and the interval is consequently thicker.

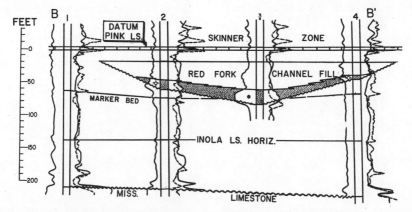

FIG. 17.—Stratigraphic profile B-B', showing Red Fork channel-fill in which over half of fill is silty shale. See Figure 14 for location of profile (modified after Scott, 1970).

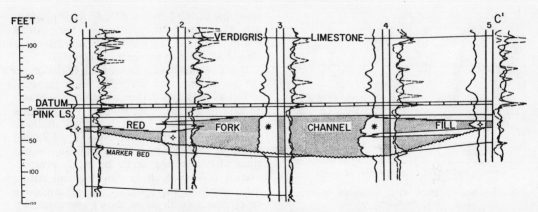

FIG. 18.—Stratigraphic profile C-C', showing Red Fork channel-fill, predominantly sandstone. See Figure 14 for location of profile (after Scott, 1970).

Figure 19 is a stratigraphic cross section of the Red Fork channel where the valley has a maximum width. As most of the fill consists of sandstone, the effects of differential compaction are minimal. Numerous additional cross sections involving the Pink limestone marker bed interval have been constructed; all show the channel aspect of the Red Fork Sandstone, with thickening at the expense of the underlying shale.

In his study Scott (1970) considered the interval from the Pink limestone to the base of the Red Fork Sandstone as constituting a GIS and, as such, reconstructed a paleodistributary system of a delta, as shown in Figure 20. To date, oil and gas production is restricted to the two principal distributaries of this delta, and it appears that all the other distributaries are primarily crevasse splays in which the sandstones generally are thinner and more argillaceous.

HYPOTHETICAL MODEL STUDY

Deltas, although quite varied in size, are seldom isolated phenomena along modern shorelines. Marginal marine conditions favorable to delta formation at the mouths of rivers generally extend for many miles along the shorelines. Most positive land areas, particularly in humid to semi-humid environments, are drained by a series of streams flowing to bordering embayments. Although present drainage into the Gulf of Mexico is dominated by the Mississippi and Rio Grande Rivers, there are, nevertheless, numerous smaller drainage courses emptying into the Gulf along the Texas-Louisiana coastline.

For a delta to form, the shoreline of an embayment must remain relatively stationary for a period of time. Thus, if a relatively stable shoreline condition favors development of a delta at the mouth of one river, it also should favor delta development at the mouths of most, if not all, rivers flowing into the same embayment.

If the shoreline is transgressing in response to cyclic subsidence of the basin area (or eustatic rises in sea level), conditions are ideal for the formation of rows of deltas at the successive stillstand positions of the shoreline. Like-

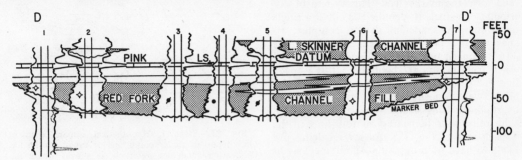

FIG. 19.—Stratigraphic profile D-D', showing Red Fork channel-fill where channel has maximum width. See Figure 14 for location of profile (after Scott, 1970).

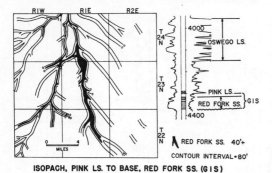

Fig. 20.—Isopach map, Pink limestone to base of Red Fork Sandstone (GIS).

along the stippled zone are likely to be prospective. Three dry holes already have been drilled within this zone, so additional information is needed in order to prospect intelligently.

It is a relatively safe assumption that a drainage course extended northwest from the apex of Delta No. 1 and that its trend was essentially normal to the paleodepositional strike. Delta No. 1 occurs where this projected drainage course intersects the strand zone. If additional drainage courses could be ascertained, similar areas of intersection would be highly prospective. A review of all existing log and testing data reveals stratigraphically higher sandstone development (with an oil show) approximately 30 mi north, in the northwest part of Figure 21. Well samples show oil stain and a drill stem test recovered oil in noncommercial quantity. This well appears to be on the edge of a pool, but which edge is not known. Four possible pool outlines are indicated with dotted circles.

At this early stage of regional analysis it is necessary to drill, or cause to be drilled, a sufficient number of step-out wells from this dry hole to find the pool and cause it to be developed.

Because Delta No. 1 produces from a distributary system deposited during one of the stillstands of a shoreline under conditions of cyclic subsidence, the sandstone of Delta No. 2 (Fig. 22) was probably deposited under similar conditions—at a later stage of cyclic subsidence. In Figure 22 the position, outline, and nature of the reservoir sandstone(s) of Delta No. 2 have

wise, conditions for delta preservation are ideal when the shoreline transgresses in a cyclic manner. This can be illustrated best by reference to a hypothetical diagram (Fig. 21), which is Stage 1 of a series of three illustrations that portray a method of prospecting for ancient deltas. Figure 21 is an isopach map of a GSS that includes the oil-productive sandstone of Delta No. 1. This GSS was picked after constructing a series of stratigraphic profiles which collectively form a correlation grid of the area. The productive sandstone(s) of Delta No. 1 occurs either at, or very near, the GSS base, which is defined by an unconformity. The isopach map of this GSS shows a fairly uniform rate of southeastward divergence. Marine transgression was from southeast to northwest, and the paleodepositional strike of the shoreline was northeast.

A detailed stratigraphic analysis of Delta No. 1 reveals that the oil-productive sandstones consist of a series of southeastward diverging channel fills. All of the channel sandstones may be either at the same stratigraphic position relative to the marker bed, at the top of the GSS, or at several closely spaced intervening levels relative to the marker bed. This system of distributary channels is identified as either a single or a composite ancient deltaic complex. Its apex occurs approximately at the position of the 475-ft isopach line (dotted), and its southeastern margin occurs at about the 510-ft isopach line. Thus, the reservoir sandstones of Delta No. 1 were deposited in a strand zone shown by stippling. This strand zone is, of necessity, subparallel with the nearest contour lines of the isopach map of the GSS. As several deltas are likely to have been deposited simultaneously at the mouths of rivers which emptied into the embayment along this strand zone, several areas

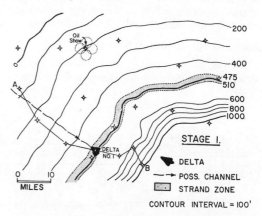

Fig. 21.—Stage 1. Hypothetical isopach map of GSS deposited under conditions of cyclic subsidence. Oil-bearing sandstone reservoir is shown where ancient drainage course discharged sediment just basinward from stillstand position of shoreline.

been determined. The drainage course of this delta is projected northwest, normal to the paleodepositional strike. Marine transgression has little or no effect on the position of a drainage course, hence it is reasonable to project the drainage through Delta No. 2 in a southeast direction. This projected drainage course intersects the strand zone on the southeast within the area labeled "Delta Prospect No. 1."

At this stage of regional analysis, Delta Prospect No. 2 assumes the same significance as Delta Prospect No. 1. It, too, lies within the area of intersection of a projected drainage course and the established second strand zone.

It is reasonable to assume that, by the time Delta Prospect Nos. 1 and 2 have been discovered and developed, numerous additional dry holes (and possibly discovery wells) will have been drilled within the area. The new data afford the means of much more detailed stratigraphic analysis and prospecting. With additional data many more pools (of various types) may be anticipated. For example, if the unconformity at the base of the GSS was fairly smooth, additional drainage channels (with deltas) should be anticipated on the east and southwest, with spacing similar to that between the two postulated channels. Additional strand zones southeast of the first strand zone, and possibly between the two established strand zones, should be anticipated.

A detailed stratigraphic profile of the GSS between Delta No. 1 and Delta Prospect No. 2 (Profile A–B, Fig. 23) will serve as a basis for determining the thickness of the shale interval

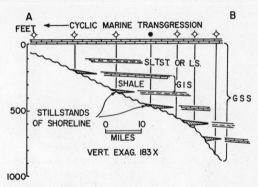

FIG. 23.—Stratigraphic profile A-B of hypothetical GSS showing sandstone development at succession of stillstand positions of shoreline.

between the pay sandstones of the two established strand zones. If this shale interval is 50 ft thick, or more, it is reasonable to postulate at least one more strand zone between them. Nonproductive sandstone developments in the dry holes between these two strand zones are tipoffs of other strand zones.

With additional well data one can determine whether the unconformity surface was fairly smooth, as well as its rate of basinward tilt at the time of cyclic marine transgression. In Figure 22 the strand zones are approximately 20 mi apart, and the amount of GSS divergence is 270 ft. A 13.5 ft/mi divergence strongly suggests the presence of one or more strand zones between the two established zones. A cyclic 25–50-ft change in sea level (or subsidence) is much more likely than 270 ft.

With the acquisition of additional drilling data, it is possible to define clearly the positions of the two drainage channels and any others present within the study area. This is accomplished by reducing the contour interval of the GSS to 50 ft or possibly 20 ft. As this GSS is, in a sense, a "cast" of the unconformity surface, the contour lines will bend upstream as illustrated in Figure 24. Selective parts of these channels may be clogged with sandstone and afford ideal conditions for oil and gas reservoirs of the simple channel type or, possibly, point bar type.

From Figure 24 it may be noted that with additional well control more strand zones have been delineated and additional deltaic reservoirs discovered and developed. The newly discovered strand zones on the southeast overlap because of the abrupt increase in the rate of divergence of the GSS on the seaward side of the hinge line. In such areas of steeper gradient

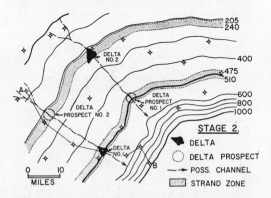

FIG. 22.—Stage 2. Hypothetical isopach map of strata deposited under conditions of cyclic subsidence. Delta No. 1 and Delta Prospect No. 1 occur where drainage channels intersect a strand zone, and Delta No. 2 and Delta Prospect No. 2 occur where headward projections of drainage channels are crossed by another strand zone.

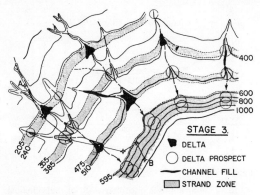

Fig. 24.—Stage 3. Hypothetical paleotopographic map, showing deltaic reservoirs and prospects in areas of intersection of strand zones and paleo-drainage zones. Individual oil pools in channel sandstone also shown.

there is less lateral shift of shoreline in response to cyclic subsidence. Consequently, a part of one delta actually might overlap the apical part of an earlier delta development. In such areas the individual strand zones are likely to be narrower than on the shelf, but the sandstones, conversely, are likely to be thicker.

Where longshore currents were operative in redistributing sediments furnished at river mouths, beach sandstones may have been deposited. Such sandstones commonly are gas-bearing, for reasons not currently understood. Examples of this are the Clinton (Lower Silurian) of east-central Ohio, Atoka (Pennsylvanian) sandstones of east-central Oklahoma, Morrow (Pennsylvanian) sandstones of northwest Oklahoma and northern Texas, and the Mesa Verde (Cretaceous) sandstones of northwestern New Mexico.

The interpretation idealized in Figure 24 presents many additional prospects for exploration drilling. In construction of the figure, all postdepositional structure has been removed. With data theoretically available for the construction of Figure 24, however, it would be a simple matter to construct a structural map, preferably with the lithologic-time marker bed at the top of the GSS as datum. Thus, the modifying effect of possible structures on accumulation of hydrocarbons could be ascertained readily. In this hypothetical example the prime emphasis is one of predicting areas for favorable reservoir sandstone development, whether of deltaic distributaries, subaerial channel sandstones, or beach deposits. At this stage the exploration geologist is ready to determine the possible modifying effects of postdepositional structure on oil and gas accumulation.

REFERENCES CITED

Barrell, J., 1912, Criteria for the recognition of ancient delta deposits: Geol. Soc. America Bull., v. 23, p. 377–446.

Busch, D. A., 1953, The significance of deltas in subsurface exploration: Tulsa Geol. Soc. Digest, v. 21, p. 71–80.

———— 1959, Prospecting for stratigraphic traps: Am. Assoc. Petroleum Geologists Bull., v. 43, no. 12, p. 2829–2843.

Capps, W. H., 1959, Stratigraphic analysis of the Missourian and lower Virgilian series in northwestern Oklahoma: M.S. thesis, Oklahoma Univ.

Gilbert, G. K., 1885, The topographic features of lake-shores: U.S. Geol. Survey 5th Ann. Rept., p. 104–108.

Halbouty, M. T., and T. D. Barber, 1961, Port Acres and Port Arthur fields, Jefferson County, Texas: Gulf Coast Assoc. Geol. Socs. Trans., v. 11, p. 225–234.

Jordan, L., 1957, Subsurface stratigraphic names of Oklahoma: Oklahoma Geol. Survey Guidebook 6, 220 p.

Khaiwka, M. H., 1968, Geometry and depositional environments of Pennsylvanian reservoir sandstones, northwestern Oklahoma: Ph.D. dissert., Oklahoma Univ.

Nanz, R. N., Jr., 1954, Genesis of Oligocene sandstone reservoir, Seeligson field, Jim Wells and Kleberg Counties, Texas: Am. Assoc. Petroleum Geologists Bull., v. 38, no. 1, p. 96–117.

Neal, E. P., 1949, Southeast Ceres field, Noble County, Oklahoma: Tulsa Geol. Soc. Digest, v. 17, p. 97–99.

———— 1951, South Ceres, Oklahoma's oddest shoestring field: World Oil, v. 133, no. 7, p. 92, 94, 98.

Rainwater, E. H., 1963, The environmental control of oil and gas occurrence in terrigenous clastic rocks: Gulf Coast Assoc. Geol. Socs. Trans., v. 13, p. 79–94.

Scott, A. J., and W. L. Fisher, 1969, Delta systems and deltaic deposition, in Delta systems in the exploration for oil and gas, a research symposium: Texas Univ. Bur. Econ. Geology, p. 10–29.

Scott, J. D., 1970, Cherokee sandstones of a portion of Noble County, Oklahoma: M.S. thesis, Oklahoma Univ.

Swann, D. H., 1964, Late Mississippian rhythmic sediments of Mississippi Valley: Am. Assoc. Petroleum Geologists Bull., v. 48, no. 5, p. 637–658.

Weimer, R. J., 1961, Upper Cretaceous delta on tectonic foreland, northern Colorado and southern Wyoming (abs.): Am. Assoc. Petroleum Geologists Bull., v. 45, no. 3, p. 417.

Winter, J. A., 1963, Computer geology uncovers secrets of Oklahoma's Endicott sand: Oil and Gas Jour., v. 61, no. 29 (July 22), p. 96–101.

Geometry of Sandstone Reservoir Bodies[1]

RUFUS J. LeBLANC[2]
Houston, Texas 77001

Abstract Natural underground reservoirs capable of containing water, petroleum, and gases include sandstones, limestones, dolomites, and fractured rocks of various types. Comprehensive research and exploration efforts by the petroleum industry have revealed much about the character and distribution of carbonate rocks (limestones and dolomites) and sandstones. Porosity and permeability of the deposits are criteria for determining their efficiency as reservoirs for fluids. Trends of certain sandstones are predictable. Furthermore, sandstone reservoirs have been less affected than carbonate reservoirs by postdepositional cementation and compaction. Fracture porosity has received less concentrated study; hence, we know less about this type of reservoir. The discussions in this paper are confined to sandstone reservoirs.

The principal sandstone generating environments are (1) fluvial environments such as alluvial fans, braided streams, and meandering streams; (2) distributary-channel and delta-front environments of various types of deltas; (3) coastal barrier islands, tidal channels, and chenier plains; (4) desert and coastal eolian plains; and (5) deeper marine environments, where the sands are distributed by both normal and density currents.

The alluvial-fan environment is characterized by flash floods and mudflows or debris flows which deposit the coarsest and most irregular sand bodies. Braided streams have numerous shallow channels separated by broad sandbars; lateral channel migration results in the deposition of thin, lenticular sand bodies. Meandering streams migrate within belts 20 times the channel width and deposit two very common types of sand bodies. The processes of bank-caving and point-bar accretion result in lateral channel migration and the formation of sand bodies (point bars) within each meander loop. Natural cut-offs and channel diversions result in the abandonment of individual meanders and long channel segments, respectively. Rapidly abandoned channels are filled with some sand but predominantly with fine-grained sediments (clay plugs), whereas gradually abandoned channels are filled mainly with sands and silts.

The most common sandstone reservoirs are of deltaic origin. They are laterally equivalent to fluvial sands and prodelta and marine clays, and they consist of two types: delta-front or fringe sands and abandoned distributary-channel sands. Fringe sands are sheetlike, and their landward margins are abrupt (against organic clays of the deltaic plain). Seaward, these sands grade into the finer prodelta and marine sediments. Distributary-channel sandstone bodies are narrow, they have abrupt basal contacts, and they decrease in grain size upward. They cut into, or completely through, the fringe sands, and also connect with the upstream fluvial sands or braided or meandering streams.

Some of the more porous and permeable sandstone reservoirs are deposited in the coastal interdeltaic realm of sedimentation. They consist of well-sorted beach and shoreface sands associated with barrier islands and tidal channels which occur between barriers. Barrier sand bodies are long and narrow, are aligned parallel with the coastline, and are characterized by an upward increase in grain size. They are flanked on the landward side by lagoonal clays and on the opposite side by marine clays. Tidal-channel sand bodies have abrupt basal contacts and range in grain size from coarse at the base to fine at the top. Laterally, they merge with barrier sands and grade into the finer sediments of tidal deltas and mud flats.

The most porous and permeable sandstone reservoirs are products of wind activity in coastal and desert regions. Wind-laid (eolian) sands are typically very well sorted and highly crossbedded, and they occur as extensive sheets.

Marine sandstones are those associated with normal-marine processes of the continental shelf, slope, and deep and those due to density-current origin (turbidites). An important type of normal-marine sand is formed during marine transgressions. Although these sands are extremely thin, they are very distinctive and widespread, have sharp updip limits, and grade seaward into marine shales. Delta-fringe and barrier-shoreface sands are two other types of shallow-marine sands.

Turbidites have been interpreted to be associated with submarine canyons. These sands are transported from nearshore environments seaward through canyons and are deposited on submarine fans in deep marine basins. Other turbidites form as a result of slumping of deltaic facies at shelf edges. Turbidite sands are usually associated with thick marine shales.

[1] Manuscript received, March 17, 1972.
[2] Shell Oil Company. This paper is based on the writer's 30 years of experience in studies of modern and ancient clastic sediments—from 1941 to 1948, with the Mississippi River Commission, under the guidance of H. N. Fisk, and, since August 1948, with Shell Development Company and Shell Oil Company.

The writer is grateful to Shell Oil Company for permission to publish this paper, and he is deeply indebted to Alan Thomson for his critical review of the manuscript; he is also grateful to Nick W. Kusakis, John Bush, Dave C. Fogt, Gil C. Flanagan, and George F. Korenek for assistance in the preparation of illustrations and reference material; to Aphrodite Mamoulides and Bernice Melde for their library assistance; to Darleen Vanderford for typing the manuscript, and to Judy Breeding for her editorial assistance.

Numerous stimulating discussions of models of clastic sedimentation and the relationship of sedimentary sequences to depositional processes were held with Hugh A. Bernard and Robert H. Nanz, Jr., during the late 1940s and 1950s, when we were closely associated with Shell's early exploration research effort. The writer is particularly indebted to these two men for their numerous contributions, many of which are included in this paper.

The writer also wishes to thank W. B. Bull, University of Arizona, for his valuable suggestions concerning the alluvial-fan model of clastic sedimentation.

INTRODUCTION

Important natural resources such as water, oil, gas, and brines are found in underground reservoirs which are composed principally of the following types of rocks: (1) porous sands, sandstones, and gravels; (2) porous limestones and dolomites; and (3) fractured rocks of various types. According to the 1971 American Petroleum Institute report on reserves of crude oil and natural gas, sandstones are the reservoirs for about 75 percent of the recoverable oil and 65 percent of the recoverable gas in the United States. It is also estimated that approximately 90 percent of our underground water supply comes from sand and gravel (Walton, 1970).

Sandstone and carbonate (limestone and dolomite) reservoirs have been intensively studied during the past 2 decades; consequently, the general characteristics and subsurface distribution of these two important types of reservoirs are relatively well known in numerous sedimentary basins. The factors which control the origin and occurrence of fracture porosity have received less attention; thus, our knowledge and understanding of this type of reservoir are more limited.

The detection of subsurface porosity trends within sedimentary basins was recognized by the petroleum industry as one of its most significant problems, and for the past 2 decades it has addressed itself to a solution through extensive research. Largely as a result of this research, which is summarized below, our ability to determine trends of porous sedimentary rocks has progressed noticeably, especially during the past 10 years.

The amount of porosity and permeability present within sedimentary rocks and the geometry of porous rock bodies are controlled mainly by two important factors: (1) the environmental conditions under which the sediments were deposited and (2) the postdepositional changes within the rocks as a result of burial, compaction, and cementation. Postdepositional diagenetic processes have less effect on the porosity and permeability of sands and sandstones than they have on carbonate sediments; consequently, porosity trends are significantly more predictable for sandstones than for limestones and dolomites.

Organization of paper—The following two parts of this paper give a brief historical summary of the early research on clastic sediments and present a classification of environments of deposition and models of clastic sedimentation.

A résumé of significant studies of modern clastic sediments—mainly by the petroleum industry, government agencies, and universities—follows. The main part of the paper concerns the sedimentary processes, sequences, and geometry of sand bodies which characterize each of the following models of clastic sedimentation: alluvial fan, braided stream, meandering stream, deltaic (birdfoot-lobate and cuspate-arcuate), coastal interdeltaic (barrier island and chenier plain), and marine (transgressive, submarine canyon, and fan).

HISTORICAL SUMMARY OF EARLY RESEARCH ON MODERN CLASTIC SEDIMENTS

Geologists are now capable of interpreting the depositional environments of ancient sedimentary facies and of predicting clastic porosity trends with a reasonable degree of accuracy (Peterson and Osmond, 1961; Potter, 1967; Rigby and Hamblin, 1972; Shelton, 1972). This capability stems from the extensive research conducted on Holocene sediments by several groups of geologists during the past 3 decades. Conditions which led to this research, and the most significant studies of clastic sedimentation which provided the models, criteria, and concepts necessary to make environmental interpretations, are summarized below.

During the late 1930s and early 1940s, petroleum geologists became aware that improved methods of stratigraphic interpretations were badly needed, and that knowledge and geologic tools necessary to explore for stratigraphic traps were inadequate. A detailed study made by the Research Committee of The American Association of Petroleum Geologists on the research needs of the industry ultimately led to the establishment of geologic research departments by major oil companies. By 1948, exploration research by the oil industry was in its early stages, and expansion proceeded rapidly thereafter.

Meanwhile, some very significant developments were occurring at Louisiana State University. H. V. Howe and R. J. Russell, together with their graduate students, had already published several Louisiana Geological Survey bulletins summarizing their pioneer work on the late Quaternary geology of southern Louisiana (Howe and Moresi, 1931, 1933; Howe et al., 1935; Russell, 1936). Their early work on the Mississippi deltaic plain and the chenier plain of southwestern Louisiana is considered to be the beginning of the modern environmental approach to stratigraphy. Fisk became fascinated with the Howe and Russell approach, and he applied results of their research to his study of Tertiary sediments. The work of Fisk (1940) in central Louisiana, which included a study of the lower Red River Valley and part of the Mississippi Valley, attracted the attention of General Max Tyler, president of the Mississippi River Commission in Vicksburg. General Tyler engaged Fisk as a consultant and provided him with a staff of geologists to conduct a geologic investigation of the lower Mississippi River alluvial valley.

The Fisk (1944) report on the Mississippi Valley, which now has become a classic geologic document, established the relations between alluvial environments, processes, and character of sediments. The AAPG, recognizing the significance of this contribution, retained Fisk as Distinguished Lecturer, and the results and significance of his work became widely known. One of his most significant contributions came when, as the petroleum industry was getting geologic research under way, he was selected by a major oil company to direct its geologic research effort in Houston.

By 1950, a few major oil companies were deeply involved in studies of recent sediments. However, the small companies did not have staff and facilities to conduct this type of research, and American Petroleum Institute Project 51 was established for the purpose of conducting research on recent sediments of the Gulf Coast. Scripps Institution of Oceanography was in charge of the project, which continued for 8 years. Results of this research were available to all companies (Shepard *et al.*, 1960).

While the petroleum industry was conducting "in-house" research and supporting the API project, some significant research was being done by the U.S. Waterways Experiment Station in Vicksburg, Mississippi, and by the new Coastal Studies Institute at Louisiana State University under the direction of R. J. Russell. These two groups conducted detailed studies of recent sediments for many years, and results were made available to the petroleum industry.

By 1955, a fairly good understanding of processes of sedimentation and character of related sediments in several depositional environments had been acquired. Although the application of this wealth of knowledge to operational problems was very difficult, some useful applications nevertheless had been made by the middle 1950s, and it was generally agreed that the initial research effort was successful.

Since 1955, geologists all over the world have become involved in studying recent sediments and applying the results to research on older rocks. Geologists with the U.S. Geological Survey and several universities have conducted studies of alluvial fans, braided streams, and eolian deposits; and the oceanographic institutions, such as Scripps, Woods Hole, and Lamont, have investigated deep-marine sediments on a worldwide basis. Publication of papers on clastic sedimentation has been increasing rapidly. The first textbook on the geology of recent sediments cites more than 700 references, 75 percent of which have appeared since 1955 (Kukal, 1971). Many of these contributions, considered to be most significant to the current understanding of clastic sediments, are cited in this paper.

Models and Environments of Clastic Sedimentation

The realm of clastic sedimentation can be divided into several conceptual models, each of which is characterized by certain depositional environments, sedimentary processes, sequences, and patterns. What are considered to be some of the most common and basic models and environments[3] of clastic sedimentation, arranged in order from the periphery to the center of a depositional basin, are listed below and are shown on Figures 1–4.

Continental
  Alluvial (fluvial) models
    Alluvial fan
    Braided stream
    Meandering stream (includes flood basins between meander belts)
  Eolian (can occur at various positions within continental and transitional models)
Transitional
  Deltaic models
    Birdfoot-lobate (fluvial dominated)
    Cuspate-arcuate (wave and current dominated)
    Estuarine (with strong tidal influence)
  Coastal-interdeltaic models
    Barrier-island model (includes barrier islands, lagoons behind barriers, tidal channels, and tidal deltas)
    Chenier-plain model (includes mud flats and cheniers)
Marine
  (Note: Sediments deposited in shallow-marine environments, such as deltas and barrier islands, are

---

[3] The classification of depositional environments presented herein was initially developed by the writer and his colleague, Hugh A. Bernard, during the early 1950s (LeBlanc and Bernard, 1954) and was recently modified (Bernard and LeBlanc, 1965). For other classifications, refer to Laporte (1968), Selley (1970), Crosby (1972), and Kukal (1971).

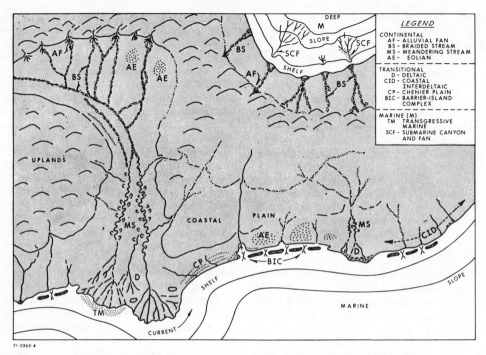

FIG. 1—Some common models of clastic sedimentation. See Figures 2–4 for details.

included under the transitional group of environments.)
Transgressive-marine model
Submarine-canyon and submarine-fan model

Résumé of Significant Studies of Modern Clastic Sedimentation

Alluvial Fans

Although much work has been done on alluvial fans, only a few papers discuss the relation of sedimentary sequences to depositional processes. Some of the more important contributions are by Rickmers (1913), Pack (1923), Blackwelder (1928), Eckis (1928), Blissenbach (1954), McKee (1957), Beaty (1963), Bull (1962, 1963, 1964, 1968, 1969, 1971), Hoppe and Ekman (1964), Winder (1965), Anstey (1965), Denny (1965, 1967), Legget et al. (1966), and Hooke (1967).

Braided Streams

Early papers on braided streams concerned channel patterns, origin of braiding, and physical characteristics of braided streams. Significant studies of this type were conducted by Lane (1957), Leopold and Wolman (1957), Chein (1961), Krigstrom (1962), Fahnestock (1963), and Brice (1964).

The relatively few papers on the relation of braided-stream deposits to depositional processes did not appear until the 1960s. Doeglas (1962) discussed braided-stream sequences of the Rhône River of France, and Ore (1963, 1965) presented some criteria for recognition of braided-stream deposits, based on the study of several braided streams in Wyoming, Colorado, and Nebraska. Fahnestock (1963) described braided streams associated with a glacial outwash plain in Washington. More recently, Williams and Rust (1969) discussed the sedimentology of a degrading braided river in the Yukon Territory, Canada. Coleman (1969) presented results of a study of the processes and sedimentary characteristics of one of the largest braided rivers, the Brahmaputra of Bangla Desh (formerly East Pakistan). N. Smith (1970) studied the Platte River from Denver, Colorado, to Omaha, Nebraska, and used the Platte model to interpret Silurian braided-stream deposits of the Appalachian region. Waechter (1970) has recently studied the braided Red River in the Texas Panhandle, and Kessler (1970, 1971) has investigated the Canadian River in Texas. Boothroyd (1970) studied braided streams associated with glacial outwash plains in Alaska.

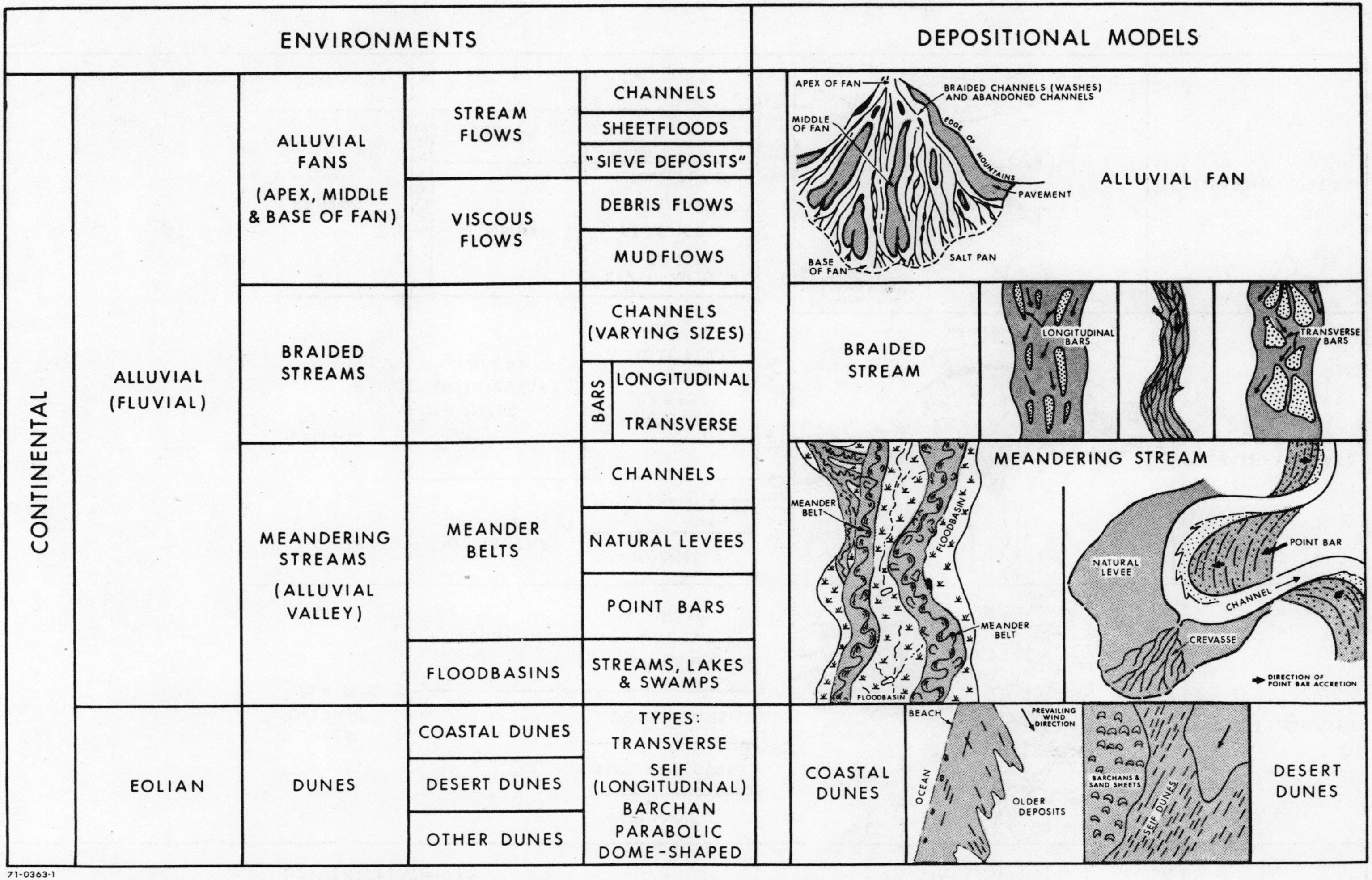

Geometry of Sandstone Reservoir Bodies

ENVIRONMENTS			DEPOSITIONAL MODELS
CONTINENTAL	ALLUVIAL (FLUVIAL)	ALLUVIAL FANS (APEX, MIDDLE & BASE OF FAN)	STREAM FLOWS — CHANNELS, SHEETFLOODS, "SIEVE DEPOSITS"
			VISCOUS FLOWS — DEBRIS FLOWS, MUDFLOWS
			ALLUVIAL FAN
		BRAIDED STREAMS	CHANNELS (VARYING SIZES); BARS — LONGITUDINAL, TRANSVERSE
			BRAIDED STREAM
		MEANDERING STREAMS (ALLUVIAL VALLEY)	CHANNELS, NATURAL LEVEES, POINT BARS, STREAMS, LAKES & SWAMPS
			MEANDER BELTS — FLOODBASINS
			MEANDERING STREAM
	EOLIAN	DUNES	COASTAL DUNES, DESERT DUNES, OTHER DUNES
			TYPES: TRANSVERSE, SEIF (LONGITUDINAL), BARCHAN, PARABOLIC, DOME-SHAPED
			COASTAL DUNES / DESERT DUNES

FIG. 2—Alluvial (fluvial) and eolian environments and models of clastic sedimentation.

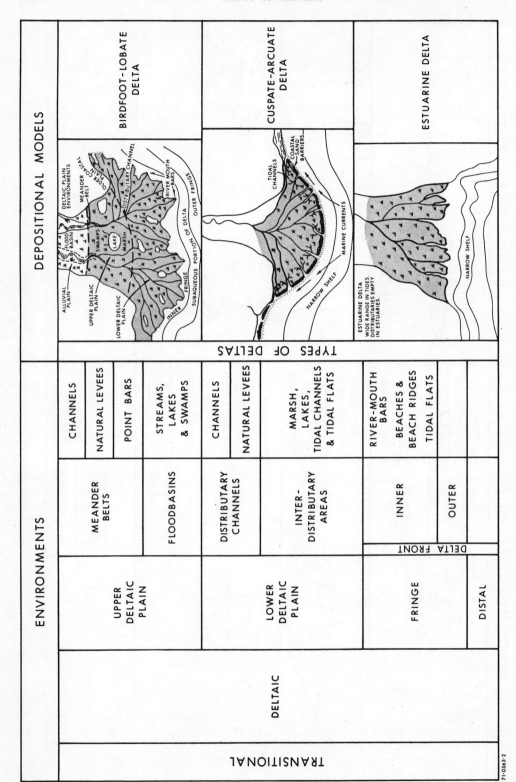

Fig. 3—Deltaic environments and models of clastic sedimentation.

Meandering Streams

H. N. Fisk's studies of the Mississippi alluvial valley, conducted for the Mississippi River Commission during the period 1941–48, represent the first significant contribution on meandering stream environments and deposits. This pioneer effort provided geologists with knowledge of the fundamental processes of alluvial-valley sedimentation. Another study of a meandering stream, the Connecticut River, and its valley was made by Jahns (1947). Important work on alluvial sediments deposited by meandering streams was also done by Sundborg (1956) in Sweden, and by Frazier and Osanik (1961), Bernard and Major (1963), and Harms et al. (1963) on the Mississippi, Brazos, and Red River point bars, respectively. Thus, by 1963 the general characteristics of point-bar sequences, and the closely related abandoned-channel and flood-basin sequences, were sufficiently well established to permit geologists to recognize this type of sedimentary deposit in outcrops and in the subsurface.

Other important contributions were made by Allen (1965a) on the origin and characteristics of alluvial sediments, by Simons et al. (1965) on the flow regime in alluvial channels, by Bernard et al. (1970) on the relation of sedimentary structures to bed form in the Brazos valley deposits, and by McGowen and Garner (1970) on coarse-grained point-bar deposits.

Deltas

The early work by W. Johnson (1921, 1922) on the Fraser delta, Russell (1936) on the Mississippi delta, Sykes (1937) on the Colorado delta, and Fisk (1944) on the Mississippi delta provided a firm basis for subsequent studies of more than 25 modern deltas during the late 1950s and the 1960s.

Fisk continued his studies of the Mississippi delta for more than 20 years. His greatest contributions were concerned with the delta framework, the origin and character of delta-front sheet sands, and the development of bar-finger sands by seaward-migrating rivermouth bars.

Scruton's (1960) paper on delta building and the deltaic sequence represents results of API Project 51 on the Mississippi delta. Additional research on Mississippi delta sedimentation, sedimentary structures, and mudlumps was reported by Welder (1959), Morgan (1961), Morgan et al. (1968), Coleman et al. (1964), Coleman (1966b), Coleman and Gagliano (1964, 1965), and also by Kolb and Van Lopik (1966). Coleman and Gagliano (1964) also discussed and illustrated processes of cyclic sedimentation. The most recent papers on the Mississippi delta are by Frazier (1967), Frazier and Osanik (1969), and Gould (1970).

Studies of three small birdfoot deltas of Texas—the Trinity, Colorado, and Guadalupe—were made by McEwen (1969), Kanes (1970), and Donaldson (1966), respectively. In addition, Donaldson et al. (1970) presented a summary paper on the Guadalupe delta. These four contributions are valuable because each one presents photographs and logs of cores of complete deltaic sequences.

European geologists associated with the petroleum industry and universities also have made valuable contributions to our understanding of deltas. Kruit (1955) and Lagaaij and Kopstein (1964) discussed their research on the Rhône delta of southern France, Allen (1965c, 1970) summarized the geology of the Niger delta of western Africa, and van Andel (1967) presented a résumé of the work done on the Orinoco delta of eastern Venezuela. More recently, the Po delta of Italy was studied by B. Nelson (1970) and the Rhône delta of southern France by Oomkens (1970).

Other recent contributions on modern deltas are by Coleman et al. (1970) on a Malaysian delta, by R. Thompson (1968) on the Colorado delta in Mexico, and by Bernard et al. (1970) on the Brazos delta of Texas.

The deltaic model is probably the most complex of the clastic models. Although additional research is needed on this aspect of sedimentation, the studies listed have provided some valuable concepts and criteria for recognition of ancient deltaic facies.

Coastal-Interdeltaic Sediments

Valuable contributions to our knowledge of this important type of sedimentation have been made by several groups of geologists. In the Gulf Coast region, the extensive Padre Island–Laguna Madre complex was studied by Fisk (1959), and the chenier plain of southwestern Louisiana was studied by Gould and McFarlan (1959) and Byrne et al. (1959). The Galveston barrier-island complex of the upper Texas coast was investigated mainly by LeBlanc and Hodgson (1959), Bernard et al. (1959, 1962), and Bernard and LeBlanc (1965).

Among the impressive studies made by Europeans during the past 15 years are those by van Straaten (1954), who presented results of very

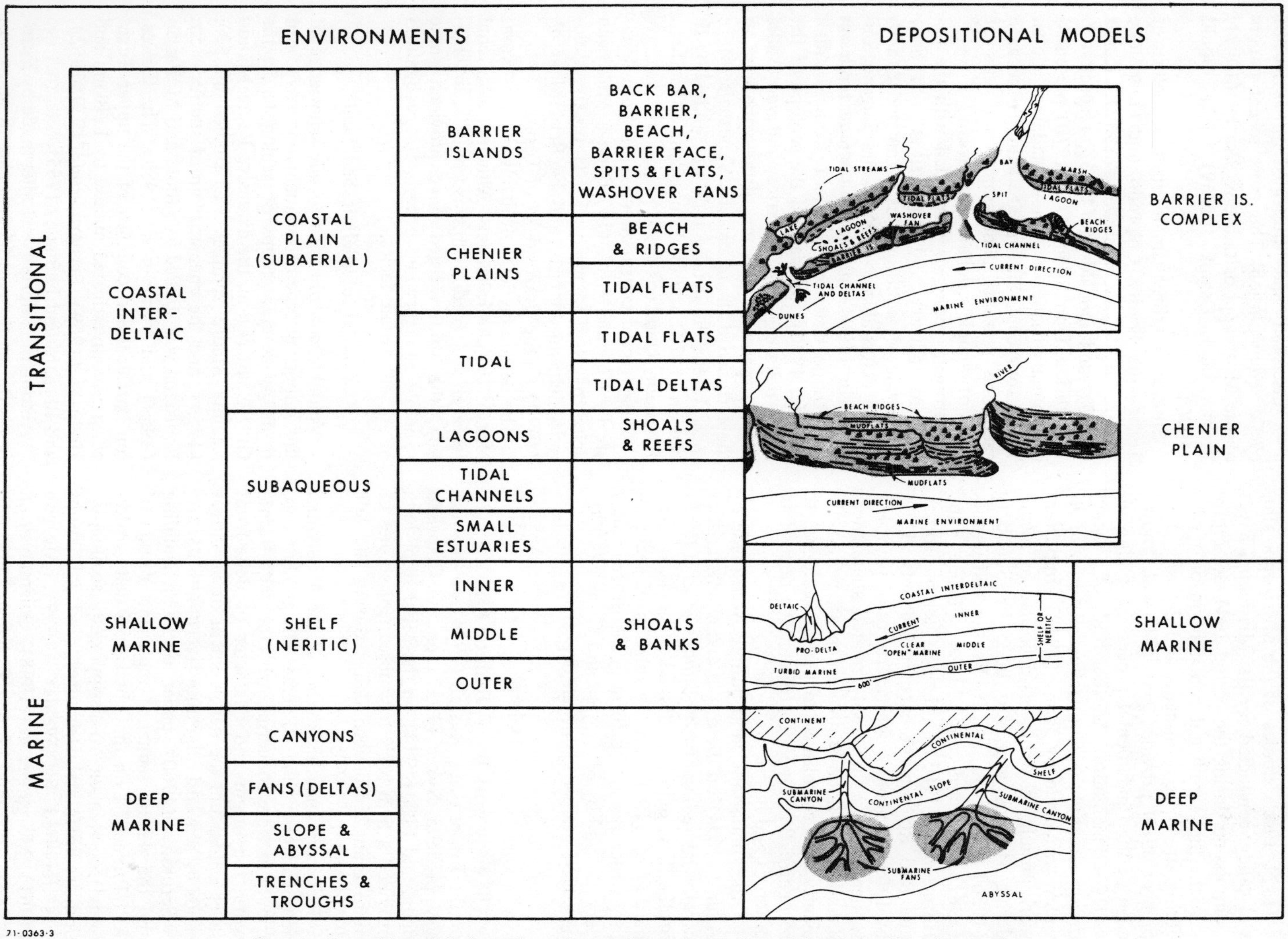

FIG. 4—Coastal-interdeltaic and marine environments and models of clastic sedimentation.

significant work on tidal flats, tidal channels, and tidal deltas of the northern Dutch coast, and by Horn (1965) and H. E. Reineck (1967), who reported on the barrier islands and tidal flats of northern Germany.

During the past several years, a group of geologists has conducted interesting research on the coastal-interdeltaic complexes which characterize much of the U.S. Atlantic Coast region. Hoyt and Henry (1965, 1967) published several papers on barriers and related features of Georgia. More recently, results of studies at the University of Massachusetts on recent coastal environments of New England were reported by Daboll (1969) and by the Coastal Research Group (1969).

In addition, Curray et al. (1969) described sediments associated with a strand-plain barrier in Mexico, and Potter (1967) summarized the characteristics of barrier-island sand bodies.

Eolian Sand Dunes

Prior to the middle 1950s, eolian depositional environments were studied principally by European geologists (Cooper, 1958). Since that time, the coastal sand dunes of the Pacific, Atlantic, and Gulf coasts of the United States, as well as the desert dunes of the United States and other countries, have been investigated by university professors and by geologists with the U.S. Geological Survey. Some of the most significant contributions, especially those concerned with dune stratification, are discussed in the section on the eolian model of clastic sedimentation.

Marine Sediments

Early work on modern marine sands, exclusive of those deposited adjacent to and related to interdeltaic and deltaic depositional environments, was conducted largely by scientists associated with Scripps, Woods Hole, and Lamont oceanographic departments. Several aspects of marine sediments were discussed by Trask et al. (1955), and the recent sands of the Pacific Ocean off California were studied by Revelle and Shepard (1939), Emery et al. (1952), and Emery (1960a). Stetson (1953) described the northwestern Gulf of Mexico sediments, and Ericson et al. (1952, 1955) and Heezen et al. (1959) investigated the Atlantic Ocean sediments. Later, Curray (1960), van Andel (1960), and van Andel and Curray (1960) reported results of the API project on the Gulf of Mexico. A few years later, results of the API project studies on the Gulf of California were reported by van Andel (1964) and van Andel and Shor (1964). Menard (1964) discussed sediments of the Pacific Ocean. For a more complete list of references to studies of recent marine sands, the reader is referred to Kuenen (1950), Guilcher (1958), Shepard et al. (1963), and Kukal (1971).

Much of the early research on modern marine environments was devoted to submarine canyons, fans, and basins considered by the investigators to be characterized mainly by turbidity-current sedimentation. Several scientists affiliated with Scripps and the University of Southern California published numerous papers on turbidites which occur in deep marine basins.

It is extremely difficult to observe the processes of turbidity-current sedimentation under natural conditions; consequently, the relations between sedimentary sequences and processes are still relatively poorly understood. Much of the research dealing with turbidity currents has been concerned with theory, laboratory models, and cores of deep-water sediments deposited by processes which have not been observed.

ALLUVIAL-FAN MODEL OF CLASTIC SEDIMENTATION

Occurrence and General Characteristics

Alluvial fans occur throughout the world, adjacent to mountain ranges or high hills. Although they form under practically all types of climatic conditions, they are more common and best developed along mountains of bold relief in arid and semi-arid regions (Figs. 5, 6).

The alluvial-fan model has the following characteristics: (1) sediment transport occurs under some of the highest energy conditions within the entire realm of clastic sedimentation, (2) deposition of clastic sediment occurs directly adjacent to the areas of erosion which provide the sediments, and (3) deposits are of maximum possible range in size of clastic particles (from the largest boulders to clays) and are commonly very poorly sorted compared with other types of alluvial sediments (Fig. 5).

The size of individual alluvial fans is controlled by drainage-basin area, slope, climate, and character of rocks within the mountain range. Individual fans range in radius from several hundred feet to several tens of miles. Coalescing fans can occur in linear belts that are hundreds of miles long. Fan deposits usually attain their maximum thicknesses and grain size near the mountain base (apex of fan) and

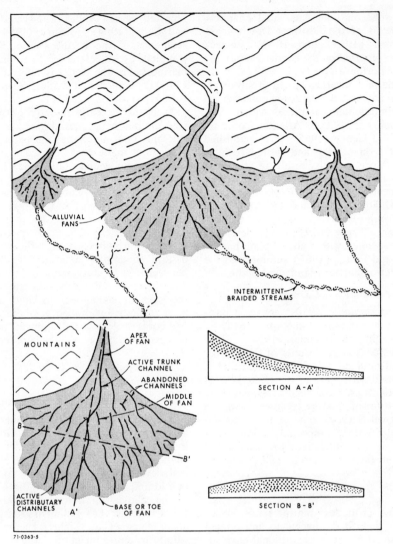

Fig. 5—Alluvial-fan model of clastic sedimentation.

gradually decrease in thickness away from the apex.

The alluvial-fan environments commonly grade downstream into braided-stream or playa-lake environments. In some areas, where mountains are adjacent to oceans or large inland lakes, alluvial fans are formed under both subaerial and submerged conditions. Such fans are now referred to as "Gilbert-type" deltas.

Alluvial-fan deposits form important reservoirs for groundwater in many areas, and adjacent groundwater basins are recharged through the fan deposits which fringe these basins.

Source, Transportation, and Deposition of Sediments

Tectonic activity and climate have a profound influence on the source, transportation, and deposition of alluvial-fan deposits. Uplift of mountain ranges results in very intensive erosion of rocks and development of a very high-gradient drainage system. The rate of weathering and production of clastic material is controlled mainly by rock characteristics and climate (temperature and rainfall).

Clastic materials are transported from source areas in mountains or high hills to alluvial fans by several types of flows: stream flows and sheetfloods and debris flows or mudflows. Sediment transport by streams is usually characteristic of large fans in regions of high to moderate rainfall. Mudflows or debris flows are more common on small fans in regions of low rainfall characterized by sudden and brief periods of heavy downpours.

Stream deposits—Streams which drain relatively small segments of steep mountain ranges have steep gradients; they may erode deep canyons and transport very large quantities of coarse debris. The typical overall stream gradient is concave upward, and the lowest gradient occurs at the toe of the fan (Fig. 5).

Hooke (1967) described a special type of stream-flow deposit, which he called "sieve deposits," on fans which are deficient in fine sediments. These gravel deposits are formed when water infiltrates completely into the fan. Bull (1969) described three types of water-laid sediments on alluvial fans: channel, sheetflood, and sieve deposits. Stream channels radiate outward from the fan apex and commonly are braided. The processes of channel migration, diversion, abandonment and filling, and development of new main channels and smaller distributary channels on the lower part of the fan surface are characteristic features. Most fan surfaces are characterized by one or a few active channels and numerous abandoned channels. Deposits on abandoned portions of gravelly and weathered fan surfaces are referred to as "pavement."

Alluvial-fan channel deposits have abrupt basal contacts and channel geometry; they are generally coarse. Bull (1972) described channel deposits as imbricated and massive or thick-bedded.

Heavy rainfall in mountainous source areas can result in floods on alluvial fans. The relatively shallow and wide fan channels are not capable of carrying the sudden influx of large volumes of water; consequently, the streams overtop their banks and flood part of the fan surface. The result is the deposition of thin layers of clastic material between channels. Bull (1969) reported sheetflood deposits to be finer grained than channel deposits, cross-bedded, and massive or thinly bedded.

Debris-flow deposits—Some workers refer to both fine-grained and coarse-grained types of plastic flowage in stream channels as mudflows, but others consider mudflows to be fine-grained debris flows. Examples of transportation and deposition of clastic sediments by mudflows

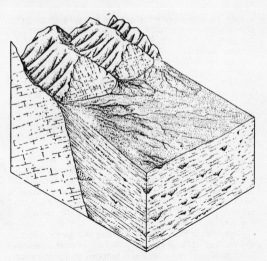

Fig. 6—Stratigraphic geometry of an alluvial fan. After Bull (1972).

were first described by Rickmers (1913) and Blackwelder (1928). The following conditions favor the development of mudflows: presence of unconsolidated material with enough clay to make it slippery when wet, steep gradients, short periods of abundant water, and sparse vegetation.

Pack (1923) discussed debris-flow deposition on alluvial-fan surfaces. Debris flows occur as a result of very sudden, severe flooding of short duration. Beaty (1963) described eye-witness accounts of debris flows on the west flank of the White Mountains of California and Nevada. Debris flows follow channels, overtop the channel banks, and form lobate tongues of debris along channels. Debris-flow deposits are very poorly sorted, fine- to coarse-grained, and unstratified; they have abrupt margins. This type of deposit is probably most common on the upper parts of the fans between the apex and midfan areas.

Summary: Character and Geometry of Alluvial-Fan Deposits

Most of the alluvial-fan studies conducted thus far have been concerned primarily with the origin and general characteristics of fans and the distribution of sediments on the surfaces of fans. An exception is Bull's excellent summary paper (Bull, 1972), which contains significant data on the geometry of channel, sheetflood, debris-flow, mudflow, and sieve deposits. The abstract of Bull's paper is quoted below:

Alluvial fans commonly are thick, oxidized, orogenic deposits whose geometry is influenced by the rate and duration of uplift of the adjacent mountains and by climatic factors.

Fans consist of water-laid sediments, debris-flow deposits, or both. Water-laid sediments occur as channel, sheetflood, or sieve deposits. Entrenched stream channels commonly are backfilled with gravel that may be imbricated, massive, or thick bedded. Braided sheets of finer-grained sediments deposited downslope from the channel may be cross-bedded, massive, laminated, or thick bedded. Sieve deposits are overlapping lobes of permeable gravel.

Debris-flow deposits generally consist of cobbles and boulders in a poorly sorted matrix. Mudflows are fine-grained debris flows. Fluid debris flows have graded bedding and horizontal orientation of tabular particles. Viscous flows have uniform particle distribution and vertical preferred orientation that may be normal to the flow direction.

Logarithmic plots of the coarsest one percentile versus median particle size may make patterns distinctive of depositional environments. Sinuous patterns indicate shallow ephemeral stream environments. Rectilinear patterns indicate debris flow environments.

Fans consist of lenticular sheets of debris (length/width ratio generally 5 to 20) and abundant channel fills near the apex. Adjacent beds commonly vary greatly in particle size, sorting, and thickness. Beds extend for long distances along radial sections and channel deposits are rare. Cross-fan sections reveal beds of limited extent that are interrupted by cut-and-fill structures.

Three longitudinal shapes are common in cross section. A fan may be lenticular, or a wedge that is either thickest, or thinnest, near the mountains.

Ancient Alluvial-Fan Deposits

Some examples of ancient alluvial-fan deposits which have been reported from the United States, Canada, Norway, and the British Isles are summarized in Table 1, together with other types of alluvial deposits.

BRAIDED-STREAM MODEL OF CLASTIC SEDIMENTATION

Occurrence and General Characteristics

Braided streams occur throughout the world under a very wide range of physiographic and climatic conditions. They are common features on extensive alluvial plains which occupy a position in the clastic realm of sedimentation between the high-gradient alluvial-fan environment at the base of mountain ranges and the low-gradient meandering-stream model of sedimentation (downstream). In physiographic provinces characterized by mountainous areas adjacent to the sea, the braided-stream environment can extend directly to the coastline and thus constitute the predominant environment of alluvial deposition. In this type of situation, meandering streams do not exist (Fig. 7). The braided stream is also a common feature of glacial outwash plains associated with the fluvio-glacial environment.

The braided-stream model is characterized by extremely variable rates of sedimentation in multiple-channel streams (Fig. 8), the patterns of which vary widely compared with meandering channels. Braided channels are usually wide and shallow; they contain numerous bars, are slightly sinuous or straight, and migrate at rapid rates. Stream gradients are high, are quite variable, and are less than those of alluvial fans but generally greater than those of meandering streams. Large fluctuations in discharge occurring over short periods of time are also common. The combination of steep gradients and high discharge rates results in the transporta-

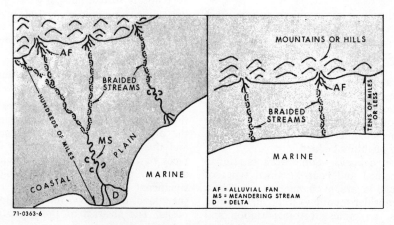

FIG. 7—Braided-stream model of clastic sedimentation.

tion and deposition of large amounts of coarse material, ranging from boulders to sand. Braided-stream deposits overall are finer than those of alluvial fans, coarser than those of meandering streams, and quite varied in stratification.

Source, Transportation, and Deposition of Sediments

Aggrading braided streams transport very large quantities of clastic material derived from a variety of sources, such as outwash plains, alluvial fans, mountainous areas, and broad plains. Unlike that of meandering streams, the bulk of the sedimentary load of most braided streams is transported as bed load. Rates of sediment transport and deposition are extremely variable, the maximum rate occurring during severe floods of short duration. High-gradient upstream segments of braided streams close to source areas are characterized by deposition of poorly sorted clastic sediments which

Table 1. Examples of Ancient Alluvial-Fan, Braided-Stream, and Meandering-Stream Deposits

Alluvial Fan	Braided Stream	Meandering Stream	Composite	Author
Arizona				Melton, 1965
California				Crowell, 1954
California			Arizona	Flemal, 1967
Colorado			California	Galehouse, 1967
			California	Boggs, 1966
			Colorado	Bolyard, 1959
		Colorado		Brady, 1969
			Colo. Plateau	Finch, 1959
			Colo. Plateau	Stokes, 1961
Colorado				Howard, 1966
Colorado				Hubert, 1960
Connecticut Valley				Klein, 1968
		Illinois		Hewitt et al., 1965
		Illinois		Shelton, 1972
			Kansas	Lins, 1950
		Kansas		Shelton, 1971
	Llano Estacado			Bretz & Horberg, 1949
	Maryland	Maryland		Hansen, 1969
Massachusetts				Wessel, 1969
Massachusetts				Stanley, 1968
			Massachusetts	Mutch, 1968
		Michigan		Shideler, 1969
		Montana		Gwinn, 1964
	Mississippi	Mississippi		Berg & Cook, 1968
Montana	Montana			Gwinn & Mutch, 1965
			Montana	Shelton, 1967
Montana				Wilson, 1967, 1970
Montana				Beaty, 1961
			Nebraska	Exum & Harms, 1968
			Nebraska	Harms, 1966
	New York	New York		Buttner, 1968
	New Jersey, New York			Smith, 1970; Shelton, 1972
			North Dakota	Royse, 1970
			Oklahoma	Visher, 1965b
		Pennsylvania		Peutner et al., 1967
	Pennsylvania			Smith, 1970
		Pennsylvania		Ryan, 1965
			Rhode Island	Mutch, 1968
S.W. USA				Bull, 1972
			Texas	Fisher & McGowen, 1969
Texas				McGowen & Groat, 1971
		Texas		McGowen & Garner, 1969; Shelton, 1971
		West Virginia		Beerbower, 1964, 1969
		Wyoming		Berg, 1968
Wyoming	Wyoming			Spearing, 1969
			Alberta	Byers, 1966
			Quebec	Dineley & Williams, 1968
Northeastern Canada				Klein, 1962
		Nova Scotia		Way, 1968
Northwest Territories	Northwest Territories	Northwest Territories		Miall, 1970
		England		Allen, 1964; Laming, 1966
Wales and Scotland				Bluck, 1965, 1967
		South Wales		Kelling, 1968
Norway				Nilsen, 1969
	Scotland			Williams, 1966, 1969
	Spain			Nagtegaal, 1966
	Spitsbergen	Spitsbergen		Moody-Stuart, 1966
		New South Wales		Conolly, 1965

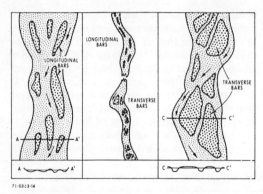

Fig. 8—Types of braided-stream channels and bars.

range in size from boulder to sand. Farther downstream, there is a gradual decrease in grain size and an increase in sorting.

The bed-load materials are transported under varying bed-form conditions, depending upon river stage. Coleman (1969) reported ripple and dune migration in the Brahmaputra River of Bangla Desh ranging from 100 ft to 2,000 ft (30–610 m) per day. Chein (1961) reported downstream movement of sandbars in the Yellow River of China to be as great as 180–360 ft (55–110 m) per day. (For comparison, the rate of bed-load movement in the meandering Mississippi is about 40 ft [12 m] per day.)

Process of channel division (braiding) by development of bars—The exact causes of channel division which results in the development of the braided pattern are not very well understood. Two methods in which channel division takes place have been described by Ore (1963) as follows:

Leopold and Wolman (1957, p. 43–44), using results of both stream-table studies and observations of natural braided streams, discuss in some detail how channel division may take place. At any time, the stream is carrying coarser fractions along the channel center than at the margins, and due to some local hydraulic condition, part of the coarsest fraction is deposited. Finer material is, in part, trapped by coarser particles, initiating a central ridge in the channel. Progressive additions to the top and downstream end of the incipient bar build the surface toward water level. As progressively more water is forced into lateral channels beside the growing bar, the channels become unstable and widen. The bar may then emerge as an island due to downcutting in lateral channels, and eventually may become stabilized by vegetation. New bars may then form by the same process in lateral channels. These authors stress that braiding is not developed by the stream's inability to move the total quantity of sediment provided to it; as incapacity leads merely to aggradation without braiding. The condition requisite to braiding is that the stream cannot move certain sizes provided; that is, the stream is incompetent to transport the coarsest fraction furnished to a given reach. Observations for the present study substantiate the braiding process of Leopold and Wolman.

Many features of streams, bars, and braided reaches result from changes in regimen (e.g., discharge, load, gradient), to a large extent representing seasonal fluctuations. Other features of bars result from normal evolution, and represent no change in regimen.

The incipient longitudinal bar formed in a channel commonly has an asymmetric, downstream-pointing, crescentic shape. This coarse part is the "nucleus" of the bar, is coarser than successive additions to the downstream end, and largely retains its position and configuration as long as any part of the bar remains. During longitudinal bar evolution downstream of this incipient bar the water and its sediment load commonly sweep from one lateral channel diagonally across the downstream end of the bar, forming a wedge of sediment with an advancing front at its downstream edge. This wedge of sediment is higher at its downstream edge, both on the longitudinal bars described here, and where found as transverse bars to be considered later. The latter build up the channel floor, independent of longitudinal bar development, simply by moving downstream.

After a certain evolutionary stage, bar height stops increasing because insufficient water for sediment transport is flowing over its surface, and deepening and widening of lateral channels slowly lower water level. From then on, the bar may be either stabilized by vegetation or dissected.

Widening of a reach after bar deposition is in some cases associated with lateral dissection of the newly formed bar. Most erosion, however, apparently occurs on the outer channel margins. If water level remains essentially constant for long periods of time, lateral dissection may establish terraces along bar margins. A compound terrace effect may be established during falling water stages. The constant tendency of the stream to establish a cross-sectional profile of equilibrium is the basic cause of lateral cutting by the stream.

Longitudinal bars which become awash during high-water stages may be dissected by small streams flowing transversely over their surfaces. In stream-table experiments, sediment added to a system eroding transverse channels on bar surfaces is first transported along lateral channels beside the bars. Eventually, these channels fill to an extent that sediment starts moving transversely over bar surfaces, and fills bar-top, transverse channels. The addition of sufficient sediment to fill lateral and bar-top channels often culminates in a transverse bar covering the whole bar surface evenly.

Another process of braiding, in addition to that described by Leopold and Wolman, takes place in well sorted sediments, and involves dissection of transverse bars. This is in opposition to construction of longitudinal bars in poorly sorted sediment, the type of braiding discussed above. Both types may occur together geographically and temporally. During extended periods of high discharge, aggradation is by large tabular bodies of sediment with laterally sinuous fronts at the angle of repose migrating downstream. Stabilization of discharge or decrease in load after establishment of these transverse bars results in their dissection by anastomosing channels; bars in this case form as residual elements of the aggradational pattern.

The transient nature of braided stream depositional surfaces is characteristic of the environment. The streams and depositional areas within the stream exhibit profound lateral-migration tendencies, especially during periods of high discharge. Channel migration takes place on several scales. Individual channels erode laterally, removing previously deposited bars. They divide and coalesce, and several are usually flowing adjacent to one another concurrently within the main channel system. The whole channel system, composed of several flowing channels with bars between, also exhibits migrating tendencies.

Braided-stream deposits—Our knowledge of modern braided-stream deposits has increased substantially during the past several years as a result of studies of several rivers in Wyoming, Colorado, and Nebraska by Ore (1963, 1965); the Brahmaputra River of Bangla Desh (formerly East Pakistan) by Coleman (1969); the Platte River of Colorado and Nebraska by N. Smith (1970); the Red River of the Texas panhandle by Waechter (1970); the Canadian River of northwest Texas by Kessler (1970, 1971); and the Copper River of Alaska by Boothroyd (1970). These studies revealed that braided-stream deposits are laid down principally in channels as longitudinal bars and transverse bars. Abandoned-channel deposits (channel fills) have been reported by Doeglas (1962) and Williams and Rust (1969).

According to Ore (1963, 1965), longitudinal-bar deposits occur mainly in upstream channel segments and transverse bars are more common in downstream segments; however, in some places these two types of bars occur together (Fig. 8). Longitudinal-bar deposits are lens-shaped and elongated in the downstream direction. Grain size decreases downstream from coarse to fine in an individual bar; deposits are poorly sorted and mainly horizontally stratified but laterally discontinuous. Transverse-bar deposits occur as long thin wedges and are highly dissected by channels. The downstream edges of transverse bars migrate to produce planar cross-stratification and some festoon crossbedding. Sediments of transverse bars are generally finer and better sorted than those of longitudinal bars.

N. Smith (1970) described some very significant relations between types of bars, stratification, and grain size in the Platte River. In the upstream segment in Colorado, the deposits consist mainly of longitudinal bars characterized by low-relief stratification, generally horizontally bedded but including some festoon crossbedding. The downstream channel segment in Nebraska is characterized by transverse-bar deposits consisting of better sorted, fine-grained sand with abundant tabular cross-stratification and some festoon crossbedding.

The Red River braided-stream sediments of West Texas consist of longitudinal-bar deposits with low-angle or horizontal stratification; they are deposited during waning flood stages (Waechter, 1970). Low-river-stage deposits consist mainly of migrating transverse-bar deposits (in channels) with tabular cross-stratification and some festoon crossbedding. The migration of very shallow channels results in stratification sets that are horizontal, tabular or lenticular, and laterally discontinuous.

Kessler (1970) reported longitudinal-bar deposits consisting mainly of fine sand in upstream reaches of the Canadian River in West Texas. Transverse-bar deposits are predominant in the downstream part of the area studied. Kessler (1971) also discussed individual flood sequences of deposits which contain parallel bedding and tabular and small-ripple cross-laminations. These sequences are covered by clay drapes and are laterally discontinuous.

Coleman (1969) presented the results of a significant study of one of the largest braided rivers of the world, the Brahmaputra in Bangla Desh. This river is 2–6 mi (3–9.5 km) wide and migrates laterally as much as 2,600 ft (790 m) per year; deposition of sediments in its channels during a single flood occurs in a definite sequence of change, ranging from ripples up to 5 ft (1.5 m) high that migrate downstream 400 ft (120 m) per day to sand waves 50 ft (15 m) high that migrate up to 2,000 ft (610 m) per day.

Williams and Rust (1969) presented results of a very detailed study of a 4-mi (6.5 km) segment of a degrading braided stream, the Donjek River of the Yukon Territory, Canada. They divided the bar and channel deposits, which range from coarse gravels to clays, into seven facies. Ninety-five percent of the bar deposits are of the longitudinal type and consist of gravel, sand, and some finer sediments. Abandoned-channel deposits consist of gradational sequences of gravels, sand, and clays that become finer upward.

Summary: Braided-Stream Deposits

Most of the sediments of modern braided streams studied during the past decade have been referred to by authors as transverse- or longitudinal-bar deposits. These sediments were deposited within braided channels during varying discharge conditions ranging from low water to flood stage. Thus, all longitudinal and transverse bars should be considered as a special type of bed form occurring within active braided channels.

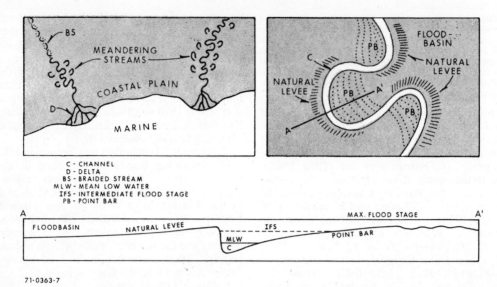

FIG. 9—Setting and general characteristics of meandering-stream model of clastic sedimentation.

Studies by Doeglas (1962) and Williams and Rust (1969) are significant because they describe abandoned-channel deposits. Doeglas discussed the methods of channel abandonment and described the channel-fill deposits as coarse grained, with channel or festoon laminations, in the upstream portions of abandoned channels, and as fine grained, silty, and rippled in the downstream portions of abandoned channels.

Ancient Braided-Stream Deposits

Some examples of ancient braided-stream deposits which have been reported from the United States, Spitsbergen, and Spain are summarized in Table 1.

MEANDERING-STREAM MODEL OF CLASTIC SEDIMENTATION

Occurrence and General Characteristics

Meandering streams generally occur in coastal-plain areas updip from deltas and downdip from the braided streams. The axis of sedimentation is usually perpendicular to the shoreline (Fig. 9).

This model is characterized by a single-channel stream which is deeper than the multichannel braided stream. Meandering streams usually have a wide range in discharge (cu ft/sec) which varies from extended periods of low-water flow to flood stages of shorter duration. Flooding can occur one or more times per year and major flooding once every several years. The meandering channel is flanked by natural levees and point bars, and it migrates within a zone (meander belt) about 15 to 20 times the channel width. Channel segments are abandoned and filled with fines as new channels develop.

Source, Transportation, and Deposition of Sediments

Sediments are derived from whatever type of deposit occurs in the drainage area. Clays and fine silts are transported in suspension (suspended load), and coarser sediments such as sand, gravel, and pebbles are transported as bed load. Sediment transport and deposition during extended low-water stages are confined to the channel and can be nil or very slow. Maximum sediment transport occurs during rising flood stage when the bed of the channel is scoured.

The maximum rate of sediment deposition occurs during falling flood stages. Grain size depends on the type of sediment available to the channel; the coarsest sediments are deposited in the deepest part of the channel, and the finest sediments accumulate in floodbasins and in some parts of the abandoned channels.

Channel migration and deposition of point-bar sediments—The most important processes of sedimentation in the meandering-stream model are related to channel migration which occurs as a result of bank caving and point-bar accretion (Fig. 10). The process of bank caving occurs most rapidly during falling flood

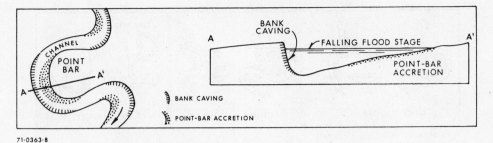

Fig. 10—Areas of bank caving and point-bar accretion along a meandering channel.

stage, when currents of maximum velocities are directed against the concave bank. Bank caving occurs at maximum rates in bends where the bed and bank materials are very sandy. Rates are much slower in areas where banks are characterized by clayey sediments (Fisk, 1947).

Deposition occurs on the convex bar (point bar) simultaneously with bank caving on the concave bank.

Bank caving and point-bar accretion result in channel migration and the development of the point-bar sequence of sediments (Fig. 11). The point bar is probably the most common and significant environment of sand deposition. The thickness of this sequence is governed by channel depths. Point-bar sequences along the Mississippi River attain thicknesses in excess of 150 ft (45 m). Medium-size rivers like the Brazos of Texas produce point-bar sequences that are 50 ft (15 m) thick (Bernard *et al.*, 1970).

Channel diversions and filling of abandoned channels—The process of channel diversion and channel abandonment is another characteristic feature of meandering streams. There are two basic types of diversion and abandonment: (1) the neck or chute cutoff of a single meander loop and (2) the abandonment of a long channel segment as a result of a major stream diversion (Fisk, 1947).

Meander loops which are abandoned as a result of neck or chute cutoffs become filled with sediment (Fig. 12A). The character of the channel fill depends on the orientation of the abandoned loop with respect to the direction of flow in the new channel. Meanders oriented with their cutoff ends pointing downstream (Fig. 12B) are filled predominantly with clays (clay plugs); those oriented with the cutoff ends pointing upstream are filled principally with sands and silts.

A major channel diversion is one which results in the abandonment of a long channel segment or meander belt, as shown in Figure 13. Channeling of flood water in a topographically low place along the bank of the active channel can rapidly erode unconsolidated sediments and create a new channel. This process can happen during a single flood or as a result of several floods. The newly established channel has a gradient advantage across the topographically lower floodbasin. A diversion can occur at any point along the channel.

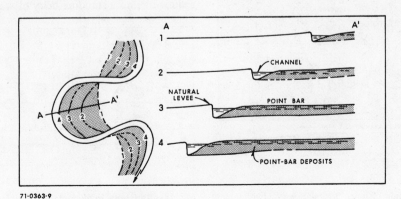

Fig. 11—Development of point-bar sequence of sediments.

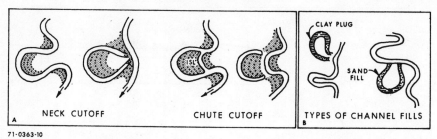

Fig. 12—Channel diversion, abandonment, and filling as a result of neck and chute cutoffs.

The character of the sediments which fill long channel segments is governed by the manner of channel diversion. Abrupt abandonment (during a single flood or a few floods) results in the very rapid filling of only the upstream end of the old channel, thus creating a long sinuous lake. These long, abandoned channels (lakes) fill very slowly with clays and silts transported by flood waters (Fig. 14, left).

Gradual channel abandonment (over a long period) results in very gradual channel deterioration. Diminishing flow transports and deposits progressively smaller amounts of finer sands and silts (Fig. 14, right).

Summary: Characteristics of Meander-Belt and Floodbasin Deposits

The meandering-stream model of sedimentation is characterized by four types of sediments: the point bar, abandoned channel, natural levee, and floodbasin. The nature of each of these four types of sediments and their interrelations are summarized in Figure 15.

Only two main types of sand bodies are associated with a meandering stream: the point-bar sands and the abandoned-channel fills. The former, which are much more abundant than the latter, occur in the lower portion of the point-bar sequence and constitute at least 75 percent of the sand deposited by a meandering stream. Coalescing point-bar sands can actually form a "blanketlike" sand body of very large regional extent. The continuity of sand is interrupted only by the "clay plugs" which occur in abandoned meander loops or in the last channel position of meander belts which have been abandoned abruptly.

Examples of ancient alluvial deposits of meandering-stream origin which have been reported in the literature are summarized in Table 1.

DELTAIC MODELS OF CLASTIC SEDIMENTATION

Occurrence and General Characteristics

Deltaic sedimentation occurs in the transitional zone between continental and marine (or inland seas and lakes) realms of deposition. Deltas are formed under subaerial and subaqueous conditions by a combination of fluvial and marine processes which prevail in an area where a fluvial system introduces land-derived sediments into a standing body of water.

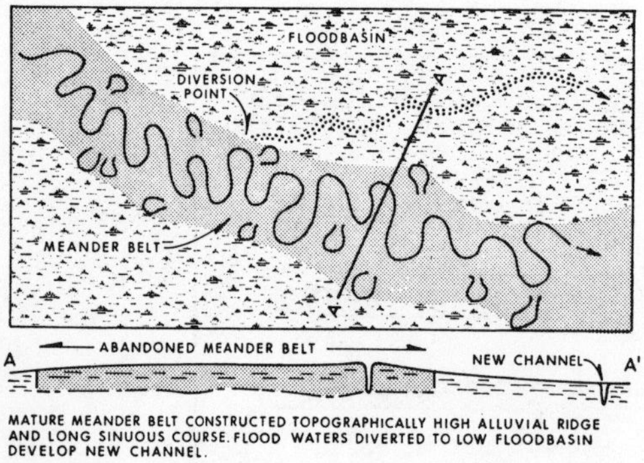

Fig. 13—Major channel diversion and abandonment of a meander belt.

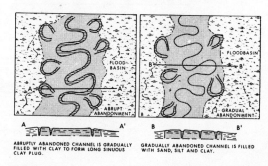

Fig. 14—Variations in character of abandoned channel fill typical of meander belts.

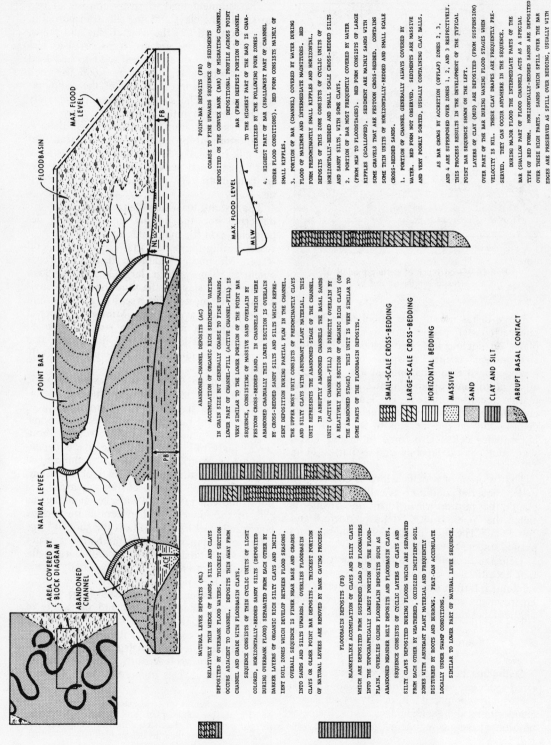

FIG. 15—Characteristics of meander-belt and floodplain deposits.

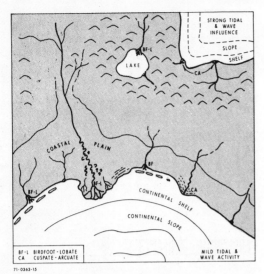

FIG. 16—Occurrence of deltaic models of clastic sedimentation.

Large deltas usually are associated with extensive coastal plains; however, all coastal plains do not include large deltas. The deltaic environment occurs downstream from the meandering-stream environment and is directly adjacent to, and updip (landward) from, the marine environment; it is flanked by the coastal-interdeltaic environment. Most large deltas occur on the margins of marine basins, but smaller deltas also form in inland lakes, seas, and coastal lagoons and estuaries (Fig. 16).

That portion of a delta which is constructed under subaerial conditions is called the "deltaic plain"; that portion which forms under water is called the "delta front," "delta platform," and "prodelta." The bulk of the deltaic mass is deposited under water.

Deltas are considered to be extremely important because they are the sites of deposition of sand much farther downdip than the interdeltaic environment, as well as being the sites where clastic deposition occurred at maximum rates.

Source and Transportation of Sediments

Sediments deposited in large deltas are derived from extensive continental regions which are usually composed of rock types of varied compositions and geologic ages. Thus, the composition of deltaic sediments can be quite varied.

The sediment load of rivers consists of two parts: (1) the clays and fine silts transported in suspension and (2) the coarser silts and sands, and in some cases gravels, transported as bed load. The ratio of suspended load to bed load varies considerably, depending upon the rock types and climatic conditions of the sediment-source areas. The suspended load is generally much greater than the bed load.

The transportation of sediment to a delta is an intermittent process. Most rivers transport the bulk of their sediments during flood stages. During extended periods of low discharge, rivers contribute very little sediment to their deltas.

The extent to which deltaic sediments are dispersed into the marine environment is dependent upon the magnitude of the marine processes during the period that a river is in flood stage. Maximum sediment dispersal occurs when a river with a large suspended load reaches flood stage at the time the marine environment is most active (season of maximum currents and wave action). Minimum dispersal occurs when a river with a small suspended load (high bed load) reaches flood stage at a time when the marine environment is relatively calm.

Size of deltas—There is an extremely wide range in the size of deltas;[4] modern deltas range in area from less than 1 sq mi (2.6 sq km) to several hundreds of square miles. Some large deltaic-plain complexes are several thousand square miles in area. Delta size is dependent upon several factors, but the three most important are the sediment load of the river; the intensity of marine currents, waves, and tides; and the rate of subsidence. For a given rate of subsidence, the ideal condition for the construction of a large delta is the sudden large influx of sediments in a calm body of water with a small tidal range. An equally large sediment influx into a highly disturbed body of water with a high tidal range results in the formation of a smaller delta, because a large amount of sediment is dispersed beyond the limits of what can reasonably be recognized as a delta. Rapid subsidence enhances the possibility for a large fluvial system to construct a large delta.

Types of deltas—A study of modern deltas of the world reveals numerous types. Bernard

---

[4] Published figures on areal extent of deltas are based on size of the deltaic plain and do not include submerged portions of the delta, which in many cases are as large as or larger than the deltaic plains.

(1965) summarized some of the factors which control delta types as follows:

Deltas and deltaic sediments are produced by the rapid deposition of stream-borne materials in relatively still-standing bodies of water. Notwithstanding the effects of subsidence and water level movements, most deltaic sediments are deposited off the delta shoreline in the proximity of the river's mouth. As these materials build upward to the level of the still-standing body of water, the remainder of deltaic sediments are deposited onshore, within the delta's flood plains, lakes, bays, and channels.

Nearly 2,500 years ago, Herodotus, using the Nile as an example, stated that the land area reclaimed from the sea by deposition of river sediments is generally deltoid in shape. The buildup and progradation of deltaic sediments produces a distinct change in stream gradient from the fluvial or alluvial plain to the deltaic plain. Near the point of gradient change the major courses of rivers generally begin to transport much finer materials, to bifurcate into major distributaries, and to form subaerial deltaic plains. The boundaries of the subaerial plain of an individual delta are the lateral-most distributaries, including their related sediments, and the coast line. Successively smaller distributaries form sub-deltas of progressively smaller magnitudes.

Deltas may be classified on the basis of the nature of their associated water bodies, such as lake, bay, inland sea, and marine deltas. Other classifications may be based on the depth of the water bodies into which they prograde, or on basin structure.

Many delta types have been described previously. Most of these have been related to the vicissitudes of sedimentary processes by which they form. Names were derived largely from the shapes of the delta shorelines. The configuration of the delta shores and many other depositional forms expressed by different sedimentary facies appear to be directly proportional to the relative relationship of the amount or rate of river sediment influx with the nature and energy of the coastal processes. The more common and better understood types, listed in order of decreasing sediment influx and increasing energy of coastal processes (waves, currents, and tides), are: birdfoot, lobate, cuspate, arcuate, and estuarine. The subdeltas of the Colorado River in Texas illustrate this relationship. During the first part of this century, the river, transporting approximately the same yearly load, built a birdfoot-lobate type delta in Matagorda Bay, a low-energy water body, and began to form a cuspate delta in the Gulf of Mexico, a comparatively high-energy water body. Many deltas are compounded; their subdeltas may be representative of two or more types of deltas, such as birdfoot, lobate, and arcuate. Less-known deltas, such as the Irrawaddy, Ganges, and Mekong, are probably mature estuarine types. Others, located very near major scarps, are referred to the "Gilbert type," which is similar to an alluvial fan.

Additional studies of modern deltas are required before a more suitable classification of delta types can be established. J. M. Coleman (personal commun.) and his associates, together with the Coastal Studies Institute at Louisiana State University, are presently conducting a comprehensive investigation of more than 40 modern deltas. Results of their studies undoubtedly will be a significant contribution toward the solution to this problem.

Only three types of deltas will be considered in this report: the birdfoot-lobate, the cuspate-arcuate, and the estuarine.

Sedimentary Processes and Deposits of the Birdfoot-Type Delta

The processes of sedimentation within a delta are much more complex and variable than those which occur in the meandering-stream and coastal-interdeltaic environment of sedimentation. It is impossible to discuss these deltaic processes in detail in a short summary paper such as this; therefore, only a brief summary of the following significant processes is presented.

1. Dispersal of sediment in the submerged parts of the delta (from river mouths seaward);
2. Formation of rivermouth bars, processes of channel bifurcation, and development of distributary channels;
3. Seaward progradation of delta, deposition of the deltaic sequence of sediments, and abandonment and filling of distributary channels; and
4. Major river diversions, abandonment of deltas, and development of new deltas.

Dispersal and deposition of sediments—Riverborne sediments which are introduced in a standing body of water (a marine body or inland lakes and seas) are transported in suspension (clays and fine silts) and as bed load (coarse silts, sands, and coarser sediments). Most of the sands and coarse silts are deposited in the immediate delta-front environment as rivermouth bars and slightly beyond the bar-front zone. The degree of sand dispersal is, of course, controlled by the level of marine energy; however, in most birdfoot deltas, sands are not transported beyond 50-ft (15 m) water depths. Fisk (1955) referred to the sands deposited around the margins of the subaerial deltaic plain as "delta-front sands," and they are called "delta-fringe sands" herein.

The finer sediments (clays and fine silts), which are transported in suspension, are dispersed over a much broader area than the fringe sands and silts. The degree of dispersal is governed by current intensity and behavior. Accumulations of clays seaward of the delta-fringe sands are referred to as "prodelta" or "distal clays" (Fig. 17).

Channel bifurcation and development of distributary channels—Some of the most significant deltaic processes are those which result in the origin and development of distributary

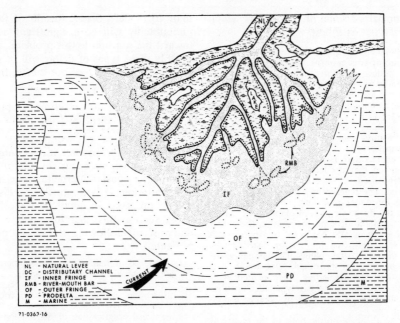

Fig. 17—Distribution of distributary-channel and fringe sands in a birdfoot-lobate delta.

channels. Welder (1959) conducted a detailed study of these processes in a part of the Mississippi delta, and Russell (1967a) summarized the origin of branching channels, as follows:

> The creation of branching channels is determined by the fact that threads of maximum turbulence and turbulent interchange (Austausch; 1.2.3; 3.5) lie deep and well toward the sides of channels, particularly if they have flat beds (typical of clay and fine sediments in many delta regions). These threads are associated with maximum scour and from them, sediment is expelled toward areas of less turbulence and Austausch. Significant load is propelled toward mid-channel, where shoals are most likely to form.

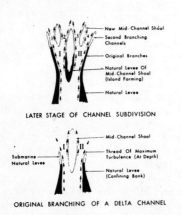

Fig. 18—Stages in development of channel bifurcation. After Russell (1967).

> At its mouth, the current of a delta channel continues forward (as a result of momentum) and creates *jet flow* into the lake or sea it enters. After leaving the confinement imposed by fixed banks, however, the current flares marginally to some extent (widening the jet, reducing its velocity, and eventually dissipating its flow energy). Near the termination of confining banks the jet flow is concentrated and moves ahead into relatively quiet water. With flaring of jet flow comes an increase in spacing between threads of most intense turbulence and exchange. There is a tendency toward scour below each thread, but the exchange process sends most of the entrained material toward marginal quiet water on both sides (Fig. 8 [Fig. 18 of this paper]). Deposition creates a submarine natural levee on the outer side of each thread. Sediment is also attracted toward and deposited in the widening area of mid-channel water, where it builds a shoal. The channel divides around the shoal, creating two distributaries, each of which develops its own marginal threads of maximum turbulence, perpetuating conditions for other divisions below each new channel mouth. If not opposed by wave erosion and longshore currents, the subdivision continues in geometric progression (2, 4, 8, 16, etc.) as the delta deposit grows forward.
> The marginal natural levees are submarine features at first and fish may swim across their crests. Later they grow upward, and for awhile become areas where logs and other flotsam accumulate and where birds walk with talons hardly submerged. Salt- or fresh-water-tolerant grasses invade the shallow water and newly created land, first along levee crests, later to widen as the levees grow larger. Salicornia and other plants become established pioneer trees such as willows, and eventually in the plant succession comes the whole complex characteristic of natural levees upstream. In tropical areas mangroves are likely to become the dominant trees.

A similar conversion exists in mid-channel, where the original shoal becomes land and either develops into a lenticular or irregular island or becomes the point of land at the head of two branching distributaries.

Progradation of delta and deposition of deltaic sequence—Fisk's discussion of the process of distributary-channel lengthening (progradation of delta seaward) is probably one of the most significant of his many contributions on deltaic sedimentation (Fisk, 1958). His description of this important aspect of delta development is presented below. (Stages in the development of a birdfoot-type delta are shown in Figure 19.)

Each of the pre-modern Mississippi River courses was initiated by an upstream diversion, similar to the one presently affecting the Mississippi as the Atchafalaya River enlarges (Fisk, 1952). Stream capture was a gradual process involving increasing flow through a diversional arm which offered a gradient advantage to the gulf. After capture was effected, each new course lengthened seaward by building a shallow-water delta and extending it gulfward. Successive stages in course lengthening are shown diagrammatically on Figure 2 [Fig. 19, this paper]. The onshore portion of the delta surface ... is composed of distributaries which are flanked by low natural levees, and interdistributary troughs holding near-sea-level marshes and shallow water bodies. Channels of the principal distributaries extend for some distance across the gently sloping offshore surface of the delta to the inner margin of the steeper delta front where the distributary-mouth bars are situated. The offshore channels are bordered by submarine levees which rise slightly above the offshore extensions of the interdistributary troughs.

In the process of course lengthening, the river occupies a succession of distributaries, each of which is favorably aligned to receive increasing flow from upstream. ... The favored distributary gradually widens and deepens to become the main stream ... ; its natural levees increase in height and width and adjacent interdistributary troughs fill, permitting marshland development. Levees along the main channel are built largely during floodstage; along the distal ends of distributaries, however, levee construction is facilitated by crevasses ... which breach the low levees and permit water and sediment to be discharged into adjacent troughs during intermediate river stages as well as during floodstage. Abnormally wide sections of the levee and of adjacent mudflats and marshes are created in this manner, and some of the crevasses continue to remain open and serve as minor distributaries while the levees increase in height. Crevasses also occur along the main stream during floodstages ... and permit tongues of sediment to extend into the swamps and marshes for considerable distances beyond the normal toe position of the levee.

Distributaries with less favorable alignment are abandoned during the course-lengthening process, and their channels are filled with sandy sediment. Abandoned distributaries associated with the development of the present course below New Orleans vein the marshlands. ... Above the birdfoot delta, the pattern is similar to that of the older courses ... ; numerous long, branching distributaries diverge at a low angle.

Stream diversions, abandonment of deltas, and development of new deltas—Deltas prograde seaward but they do not migrate laterally, as a point bar does, for example. A delta *shifts position* laterally if a major stream diversion occurs upstream in the alluvial environment or in the upper deltaic-plain region (Fig. 20). Channel diversions were discussed in the section on the meandering-stream model.

Deltas, like meander belts, can be abandoned abruptly or gradually, depending upon the time required for channel diversion to occur. Once a delta is completely abandoned, all processes of deltaic sedimentation cease to exist in that particular delta. With a standing sea level, the sediments of the abandoned delta compact, and subsidence probably continues. The net result is the encroachment of the marine environment over the abandoned delta. This process has erroneously been referred to by some authors as "the destructive phase of deltaic sedimentation." The author maintains that the proper terminology for this process is "transgressive marine sedimentation." The two processes and their related sediments are significantly different, as the discussion of the transgressive marine model of sedimentation demonstrates (see the succeeding section on this model).

As the marine environment advances landward over an abandoned subaqueous delta front and the margins of the deltaic plain, the upper portion of the deltaic sequence of sediment is removed by wave action. The amount of sediment removed depends on the inland extent of the transgression and on the rate of subsidence. The front of the transgression is usually characterized by deposition of thin marine sand units. Seaward, sediments become finer and grade into clays. Thus, local marine transgressions which occur because of delta shifts result in the deposition of a very distinctive marine sedimentary sequence which is easily distinguished from the underlying deltaic sequence.

Concurrent with marine transgression over an abandoned delta, a new delta will develop on the flanks of the abandoned delta. Sedimentary processes in the new delta are similar to those described under the discussion of progradation of deltas.

Repeated occurrences of river diversions result in the deposition of several discrete deltaic masses which are separated by thin transgressive marine sequences (Fig. 20). Under ideal conditions, deltaic facies can attain thicknesses

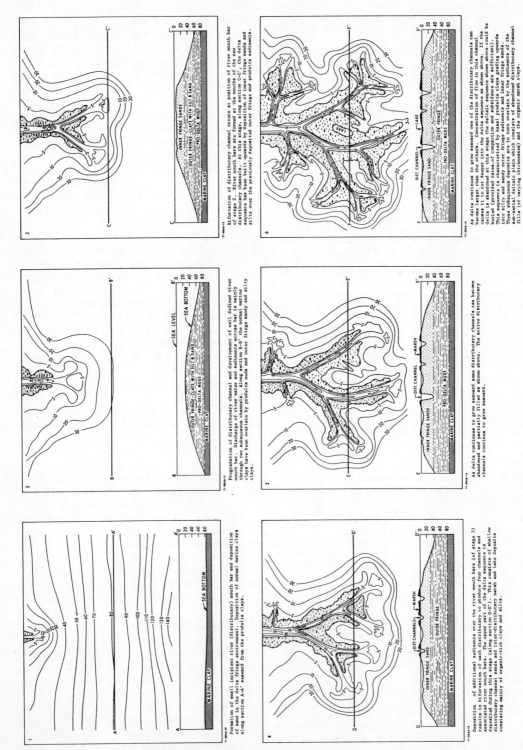

FIG. 19—Stages in development of birdfoot delta and deposition of deltaic sequence.

Geometry of Sandstone Reservoir Bodies

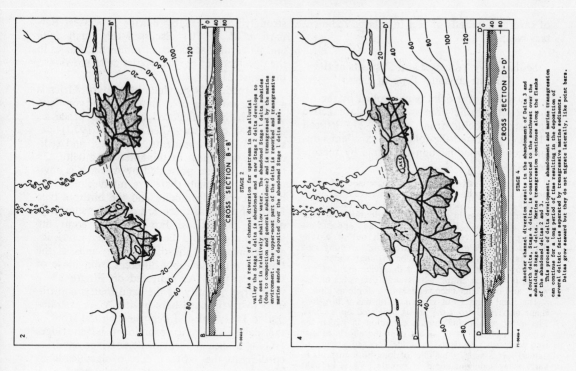

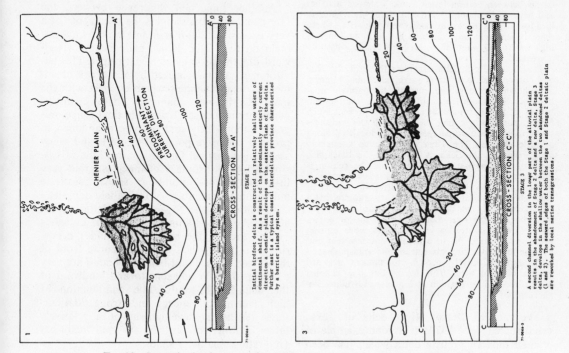

Fig. 20—Stages in development of a deltaic-plain complex and related deltaic facies.

of several hundreds of feet in a large sedimentary basin.

Sedimentary Processes and Deposits of the Cuspate-Arcuate Type of Delta

The shape of a delta is controlled by the influence of marine processes which are active against the delta front (Table 2). Russell (1967a) presented the following excellent summary of the modification of deltas by marine processes.

> The depositional processes characteristic of river mouths are opposed by marine processes that work toward removal of deposits. In a quiet sea or lake the geometric increase in number of distributaries is most closely approached. Below the most inland and earliest forking of the river, the delta builds out as a fan-shaped accumulation, with distributaries creating ribs with natural levees separating basins that widen and open toward the sea. The point deserving greatest emphasis is that the entire delta system originates underwater and only later becomes features visible as land.
>
> The ideal delta front is *arcuate* or has a *bird-foot* shape as viewed from the air or indicated on a chart. The latter pattern indicates a condition in which the deposition of load is dominant over the efforts of marine processes. It results from the forward growth of natural levees and the inability of longshore currents to carry away sediment about as rapidly as it is brought to the river mouth. The delta of the Mississippi is the largest and most typically cited example. Some talons of the foot extend out more than 20 miles and the basins between natural levees flank V-shaped marshes and bays up to about 1.5 fathoms deep. Many smaller bird-foot deltas occur in lakes and estuaries, where there is relatively little distance for fetch to generate high waves and where there are only feeble longshore currents.
>
> The arcuate-front deltas, such as those of the Nile and Niger, indicate sufficient wave action and removal of sediment by longshore currents to maintain relatively stable, smooth fronts. In some cases the momentum of jet flow is apparently sufficient to prevent much flaring, and a single pair of natural levees advances seaward to form a *cuspate* delta front, localized along a single channel. The Tiber, Italy, is the commonly cited example. The Sakayra River, on the Black Sea coast of Turkey has such a delta, but the reason is dominance of wave action. Ahead of it is a large area of shoal water with an extremely irregular system of channels and natural levees (changing so rapidly that a pilot keeps daily watch over them in order to guide boats back to the river mouth). Levees are prevented from growing up to sea level because wave erosion keeps them planed off to a depth of a few feet and because longshore currents entrain and transport sediment away effectively enough to prevent seaward growth of land area. This leaves but one channel mouth in a central position as a gently protruding single cusp.

The modern Brazos River delta of Texas (constructed since 1929) is a good example of a small modern arcuate delta which has been strongly influenced by marine processes. Bernard *et al.* (1970) discussed this small delta and its vertical sequence of sediments. Stages in the development of this type of delta are shown in Figure 21.

The modern Niger delta of western Africa is a classic example of a large arcuate-type delta that is highly influenced by marine processes and tidal currents. Allen (1965c, 1970) described the environments, processes, and sedimentary sequences of this interesting delta. On the basis of data presented by Allen, it is obvious that, although there are many similarities between the Niger delta and the birdfoot-type Mississippi delta, there are certainly some significant differences. For example, from the standpoint of sand bodies, the characteristics and geometry of the delta-fringe sands of the Niger are considerably different from those of the Mississippi. As indicated on Figure 22, a very large quantity of the sand that is contributed to rivermouth bars by the Niger is transported landward and deposited on the front of the deltaic plain as prominent beach ridges (this is a special form of delta-fringe sand according to the writer's deltaic classification). This process results in the development of a thick body of clean sands along the entire front of the deltaic plain.

Another important difference between the Niger and the Mississippi is the occurrence of a very extensive tidal-marsh and swamp environment on the Niger deltaic plain behind (landward of) the prominent beach ridges. This environment is characterized by a network of numerous small channels which connect with the main distributary channels. These channels, which are influenced by a wide tidal range, migrate rather freely and become abandoned to produce extensive point-bar deposits and many abandoned channel fills. In contrast, the Mississippi River distributary channels migrate very little and, hence, point-bar sands constitute only a small percentage of the deltaic deposits.

In summary, the Niger arcuate delta is characterized by prominent delta-fringe sands (which include the beach-ridge sands) occurring as a narrow belt along the entire front of the deltaic plain. Point-bar sand bodies are very common directly adjacent to and landward of the delta-fringe sands. The combination of high-level marine energy and strong tidal currents results in development of a relatively large quantity of distributary and point-bar (migrating channels) sands.

Table 2. Factors Which Influence Characteristics of Deltaic Deposits

Factors Which Influence Deltaic Sedimentation

Characteristics of Deltaic Deposits		Magnitude of Fluvial Processes (Sediment Influx)	Magnitude of Marine Processes (Wave Energy, Range in Tide & Currents)	Depth of Water	Subsidence Rate	Sea Level	Salinity of Water	Climate and Vegetation
Thickness of Deltaic Sequence		Large sediment influx required to produce thick sequence.	High wave and current energy tends to disperse sediments and reduce rate of sedimentation.	Water depths in basin controls max. thickness possible. Subsidence permits development of sequence thicker than depth.	Rapid subsidence produces thick sequence. Thin sequences occur in stable areas.	Rising sea level results in thicker sequences.		Abundant vegetation promotes development of thick organic rich layer in upper part of sequence.
Sand Content	Distributary Channel Fills	Percent of sand in channels depends on rapid or gradual diversion. Gradual abandonment produces more sand in fills.	Large tidal range can transport sand upstream near river mouths and increase sand content in fills.					
	Fringe	Amount of sand in fringe depends on sand load of river. Small for muddy rivers; large for rivers with high sand load.	Strong currents may transport sand along coast away from delta thus reducing sand content of fringe.					
Porosity and Permeability	Distributary Channel Fills		Large tidal range results in deposition of clays and salts in lower reaches of channels thus lowering porosity and permeability.					
	Fringe	High sand content. Rivers generally have best developed fringe sands.	Strong wave and current winnows fines and deposition of clean sand occurs. Dirty sands are deposited in areas of low marine energy.					
Geometry of Deltaic Mass			Strong marine currents can disperse sediment over extensive area to form broad subaqueous fringe and pro-delta masses.	Relatively thin blanket-like deltaic masses form in shallow water. Geometry influenced by bottom topography.	Thin blanket-like delta masses occur in stable areas. Local downwarps result in ladle-shape delta masses.	Rising sea level will produce on-lap fringe over marsh facies. Falling sea level causes strong off-lap of marsh over fringe.	Fine sediments in suspension flow over denser sea water and may be dispersed over broad areas by currents.	
Influence of Organisms on Sediments		Rapid influx of sediment and high turbidity greatly reduces organism population and degree of bioturbation.	Organisms have little influence on sediments in near shore zone with large wave energy.					
Number of Distributary Channel Fills			Strong wave and current action prevents normal development of river mouth bars necessary for bifurcation of channel.	Deltas constructed in shoal water have largest number of distributary channels.				
Thickness (Depth) of Dist. Channel Fills					Rapidly subsiding deltaic plain may result in thicker dist. channel fills.	Falling sea level may cause deepening of channels. Rising sea level causes shoaling.		
Rate of Sedimentation		Rates of sedimentation generally greatest in deltas of rivers with large sediment.	Strong currents can disperse sediments and reduce deposition rates.					Rapid growth of vegetation traps sediments and increases rate of deposition.
Thickness of Organic Rich Clays and Peats (Conc) in Upper Part of Sequence					Thick section of peat and organic clays can accumulate in a near sea level environment (with subsidence).	A rising sea level will result in thick section of organic material in upper portion of sequence.		Warm, wet deltas with dense vegetation are characterized by thick peats and organic rich clays.

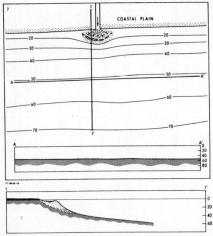

River diversion (at a point within alluvial valley or deltaic plain) results in the development of a new channel. During the initial stage of delta development a relatively small river mouth bar is formed in the marine environment. The distance of this bar from the original shoreline is approximately four times the width of the channel. This type of river mouth bar is generally breached by two or more subaqueous channels which radiate from the river mouth. The bar crest is characterized by sand and the bar-back area (which acts as a stilling basin) is characterized by clays and silts with minor amounts of sand. Seaward from the bar crest (the bar-front) the sediments grade from sands to silts and clays. Deposition along section A-A' consists of normal marine muds.

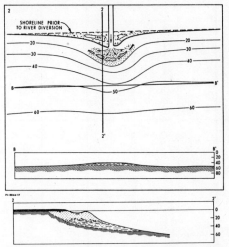

The initial river mouth bar (shown in stage 1) continues to grow (both vertically and seaward by accretion. If major flood occurs when marine environment is relatively quiet (non-storm period) the river mouth bar is built upwards to an elevation slightly above mean sea-level. Vegetation on newly formed subaerial bar traps fine-grained sediments and the bar becomes attached to the mainland. Thus the initial river mouth bar shown in Stage 1 becomes part of the deltaic plain. The river channel cuts across the old bar (of stage 1) and a new river mouth bar develops offshore. Deposition along section 2-2 consists of prodelta clays.

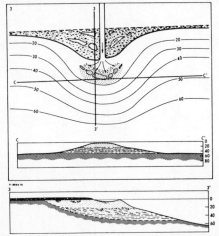

The stage 2 river mouth bar continues to grow seaward by accretion and vertically during flood stages. When bar crest reaches elevation slightly above mean sea-level it becomes attached to the stage 2 deltaic plain. The area of the stage 2 bar is separated from the stage 1 bar by shallow lakes. A new river mouth bar is constructed in the middle part of section 3-3, and delta fringe sands are deposited over the older outer fringe and prodelta sediments.

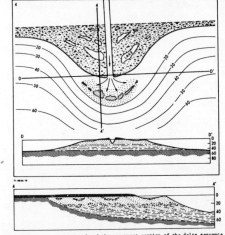

Delta continues to grow seaward and the uppermost portion of the delta sequence is deposited as the subaerial deltaic plain is constructed over the area along the center of section 4-4'. The rate at which this type of delta advances seaward is governed by the frequency and magnitude of river floods and storms. If major floods occur frequently during periods characterized by relatively low marine energy (non-storm) the delta will advance seaward rapidly. Severe current and wave action associated with major storms, especially during non-flood stages of the river, have a tendency to partially destroy the river mouth bars and thereby reduce the rate of seaward delta advance.

FIG. 21—Stages in development of a cuspate delta.

Sedimentary Processes and Deposits of Estuarine-Type Delta

Large deltas such as the Ganges, Amazon, and Colorado (in Gulf of California) are considered to be examples of estuarine-type deltas. Although our knowledge of these deltas is extremely limited, it is now reasonably well established that they are associated with extreme tidal conditions (up to 25 ft [8 m] at the mouth of the Colorado River). It is apparent that very strong tidal currents have a profound influence on the distribution of sediments. Sands are known to be transported for great distances in front of these deltas; however, the geometry of these sand bodies is unknown. Additional studies of this type delta are badly needed.

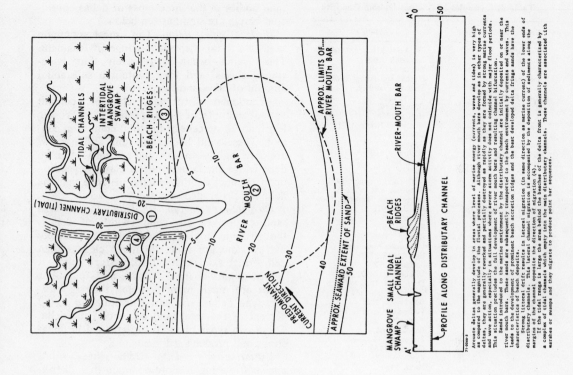

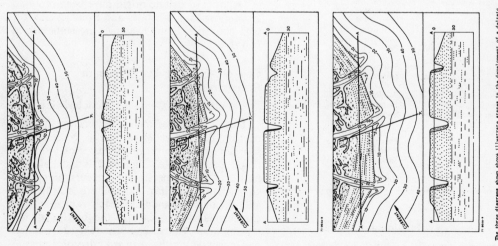

FIG. 22—Processes of deltaic sedimentation associated with an arcuate delta.

Table 3. Examples of Ancient Deltaic Deposits

Geographic Occurrence	Author
California	Todd and Monroe, 1968
Illinois	Lineback, 1968
	Swann et al., 1965
Indiana	Hrabar and Potter, 1969
	Wier and Girdley, 1963
Iowa & Illinois	Laury, 1968
Kansas	Brown, 1967
	Hattin, 1965
Louisiana	Clark and Rouse, 1971
	Curtis, 1970
Michigan	Asseez, 1969
Miss., La. & Ala.	Galloway, 1968
Montana	Sims, 1967
Nebraska	Shelton, 1972
New Mexico	Schlee and Moench, 1961
New York	Friedman and Johnson, 1966
	Lumsden and Pelletier, 1969
New York & Ontario	Martini, 1971
New York	Wolff, 1967
North Dakota	Shelton, 1972
Ohio	Knight, 1969
	Lené and Owen, 1969
Oklahoma	Busch, 1953. 1971
	Shelton, 1972
	Visher et al., 1971
Oregon	Dott, 1964, 1966
	Snavely et al., 1964
Pa., W. Va., Ohio	Beerbower, 1961
	Ferm and Cavaroc, 1969
South Dakota	Pettyjohn, 1967
Texas	Brown, 1969
	Fisher and McGowen, 1969
	Gregory, 1966
	LeBlanc, 1971
	Nanz, 1954
	Shannon and Dahl, 1971
	Wermund and Jenkins, 1970
	Shelton, 1972
W. Va., Pa., Ohio	Donaldson, 1969
Wyoming	Barlow and Haun, 1966
	Dondanville, 1963
	Hale, 1961
	Paull, 1962
	Weimer, 1961b
Wyoming & Colorado	Weimer, 1965
Several states, U.S.A.	Fisher et al., 1969
N. Appalachians	Ferm, 1970
Central Appalachians	Horowitz, 1966
Central Appalachians	Dennison, 1971
Upper Miss. embayment & Illinois basin	Pryor, 1960, 1961
Upper Miss. Valley	Swann, 1964
Okla., Iowa, Mo., Kans., Ill., Ind., Ky.	Manos, 1967
Okla. to Penn.	Wanless et al., 1970
Central Gulf Coast	Mann and Thomas, 1968
Alberta, Canada	Carrigy, 1971
	Shawa, 1969
	Shepheard and Hills, 1970
	Thachuk, 1968
England	Allen, 1962
	Taylor, 1963
Ireland	Hubbard, 1967
Scotland	Greensmith, 1966

Summary: Deltaic Sand Bodies

There are three basic types of deltaic sand bodies: delta-fringe, abandoned distributary-channel, and point-bar sands. The relative abundance and general characteristics of these sand bodies in the three types of deltas considered herein are summarized below.

Birdfoot-type delta—The most common sands are those of the delta-fringe environment. These sands occur as relatively thin, widespread sheets, and they contain a substantial amount of clays and silts.

Abandoned distributary channels contain varied amounts of sand, probably composing less than 20 percent of the total delta sand content. These sand bodies are long and narrow, are only slightly sinuous, and are encased in the delta-fringe sands or prodelta clays, depending upon channel depths and the distance that the delta has prograded seaward.

Cuspate-arcuate type of delta—Delta-fringe sand complexes are wide (width of delta), though individual sand bodies are relatively narrow, and are generally much cleaner than delta-fringe sands of the birdfoot-type delta.

Distributary-channel sands and point-bar sands are much more common than in birdfoot-type deltas and can constitute up to 50 percent of the total sand content of the delta. These two types of sands are encased in delta-fringe and prodelta sediments.

Estuarine-type delta—Delta-fringe sands appear to be much more common than distributary and point-bar sands. They probably extend for great distances within the marine environment in front of the delta; however, their geometry remains unknown.

Ancient Deltaic Deposits

Deposits of deltaic origin have been reported from more than 40 states and from several foreign countries. Some examples are summarized in Table 3.

COASTAL-INTERDELTAIC MODEL OF SEDIMENTATION

Setting and General Characteristics

This type of sedimentation occurs in long, narrow belts parallel with the coast where shoreline and nearshore processes of sedimentation predominate. The ideal interdeltaic deposit, as the name implies, occurs along the coast between deltas and comprises mud flats and cheniers (abandoned beach ridges) of the chenier-plain complex and the barrier-island–lagoon–tidal-channel complex (Fig. 23). It can also occur along the seaward edge of a coastal plain which is drained by numerous small streams and rivers but is devoid of any sizable deltas at the marine shoreline.

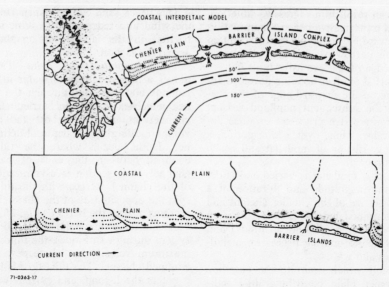

FIG. 23—General setting and characteristics of coastal-interdeltaic model of clastic sedimentation.

Source and Transportation of Sediments

Most of the sediments deposited are derived from land, but minor amounts come from the marine environment. A portion of the sediment transported to the marine shoreline by rivers and smaller streams is dispersed laterally by marine currents for great distances along the coast. Clays and fine silts are carried in suspension, and sand is transported mainly as bed load or by wave action in the beach and nearshore zone. The suspended silt and clay load is dispersed at a rapid rate and is most significant in the development of the mud flats of the chenier plain. Lateral movement of the sand bed load occurs at a relatively slow rate and is most significant in the development of the cheniers and the barrier-island complex.

A minor amount of sediment can also be derived from adjacent continental-shelf areas if erosion occurs in the marine environment.

Sedimentary processes and deposits of chenier plain—Major floods result in the sudden large influx of sediments at river mouths. Much of the suspended load introduced to the coastal-marine environment is rapidly dispersed laterally along the coast by the predominant longshore drift. A considerable portion of this suspended load is deposited along the shoreline (on the delta flank) as extensive mud flats. This period of regressive sedimentation (progradation or offlap) occurs in a relatively short period when rivers are at flood stages (Fig. 24).

During long periods when rivers are not flooding, the supply of sediment to the coast is reduced considerably or is nil. Coastal-marine currents and wave action rework the seaward edge of the newly formed mud flat, and a transgressive situation develops. A slight increase in

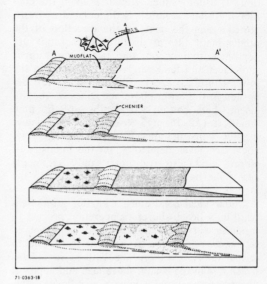

FIG. 24—Stages in development of a chenier plain.

sand supply can result in a regressive situation, and the initial transgressive beach accumulation will grow seaward by regressive beach accretion to form a long, narrow, well-defined chenier on the seaward edge of the extensive mud flat.

Another period of river flooding develops another mud flat on the seaward edge of the chenier. During the subsequent nonflood season, the coastal-transgressive processes produce another beach ridge. Thus, over a long period, a chenier plain consisting of mud flat and beach ridges is constructed.

The width of a mud flat is varied and is dependent on the magnitude and duration of a river flood. The size of the chenier (height and width) is determined by two factors: duration of the nonflood season (absence of muds) and magnitude of coastal-marine processes, including storm tides and waves.

Small streams which drain to the coastline across a chenier plain contribute little sediment to the chenier-plain environment. The mouths of these streams are generally deflected in the direction of the littoral drift.

Sedimentary processes and deposits of barrier-island complex—The typical barrier-island complex comprises three different but related depositional environments: the barrier island, the lagoon behind the barrier, and the tidal channel–tidal deltas between the barriers.

The seaward face of a barrier island is primarily an environment of sand deposition. Coastal-marine energy (currents and wave action) is usually much greater than in the chenier–mud-flat regions. Sediments are transported along the coast in the direction of the predominant littoral drift. Coarser sands are deposited mainly on the beach and upper shoreface, and finer sands are deposited in the lower shoreface areas. Silts and clays are deposited in the lower shoreface zones on the adjacent shelf bottom—at depths greater than 40–50 ft (12–15 m). Storm tides and waves usually construct beach ridges several feet above sea level, depending on the intensity of storms, and also transport sandy sediments across the barrier from the beach zone to the lagoon.

Under ideal conditions, a barrier grows seaward by a beach-shoreface accretion process to produce a typical barrier-island sequence of sediments which grades upward from fine to coarse (Figs. 25, 27). The various organisms which live in the beach, shoreface, and adjacent offshore areas usually have a significant influence on the character of sedimentary structures.

Dry beach sand can be transported inland by the wind and redeposited as dune sand on the barrier, in the lagoon, or on the mainland across the lagoon.

Tidal channel–tidal delta—Tidal action moves a large quantity of water in and out of lagoons and estuaries through the tidal channels which exist between barrier islands. These channels are relatively short and narrow and vary considerably in depth. Maximum channel depths occur where the tidal flow is confined between the ends of barriers. The channel cross section is asymmetric: one side of the channel merges with the tidal flats and spit; the opposite side of the channel has abrupt margins against the barrier (Fig. 26).

As marine waters enter the lagoon or estuary system during rising tides, the inflow attains its maximum velocity in the deepest part of the confined channel. The tidal flow is dispersed as it enters the lagoon, and current velocities are greatly reduced. The result is the deposition of sediment in the form of a tidal delta which consists of a shallow distributary channel separated by sand or silt shoals. Similar tidal deltas are also formed on the marine side of the system by similar processes associated with the falling or outgoing tide.

The depth of tidal channels and the extent of tidal deltas are dependent on the magnitude of the tidal currents. The deepest channels and the largest deltas are associated with large lagoons and estuaries affected by extreme tidal ranges.

Tidal channels migrate laterally in the direction of littoral drift by eroding the barrier head adjacent to the deep side of the channel and by spit and tidal-flat accretion on the opposite side.

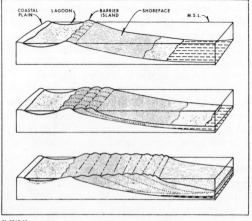

Fig. 25—Stages in development of a barrier island.

Lateral migration of the tidal system results in the deposition of the tidal-channel and tidal-delta sequences of sediments.

Summary: Characteristics of Coastal-Interdeltaic Deposits

The coastal-interdeltaic model of sedimentation is characterized by six distinct but related types of deposits: mud flat, chenier, barrier island, lagoon, tidal channel, and tidal delta. Characteristics of these deposits are summarized in Figure 27.

Three main types of sand bodies are associated with this model: barrier island, chenier, and tidal channel–tidal delta. The barrier-island sand body, which is the largest and most significant of the three, is long (usually tens of miles) and narrow (2–6 mi or 3–10 km), is oriented parallel with the coastline, and attains maximum thicknesses of 50–60 ft (15–18 m). The chenier sand bodies are very similar to those of the barriers; however, they are generally only about a third as thick. Tidal-channel sand bodies are oriented perpendicular to the barrier sands, and their thickness can vary considerably (less than, equal to, or greater than that of the barrier sands), depending on the depth of tidal channels.

Ancient Coastal-Interdeltaic Deposits

Examples of ancient coastal-interdeltaic deposits reported from 13 states are summarized in Table 4.

Table 4. Examples of Ancient Coastal-Interdeltaic Deposits

Geographic Occurrence	Author
Colorado	Griffith, 1966
Florida	Gremillion et al., 1964
Georgia	Hails and Hoyt, 1969
	MacNeil, 1950
Illinois	Rusnak, 1957
Louisiana	Sloane, 1958
Louisiana & Arkansas	Thomas and Mann, 1966
	Berg and Davies, 1968
Montana	Cannon, 1966
	Davies et al., 1971
	Shelton, 1965
New Mexico	Sabins, 1963
New York	McCave, 1969
Oklahoma & Kansas	Bass et al., 1937
	Boyd and Dyer, 1966
	Dodge, 1965
Texas	Fisher and McGowen, 1969
	Fisher et al., 1970
	Shelton, 1972
	Harms et al., 1965
	Jacka, 1965
Wyoming	Miller, 1962
	Paull, 1962
	Scruton, 1961
	Weimer, 1961a

EOLIAN MODEL OF SAND DEPOSITION

Occurrence and General Characteristics

A very common process of sedimentation is transportation and deposition of sand by the wind. Two basic conditions are necessary for the formation of windblown sand deposits: a large supply of dry sand and a sufficient wind velocity. These conditions are commonly present along coastlines characterized by sandy beaches and also in semiarid regions and deserts, where weathering and fluvial sedimentation produce a large quantitiy of sand (Fig. 28).

Under certain conditions, sands on the downstream parts of alluvial fans and along braided streams are transported and redeposited by the wind (Glennie, 1970). Sands originally deposited on point bars of meandering streams and along distributary channels of some deltas are also picked up by the wind and redeposited locally as dune sand. Similarly, sands deposited along beaches of the coastal-interdeltaic environments are redeposited by onshore winds as sand dunes on barrier islands or on the mainland. Thus, the eolian process of sand deposition is likely to occur within all models of clastic sedimentation discussed in the preceding sections.

Eolian Transport and Sedimentation

The complex processes of sand transport and deposition by the wind were studied and de-

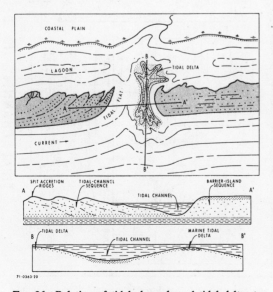

FIG. 26—Relation of tidal channels and tidal deltas to barrier islands.

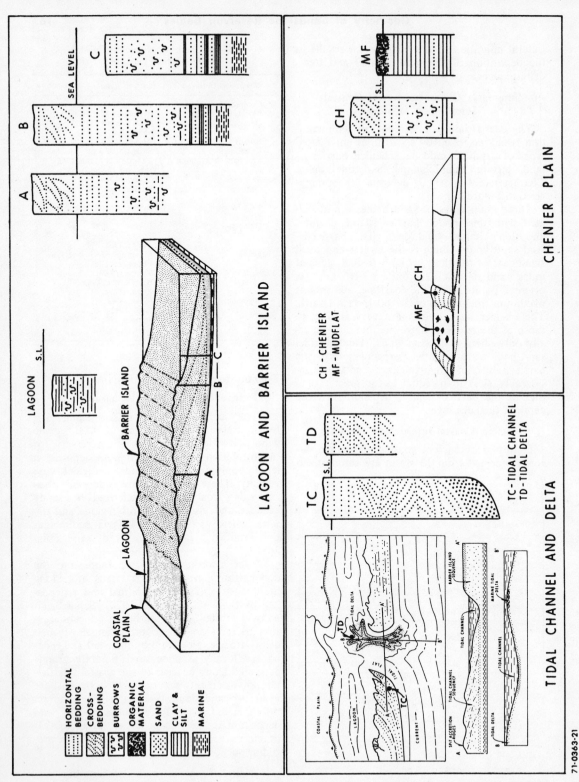

FIG. 27—Sedimentary sequences typical of chenier-plain and barrier-island complexes of coastal-interdeltaic model of clastic sedimentation.

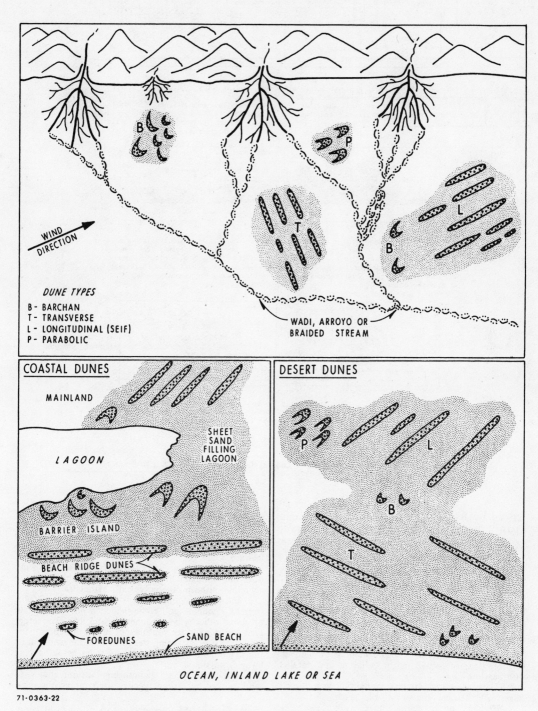

Fig. 28—Occurrences of eolian sands in coastal and desert regions.

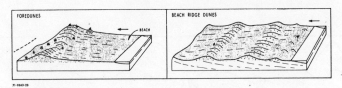

Low mounds and small elongate ridges a few feet high occurring adjacent and parallel to sand beach and shoreline, usually partly stabilized by vegetation.

Dunes against vegetation coalesce to form long, slightly sinuous ridge or series of ridges parallel to coastline. Closely associated with beach accretion ridges formed by wave action. Characteristic of barrier islands and shorelines on flanks of deltas.

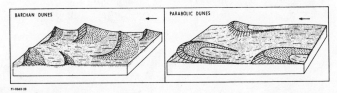

Concentric with steep slope on concave (leeward) side facing away from beach. Horns extend downwind. Can occur as scattered isolated dunes or several barchans can join to form sinuous ridge which resembles transverse dunes.

U-shaped with open end toward beach (windward) and steep side away from beach. Results from sand blowouts. Middle part moves forward (downwind) with respect to sides. Long arms usually anchored by vegetation.

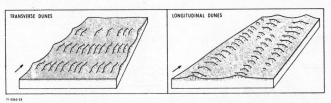

Dunes or ridges occur parallel or slightly oblique to coastline and elongated in direction perpendicular to effective wind direction. Generally symmetrical in cross section. Leeward side steep and windward side has very gentle slope.

Elongated parallel to wind direction and usually oblique or perpendicular to coastline. Cross section symmetrical. Separated from each other by flat areas. Seif dunes are special type of longitudinal dunes.

FIG. 29—Some common types of coastal dunes which also occur in deserts.

scribed by Bagnold (1941). Recently, Glennie (1970) summarized this type of sedimentation as observed under desert conditions.

The most common method of sand deposition is in the form of sand dunes. Many types of dunes have been recognized and described by numerous authors (Fig. 29). H. Smith (1954) presented the following classification and description of coastal dunes which can occur either under active or stabilized conditions.

1. Foredune ridges, or elongate mounds of sand up to a few tens of feet in height, adjacent and parallel with beaches.
2. U-shaped dunes, arcuate to hairpin-shaped sand ridges with the open end toward the beach.
3. Barchans, or crescentic dunes, with a steep lee slope on the concave side, which faces away from the beach.
4. Transverse dune ridges, trending parallel with or oblique to the shore, and elongated in a direction essentially perpendicular to the dominant winds. These dunes are asymmetric in cross profile, having a gentle slope on the windward side and a steep slope on the leeward side.
5. Longitudinal dunes, elongated parallel with wind direction and extending perpendicular or oblique to the shoreline; cross profile is typically symmetric.
6. Blowouts, comprising a wide variety of pits, troughs, channels, and chute-shaped forms cutting into or across other types of dunes or sand hills. The larger ones are marked by conspicuous heaps of sand on the landward side, assuming the form of a fan, mound, or ridge, commonly with a slope as steep as 32° facing away from the shore.
7. Attached dunes, comprising accumulations of sand trapped by various types of topographic obstacles.

McKee (1966) described an additional type, the dome-shaped dune, from White Sands National Monument, and Glennie (1970) described the seif dune of Oman, which is a special type of longitudinal dune. Many other dune types have been described; however, the above types appear to be the most common.

Studies of modern eolian sand bodies—Cooper (1958) reviewed the early studies of sand dunes, mainly by Europeans, and summarized the status of dune reseach in North America. Additional sand-dune studies in the United States since 1959 were made in Alaska by Black (1961), on the Texas coast by McBride and Hayes (1962), on the Georgia coast by Land (1964), in the Imperial Valley of California by Norris (1966), in coastal California by Cooper (1967), and in the San Luis Valley of Colorado by R. Johnson (1967). Additional studies outside the United States were made in southern Peru by Finkel (1959), in Baja California by Inman *et al.* (1966), in Libya by McKee and Tibbitts (1964), in Russia by Zenkovich (1967), and in Australia by Folk (1971).

During the past several years, some very important studies on eolian sands, which included detailed observations of internal dune structure and stratification in deep trenches cut through dunes, were made along the Texas coast by McBride and Hayes (1962), in White Sands National Monument by McKee (1966), along the Dutch coast by Jelgersma *et al.* (1970), and in the deserts of the Middle East by Glennie (1970). These authors presented photographs and sketches of various types of sedimentary structures exposed in trench walls and described their relations to dune types, wind regime, and grain-size distribution. These studies have provided some badly needed criteria for recognition of ancient eolian sands. The following summary of the geometry and general characteristics of modern eolian sand bodies was prepared largely from the references cited above.

Summary: Coastal Eolian Sand Bodies

Coastal eolian sand bodies, consisting of several types of dunes, are very long and quite narrow; they range in thickness from a few feet to a few hundreds of feet, and are aligned parallel with or oblique to the coastline. Because these sands are derived from beach deposits and form in vegetated areas, they commonly contain fragments of both shells and plants. They are characterized by high-angle crossbedding and are usually well sorted. The adjacent and laterally equivalent beach deposits are generally horizontally bedded and have some low-angle crossbedding.

Summary: Eolian Sand Bodies of Desert Regions

Desert eolian sand bodies differ from coastal eolian sands mainly in their distribution. The internal sedimentary structures and their relations to dune types are similar (Bigarella *et al.*, 1969). Seif dunes are products of two wind directions and appear to occur more commonly in desert areas. These dunes are characterized by high-angle crossbedding in two directions.

Ancient Eolian Deposits

Ancient eolian deposits have been reported from the Colorado Plateau by Baars (1961) and Stokes (1961, 1964, 1968), from the southwestern United States by McKee (1934), from England by Laming (1966), and from Brazil and Uruguay by Bigarella and Salamuni (1961). Criteria for recognition of eolian deposits have been summarized by Bigarella (1972).

MARINE CLASTIC SEDIMENTATION

Transportation and deposition of sand in the marine environment occur under a wide range of geologic and hydrologic conditions, ranging from those of the coastal shallow-marine environments to the deeper water environments of the outer continental shelves, the slopes, and the abyssal plains (Fig. 30).

As indicated in the Introduction, sands deposited under regressive (progradational) conditions within the coastal shallow-marine environments are considered herein as products of either the coastal-interdeltaic or the deltaic model of sedimentation. Other important shallow-marine sand bodies are produced as a result of marine transgressions.

During the past several years, studies made principally by the major oceanographic institutions on the modern deep-marine environments and research by petroleum geologists, university professors, and graduate students on ancient clastic sediments of various geologic ages have revealed that sand bodies of deep-marine origin are rather common throughout the world. Although most geologists now accept the fact that some sands are of deep-marine origin, our understanding of the various geologic processes which produce these sand bodies is relatively poor. The writer's personal experience with this type of clastic sedimentation is limited; however, on the basis of familiarity with the literature on marine sediments, it appears that most deep-marine sands are depos-

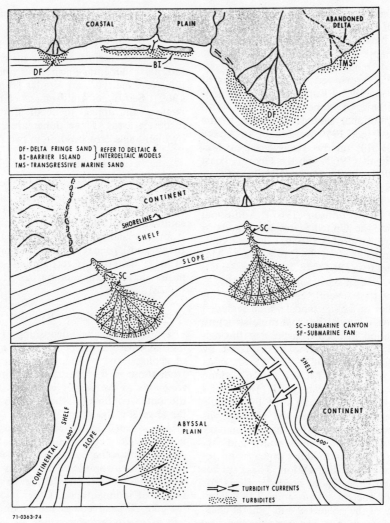

Fig. 30—Deposition of sand in marine environments.

ited under three principal types of environmental conditions: (1) on the outer shelf, the slope, and the continental rise, as a result of slumping, sliding, and tectonic activity such as earthquakes; (2) in abyssal plains, by density (turbidity) and bottom currents; and (3) in submarine canyons, fan valleys, and fans, by both bottom and density currents.

Only two of these several types of marine sands are discussed in this paper: (1) the shallow-marine sands deposited as a result of transgressive-marine sedimentation associated with the shifting of deltas and (2) the deep-marine sands deposited in submarine canyons, fan valleys, and fans.

Transgressive-Marine Model of Clastic Sedimentation

Setting and general characteristics—Deposition of clastic sediments during periods of marine transgressions (onlap) is a common process of sedimentation in most basins. There are two basic types of marine transgression: that which is associated with the shifting of deltas as a result of major river diversions during a period of standing sea level, and that which occurs as a result of a relative rise in sea level (due to subsidence of a coastal plain or eustatic rise in sea level). The inland and lateral extents of marine transgressions resulting from delta shifts are limited in size, depending on the dimensions of the abandoned deltaic plains. Marine transgressions resulting from relative changes in sea level extend over much broader regions and are commonly referred to as regional transgressions. Their dimensions are governed mainly by the topography of the coastal plain being transgressed and by the amount of relative rise in sea level. Thus, transgressive-marine deposition can occur locally over abandoned deltas or regionally over eolian, alluvial, interdeltaic, and deltaic deposits of a large part of a coastal plain.

Modern marine transgressions resulting from major changes in drainage and delta shifts have been described by several authors: Russell, 1936; Russell and Russell, 1939; Kruit, 1955; van Straaten, 1959; Scruton, 1960; Curray, 1964; Coleman and Gagliano, 1964; Rainwater, 1964; Coleman, 1966b; Scott and Fisher, 1969; L. Brown, 1969; and Oomkens, 1970.

Sources, transportation, and deposition of sediments—After a delta is abandoned because of upstream channel diversion, a very significant change occurs in conditions of sedimentation. The abandoned deltaic plain and subaqueous delta front no longer receive sediment and gradually subside owing to the compaction of the deltaic deposits. The seaward edge of the abandoned delta is attacked by marine wave and current action and recedes landward at relatively slow rates. As marine processes erode the upper part of the deltaic sequence, the sandy sediments within the sequence are winnowed and deposited along the advancing shoreline as barrier islands, beaches, and shallow-marine sands; finer sediments are deposited farther offshore. Thus, the transgressive-marine depositional profile is characterized by sands and shell material nearshore and by progressively finer sediments offshore. Over a period of time, as the transgression proceeds inland, the thin veneer of shallow-marine sands which are deposited over the underlying delta sediments is in turn overlain by marine silts and clays. Stages in the development of such a trangressive-marine sand body are illustrated in Figure 31.

Character of sediments—This type of sedimentation, although largely restricted in extent to abandoned deltas and adjacent and laterally

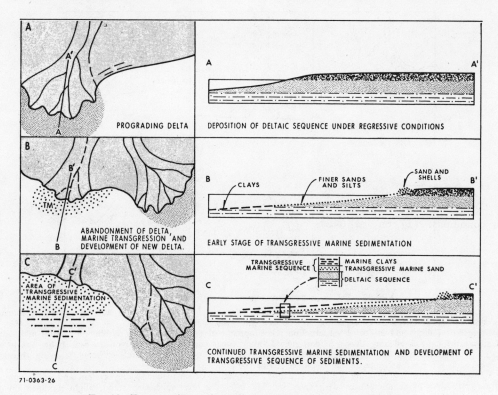

Fig. 31—Transgressive-marine sedimentation resulting from delta shifts.

equivalent interdeltaic and offshore-marine environments, is significant because it produces very diagnostic blanketlike layers of marine sediments (thin shallow-marine sand overlain by clays) which separate the individual deltaic units (Fig. 31). These layers usually provide the only good correlations within thick deltaic facies, and the marine shales act as impervious seals between deltaic sand bodies. The transgressive-marine sands containing calcareous shell material usually become cemented and thus do not form very efficient reservoirs.

Submarine Canyon-Fan Model of Clastic Sedimentation

Occurrence and general characteristics—The occurrence of modern and Pleistocene sands in deep-marine environments of the world is well documented as a result of numerous deep-sea investigations by oceanographic institutions during the past 20 years. Although there is much controversy regarding the origin of these sands, it is certain that such sands do exist. An analysis of the literature reveals that some of the most common deep-sea sands are those associated with submarine canyons and fans. (For a discussion of types of submarine canyons, troughs, and valleys, the reader is referred to Shepard and Dill, 1966.)

Submarine canyons and fans are common features associated with continental shelves, slopes, and rises. The canyons and fans off the Pacific Coast of the United States and Canada have received the most attention. Significant papers on these features off the coasts of Washington, Oregon, California, and Baja California, and off the Gulf and Atlantic coasts, are listed in the selected references. Also included are references to papers on canyons and fans in the Mediterranean, the Atlantic Ocean off Africa, and the Indian Ocean off Pakistan.

Characteristics and origin of submarine canyons have been discussed by numerous authors (for summary, see Shepard and Dill, 1966). Although it is still uncertain how some deep-sea canyons and valleys originated, it is now reasonably well established that a large number of canyons and fans are related to rivers, and that they were formed during stages of low sea level of the Quaternary Period (Figs. 32, 33). For

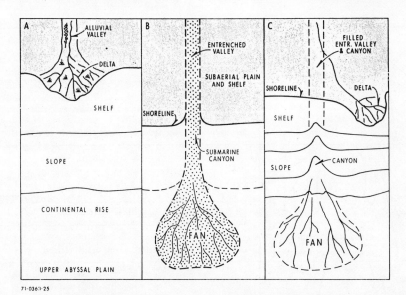

FIG. 32—Stages in development of submarine canyon and fan.

Stage A: Standing sea-level situation. Development of alluvial valley and delta and deposition of marine clays on shelf and slope. Base of aggrading river is well below sea level.

Stage B: Falling and low-sea-level situation. Development of entrenched-valley system on coastal plain and of submarine canyon offshore. Bases of entrenched valley (near coast) and of canyon are well below sea level. Rates of sedimentation are very high. Material removed by canyon-cutting and sediments flowing through canyon while it formed are deposited as extensive submarine fan.

Stage C: Rising and standing sea-level situation. Alluviation of entrenched-valley system and partial filling of canyon. Rates of sedimentation are greatly reduced after sea level reaches a stand. Slight modification of fan by normal-marine processes occurs.

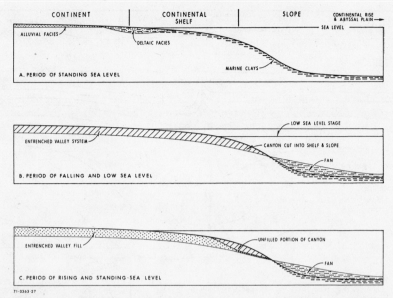

FIG. 33—Relation of submarine-fan deposits to submarine-canyon and entrenched-valley system.

example, the Mississippi Canyon off southern Louisiana is a continuation of the late Pleistocene Mississippi entrenched valley system (Fisk, 1944; Osterhoudt, 1946; Fish and McFarlan, 1955; and Bergantino, 1971). Also, the Astoria canyon and fan off Oregon are related to the Columbia River (Duncan and Kulm, 1970); the Newport Canyon is related to the Santa Ana River of California (Felix and Gorsline, 1971); the Congo Canyon connects with the Congo River (Heezen et al., 1964); the Monterey and Soquel canyons and fans occur off the Great Valley of California (Martin and Emery, 1967); the Bengal deep-sea fan and the "Swatch-of-No-Ground" canyon occur off the Ganges River delta (Curray and Moore, 1971); and the Inguri canyon is related to a river flowing in the Caspian Sea (Trimonis and Shimkus, 1970). The *National Geographic* magazine maps of the Indian and Atlantic Ocean floors (Heezen and Tharp, 1967, 1969) show large fans off the Indus and Amazon Rivers and also off the Laurentian Trough, and the Hudson Canyon is associated with the Hudson River. Seismic reflection surveys between canyon heads (on shelves) and the coastline most probably will reveal more examples of canyons related to entrenched river valleys on land.

There is an extremely wide variation in the size of submarine canyon-fan systems. Some of the small ones off California studied by Gorsline and Emery (1959) include short canyons 5–10 mi (8–16 km) long and fan areas of about 50 sq mi (130 sq km). The largest canyon-fan systems studied thus far are those of the Congo, Ganges, and Rhône Rivers. The Bengal fan is 2,600 km long and 1,100 km wide; the Congo fan is more than 520 km long and 185 km wide; and one of the largest fans off the Pacific coast of the United States, the Delgade fan, is 300 km long and 330 km wide (Normark, 1970). Menard (1960) discussed the dimensions of several other fans.

Some very significant studies of deep-sea sands associated with canyons and fans—based on core, seismic reflection, and bathymetric data, and bottom observations and photography by divers—have been made during the past 3 years (Winterer et al., 1968; Carlson and Nelson, 1969; Shepard et al., 1969; Curray and Moore, 1971; Normark, 1970; Nelson et al., 1970; Piper, 1970; Duncan and Kulm, 1970; and Felix and Gorsline, 1971).

Physiographic features.—Detailed bathymetric surveys over several canyons and fans of various sizes have revealed that these submarine features are characterized by physiographic features very similar to those of subaerial alluvial fans. The canyons are V-shaped and have steep walls and gradients. The surfaces of the fans are characterized by lower gradient distributary channels with natural levees and by topographically low interchannel areas. Some fans are crossed by relatively large fan valleys

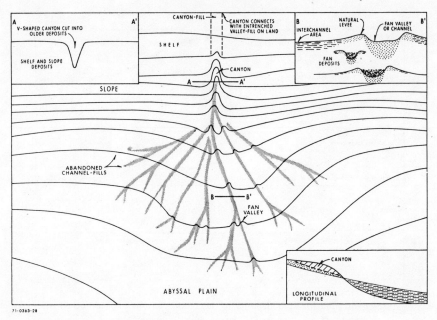

FIG. 34—Submarine-canyon and fan model of clastic sedimentation.

which also have natural levees. The principal physiographic features of a typical canyon and fan are illustrated in a generalized fashion in Figure 34.

The overall shape of a submarine fan can either be symmetrical or asymmetrical, depending on strong current directions and on the presence of high topographic features on the abyssal plains. Sizes of the distributary channels and natural levees are widely varied; the larger channels usually have the highest and broadest natural levees. The lower parts of fans merge with the abyssal plains. Channels on fan surfaces were probably formed by depositional processes. Erosional channels that have been reported probably represent an entrenchment stage, as is the case with subaerial alluvial fans.

Many canyons were cut across continental shelves and slopes, and the fans were constructed at the base of the slopes or on the continental rises. Some canyons presently do not extend landward across the continental shelves (*e.g.*, the Mississippi Canyon) because they have been filled with sediments. Seismic surveys reveal that this type of canyon was once connected with inland entrenched valley systems.

Longitudinal profiles of canyons and fans are concave upward. The steepest gradients occur in the upper (landward) portions of canyons, and the lowest gradients occur on the outer or lower portions of fans.

Depositional processes and character of sediments—It is absolutely certain that large quantities of sediment, including a significant amount of sand, have been transported through submarine canyons and deposited as submarine fans in deep-sea environments. The manner in which these sediments were transported, especially the sands, is much less certain. Nearly 2 decades ago, some very strong statements were made by oceanographers regarding the turbidity-current origin of both the canyons and the fan deposits. Although no one had actually seen or measured a turbidity current in a canyon or over a submarine fan, the turbidity-current concept was very popular with most oceanographers during the early 1950s. During the past 20 years, numerous additional observations have been made, but no one has yet seen a live turbidity current in a natural marine environment. On the basis of direct observations of the ocean bottom and sedimentary structures in cores, many oceanographers now believe that some submarine-fan sand deposits were transported mainly by normal bottom currents, especially during low stages of sea level of the Pleistocene. A typical example is the origin of the sand associated with the Mississippi cone in

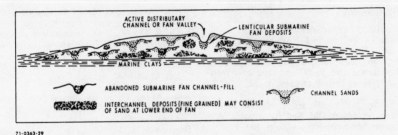

FIG. 35—Generalized distribution of submarine-fan deposits.

the Gulf of Mexico off southern Louisiana. Greenman and LeBlanc (1956) did not consider these sands to be of turbidity-current origin, but Ewing et al. (1958) were certain that the sands were transported and deposited by turbidity currents. Twelve years later, Huang and Goodell (1970) concluded, on the basis of detailed studies of sedimentary structures observed in numerous cores, that the sands are not of turbidity-current origin, but that the mechanisms of transport are bottom currents, differential pelagic settling, and mass movement by sliding and slumping. Walker and Massingill (1970) reported that part of the Mississippi cone sediments were recently involved in large-scale slumps. They presented evidence that one slump moved from near the mouth of the Mississippi Canyon southeastward for at least 160 n. mi. Thus, the origin of these deep-sea sands and many others remains a problem.

Regardless of the mechanisms of sediment transport through submarine canyons and of deposition of fans, the general nature and distribution of fan deposits have been determined for several fans. The coarsest and most poorly sorted sediments occur in canyons. Sands are common in distributary channels and fan valleys and on the lower parts of the open fan. Sandy sediments also occur on natural levees, but the interchannel areas are characterized by fine-grained sediments (Fig. 35). Core data from several fans indicate that sand bodies are usually thin and very lenticular, and are interbedded with fine-grained sediments.

For details concerning the sedimentary structures which characterize submarine-canyon and fan deposits, the reader is referred to Carlson and Nelson (1969); Shepard et al. (1969); Stanley (1969); Huang and Goodell (1970); and Haner (1971).

Horn et al. (1971) described the characteristics of sediments related to submarine canyons, fans, and adjacent abyssal plains of the northeast Pacific Ocean off Alaska, Canada, Washington, Oregon, and northern California. They interpreted sediments with a wide range in layer thickness, with graded and nongraded layers, and with sand in the basal parts of graded units to be proximal turbidites related to main routes followed by turbidity currents (probably channels). The finer grained sediments, mainly graded silts and clays, were interpreted as distal turbidites deposited beyond the main avenues of turbidite flows.

It is the opinion of the writer that many of the submarine canyons and related fans which now are found off rivers are the products of entrenchment (canyons) and deposition (fans) during stages of low sea level of the Pleistocene. Oceanographers who have studied several of these fan deposits have concluded that they are of Miocene to Pleistocene age. The geologic-age determinations were made on the basis of present sediment load of the related rivers and known thickness of fan deposits. This writer suggests that rates of sedimentation were probably several times greater during Pleistocene low-sea-level stages than at the present time (period of higher and standing sea level) and, consequently, that the fan deposits are probably chiefly of Pleistocene age.

Ancient examples of submarine canyon and fan deposits—Some examples of ancient deposits of submarine canyon and fan origin have been described from the Gulf coast by Osterhoudt (1946), Bornhauser (1948, 1960), Hoyt (1959), Paine (1966), and Sabate (1968); from California by Sullwold (1960), Martin (1963), Bartow (1966), Dickas and Payne (1967), Normark and Piper (1969), Piper and Normark (1971), Davis (1971), Fischer (1971), and Shelton (1972); from Canada by Hubert et al. (1970); from Europe by Walker (1966), Stanley (1967, 1969), and Kelling and Woollands (1969); and from Australia by Conolly (1968).

Selected References

Al-Habeeb, K. H., 1970, Sedimentology of the floodplain sediments of the middle Euphrates River, Iraq: Baghdad Univ., MS thesis.

Allen, J. R. L., 1962, Petrology, origin, and deposition of the highest lower Old Red Sandstone of Shropshire, England: Jour. Sed. Petrology, v. 32, p. 657–697.

―――― 1964, Studies in fluviatile sedimentation: six cyclothems from the lower Old Red Sandstones, Anglo-Welsh basin: Sedimentology, v. 3, p. 163–198.

―――― 1965a, A review of the origin and characteristics of Recent alluvial sediments: Sedimentology, v. 5, p. 89–191.

―――― 1965b, Coastal geomorphology of eastern Nigeria: beach-ridge barrier-islands and vegetated tidal flats: Geologie en Mijnbouw, v. 44, p. 1–21.

―――― 1965c, Late Quaternary Niger delta and adjacent areas: sedimentary environments and lithofacies: Am. Assoc. Petroleum Geologists Bull., v. 49, no. 5, p. 547–600.

―――― 1970, Sediments of the modern Niger delta, *in* Deltaic sedimentation—modern and ancient: Soc. Econ. Paleontologists and Mineralogists Spec. Pub. 15, p. 138–151.

American Petroleum Institute, 1971: Petroleum facts and figures.

Anderson, G. S., and K. M. Hussey, 1962, Alluvial fan development at Franklin Bluff, Alaska: Iowa Acad. Sci. Proc., v. 69, p. 310–322.

Andrews, P. B., 1970, Facies and genesis of a hurricane washover fan, St. Joseph Island, central Texas: Texas Univ. Bur. Econ. Geology Rept. Inv. 67, 147 p.

Anstey, R. L., 1965, Physical characteristics of alluvial fans: U.S. Army Quartermaster Research and Eng. Command, Earth Sci. Div. Tech. Rept. ES-20, 109 p.

Arnborg, L. E., 1948, The delta of the Angerman River: Geog. Annaler, v. 30, p. 673–690.

Asseez, L. O., 1969, Paleogeography of Lower Mississippian rocks of Michigan basin: Am. Assoc. Petroleum Geologists Bull., v. 53, no. 1, p. 127–135.

Attia, M. I., 1954, Deposits in the Nile Valley and the delta: Cairo, Egypt Geol. Survey, 356 p.

Axelson, V., 1967, The Laitaure delta, a study of deltaic morphology and processes: Geog. Annaler, v. 49, p. 1–127.

Baars, D. L., 1961, Permian blanket sandstones of Colorado Plateau, *in* J. A. Peterson and J. C. Osmond, eds., Geometry of sandstone bodies: Am. Assoc. Petroleum Geologists, p. 179–219.

Bagchi, K., 1944, The Ganges delta: Calcutta, Calcutta Univ. Press, 157 p.

Bagnold, R. A., 1941, The physics of blown sand and desert dunes: New York, W. Morrow and Co., 265 p.

Bakker, W. T. J. N. P., and T. Edelman, 1965, The coast line of river-deltas: 9th Conf. on Coastal Engineering Proc., p. 199–218.

Barlow, J. A., Jr., and J. D. Haun, 1966, Regional stratigraphy of Frontier Formation and relation to Salt Creek field, Wyoming: Am. Assoc. Petroleum Geologists Bull., v. 50, p. 2185–2196.

Bartow, J. A., 1966, Deep submarine channel in upper Miocene, Orange County, California: Jour. Sed. Petrology, v. 36, no. 3, p. 700–705.

Bass, N. W., *et al.*, 1937, Origin and distribution of Bartlesville and Burbank shoestring oil sands in parts of Oklahoma and Kansas: Am. Assoc. Petroleum Geologists Bull., v. 21, p. 30–66.

Bates, C. C., 1953, Rational theory of delta formation: Am. Assoc. Petroleum Geologists Bull., v. 37, p. 2119–2162.

Beall, A. O., Jr., 1968, Sedimentary processes operative along the western Louisiana shoreline: Jour. Sed. Petrology, v. 38, p. 869–877.

Beaty, C. B., 1961, Boulder deposit in Flint Creek Valley, western Montana: Geol. Soc. America Bull., v. 72, p. 1015–1020.

―――― 1963, Origin of alluvial fans, White Mountains, California and Nevada: Assoc. Am. Geographers Annals, v. 53, p. 516–535.

Beerbower, J. R., 1961, Origin of cyclothems of the Dunkard Group (Upper Pennsylvanian-Lower Permian) in Pennsylvania, West Virginia, and Ohio: Geol. Soc. America Bull., v. 72, p. 1029–1050.

―――― 1964, Cyclothems and cyclic depositional mechanisms in alluvial plain sedimentation: Kansas Geol. Survey Bull. 169, v. 1, p. 31–42.

―――― 1969, Interpretation of cyclic Permo-Carboniferous deposition in alluvial plain sediments in West Virginia: Geol. Soc. America Bull., v. 80, p. 1843–1848.

Berg, R. R., 1968, Point bar origin of Fall River sandstone reservoirs, northeastern Wyoming: Am. Assoc. Petroleum Geologists Bull., v. 52, no. 11, p. 2116–2122.

―――― and B. C. Cook, 1968, Petrography and origin of lower Tuscaloosa sandstones, Mallalieu field, Lincoln County, Mississippi: Gulf Coast Assoc. Geol. Socs. Trans., v. 18, p. 242–255.

―――― and D. K. Davies, 1968, Origin of Lower Cretaceous Muddy Sandstone at Bell Creek field, Montana: Am. Assoc. Petroleum Geologists Bull., v. 52, no. 10, p. 1888–1898.

Bergantino, R. N., 1971, Submarine regional geomorphology of the Gulf of Mexico: Geol. Soc. America Bull., v. 82, p. 741–752.

Bernard, H. A., 1965, A resume of river delta types: Am. Assoc. Petroleum Geologists Bull., v. 49, p. 334–335.

―――― and R. J. LeBlanc, 1965, Resume of Quaternary geology of the northwestern Gulf of Mexico province, *in* Quaternary of the United States, Review volume for 7th Cong., Internat. Assoc. for Quaternary Research: Princeton, New Jersey, Princeton Univ. Press, p. 137–185.

―――― and C. F. Major, 1963, Recent meander belt deposits of the Brazos River: an alluvial "sand" model: Am. Assoc. Petroleum Geologists Bull., v. 47, no. 2, p. 350.

―――― R. J. LeBlanc, and C. F. Major, 1962, Recent and Pleistocene geology of southeast Texas, *in* Geology of the Gulf Coast and central Texas and guidebook of excursions, Geol. Soc. America 1962 Ann. Mtg.: Houston Geol. Soc., p. 175–224.

―――― C. F. Major, and B. S. Parrott, 1959, The Galveston barrier island and environs—a model for predicting reservoir occurrence and trend: Gulf Coast Assoc. Geol. Socs. Trans., v. 9, p. 221–224.

―――― *et al.*, 1970, Recent sediments of southeast Texas; a field guide to the Brazos alluvial and deltaic plains and the Galveston barrier island complex: Texas Univ. Bur. Econ. Geology Guidebook No. 11.

Beutner, E. C., *et al.*, 1967, Bedding geometry in a Pennsylvanian channel sandstone: Geol. Soc. America Bull., v. 78, no. 7, p. 911–916.

Bigarella, J. J., 1972, Eolian deposits—their characteristics, recognition, and importance, *in* J. K. Rigby and W. K. Hamblin, eds., Recognition of ancient sedimentary environments: Soc. Econ. Paleontologists and Mineralogists Spec. Pub. 16, 340 p.

———— and R. Salamuni, 1961, Early Mesozoic wind patterns as suggested by dune bedding in the Botucatu Sandstone of Brazil and Uruguay: Geol. Soc. America Bull., v. 72, p. 1089–1106.

———— et al., 1969, Coastal dune structure from Parana, Brazil: Marine Geology, v. 7, p. 5–55.

Bird, E. C. F., 1961, The coastal barriers of East Gippsland, Australia: Jour. Geology, v. 127, p. 460–468.

———— 1962, The river deltas of the Gippsland lakes: Royal Soc. Victoria Proc., v. 75, pt. 1, p. 65–74.

Black, R. F., 1961, Eolian deposits of Alaska: Arctic, v. 4, no. 2, p. 89–111.

Blackwelder, E., 1928, Mudflow as a geologic agent in semi-arid mountains: Geol. Soc. America Bull., v. 39, p. 465–483.

———— 1931, Desert plains: Jour. Geology, v. 39, p. 133–140.

Blissenbach, E., 1952, Relation of surface angle distribution to particle size distribution on alluvial fans: Jour. Sed. Petrology, v. 22, p. 25–28.

———— 1954, Geology of alluvial fans in semi-arid regions: Geol. Soc. America Bull., v. 65, p. 175–189.

Bluck, B. J., 1964, Sedimentation of an alluvial fan in southern Nevada: Jour. Sed. Petrology, v. 34, no. 2, p. 395–400.

———— 1965, The sedimentary history of some Triassic conglomerates in The Vale of Glamorgan, South Wales: Sedimentology, v. 4, p. 225–246.

———— 1967, Deposition of some upper Old Red Sandstone conglomerates in the Clyde area: a study in the significance of bedding: Scottish Jour. Geology, v. 3, no. 8, pt. 2, p. 139–167.

Boggs, S., Jr., 1966, Petrology of Minturn Formation, east-central Eagle County, Colorado: Am. Assoc. Petroleum Geologists Bull., v. 50, no. 7, p. 1399–1422.

Bolyard, D. W., 1959, Pennsylvanian and Permian stratigraphy in Sangre De Cristo Mountains between La Veta Pass and Westcliffe, Colorado: Am. Assoc. Petroleum Geologists Bull., v. 43, p. 1896–1939.

Bonham-Carter, G. F., and A. J. Sutherland, 1967, Diffusion and settling of sediments at river mouths: a computer simulation model: Gulf Coast Assoc. Geol. Socs. Trans., v. 17, p. 326–338.

Boothroyd, J. C., 1970, Recent braided stream sedimentation, south-central Alaska (abs.): Am. Assoc. Petroleum Geologists Bull., v. 54, no. 5, p. 836.

Bornhauser, M., 1948, Possible ancient submarine canyon in southwestern Louisiana: Am. Assoc. Petroleum Geologists Bull., v. 32, p. 2287–2290.

———— 1960, Depositional and structural history of northwest Hartburg field, Newton County, Texas: Am. Assoc. Petroleum Geologists Bull., v. 44, p. 458–470.

Bouma, A. H., 1965, Sedimentary characteristics of samples collected from some submarine canyons: Marine Geology, v. 3, no. 4, p. 291–320.

Boyd, D. R., and B. F. Dyer, 1966, Frio barrier bar system of south Texas: Am. Assoc. Petroleum Geologists Bull., v. 50, p. 170–178.

Brady, L. L., 1969, Stratigraphy and petrology of the Morrison Formation (Jurassic) of the Canon City, Colorado area: Jour. Sed. Petrology, v. 39, no. 2, p. 632–648.

Bretz, J. H., and L. Horberg, 1949, The Ogallala Formation west of the Llano Estacado: Jour. Geology, v. 57, no. 5, p. 477–490.

Brice, J. C., 1964, Channel pattern and terraces of the Loup Rivers in Nebraska: U.S. Geol. Survey Prof. Paper 422-D, 41 p.

Brickman, R., and L. J. Pons, 1968, A pedo-geomorphological classification and map of the Holocene sediments in the coastal plain of the three Guianas: Wageningen, Netherlands Soil Survey Institute, 40 p.

Brown, L. F., Jr., 1969, North Texas (eastern shelf) Pennsylvanian delta systems, in Delta systems in exploration for oil and gas: Texas Univ. Bur. Econ. Geology, p. 40–53.

Brown, S. L., 1967, Stratigraphy and depositional environment of the Elgin Sandstone (Pennsylvanian) in south-central Kansas: Kansas Geol. Survey Bull. 187, pt. 3, 9 p.

Buffington, E. C., 1952, Submarine "natural levees": Jour. Geology, v. 60, p. 473–479.

Bull, W. B., 1960, Types of deposition on alluvial fans in western Fresno County, California (abs.): Geol. Soc. America Bull., v. 71, p. 2052.

———— 1962, Relation of textural (cm) patterns to depositional environment of alluvial fan deposits: Jour. Sed. Petrology, v. 32, p. 211–216.

———— 1963, Alluvial fan deposits in western Fresno County, California: Jour. Geology, v. 71, p. 243–251.

———— 1964, Geomorphology of segmented alluvial fans in western Fresno County, California: U.S. Geol. Survey Prof. Paper 352-E, p. 59–129.

———— 1968, Alluvial fans: Jour. Geol. Education, v. 16, p. 101–106.

———— 1969, Recognition of alluvial-fan environments in stratigraphic record (abs.): Am. Assoc. Petroleum Geologists Bull., v. 53, p. 710.

———— 1972, Recognition of alluvial-fan deposits in the stratigraphic record, in J. K. Rigby and W. K. Hamblin, eds., Recognition of ancient sedimentary environments: Soc. Econ. Paleontologists and Mineralogists Spec. Pub. 16, 340 p.

Busch, D. A., 1953, The significance of deltas in subsurface exploration: Tulsa Geol. Soc. Digest, v. 21, p. 71–80.

———— 1971, Genetic units in delta prospecting: Am. Assoc. Petroleum Geologists Bull., v. 55, p. 1137–1154.

Buttner, P. J. R., 1968, Proximal continental rhythmic sequences in the Genesee Group (lower Upper Devonian): Geol. Soc. America Spec. Paper 106, p. 109–126.

Byers, P. R., 1966, Mineralogy and origin of the Eastend and Whitemud Formations of south-central and southwestern Saskatchewan and southeastern Alberta: Kingston, Ontario, Queen's University, Master's thesis.

Byrne, J. V., et al., 1959, The chenier plain and its stratigraphy, southwestern Louisiana: Gulf Coast Assoc. Geol. Socs. Trans., v. 9, p. 237–260.

Cadwell, D. H., and J. H. Moss, 1969, Floodplain sediments and their movement during a high discharge, near Manheim, Pennsylvania: Pennsylvania Acad. Sci. Proc., v. 43, p. 185–186.

Cannon, J. L., 1966, Outcrop examination and interpretation of paleocurrent patterns of the Blackleaf Formation near Great Falls, Montana: Billings Geol. Soc. 17th Field Conf. Guidebook, p. 71–111.

Carlson, P. R., and C. H. Nelson, 1969, Sediments and sedimentary structures of the Astoria submarine canyon-fan system, northeast Pacific: Jour. Sed. Petrology, v. 39, p. 1269–1282.

Carrigy, M. A., 1971, Deltaic sedimentation in Athabasca tar sands: Am. Assoc. Petroleum Geologists Bull., v. 55, p. 1155–1169.

Casey, S. R., and R. B. Cantrell, 1941, Davis sand lens, Hardin field, Liberty County, Texas, in A. I. Levorsen, ed., Stratigraphic type oil fields: Am. Assoc. Petroleum Geologists, p. 564–599.

Cavaroc, V. V., 1969, Delineation of some outer deltaic plain subenvironments based on sedimentary properties in vertical section (abs.): Am. Assoc. Petroleum Geologists Bull., v. 53, no. 3, p. 710–711.

Challinor, J., 1946, Two contrasted types of alluvial deposits: Geol. Mag., v. 83, p. 162–164.

Chamberlin, T. K., 1964, Mass transport of sediment in the heads of Scripps submarine canyon, California, in R. L. Miller, ed., Papers in marine geology: New York, The Macmillan Co., p. 42–64.

Chawner, W. D., 1935, Alluvial fan flooding: the Montrose, California flood of 1934: Geog. Rev., v. 25, p. 255–263.

Chein, C., 1961, The braided river of the lower Yellow River: Sci. Sinica, v. 10, no. 6, p. 734–754.

Clark, R. H., and J. T. Rouse, 1971, A closed system for generation and entrapment of hydrocarbons in Cenozoic deltas, Louisiana Gulf Coast: Am. Assoc. Petroleum Geologists Bull., v. 55, p. 1170–1178.

Coastal Research Group, 1969, Field trip guidebook, coastal environments of northeastern Massachusetts and New Hampshire: Massachusetts Univ. Geol. Dept. Contr. 1-CRG, 462 p.

Coleman, J. M., 1966a, Ecological changes in the massive fresh-water clay sequence: Gulf Coast Assoc. Geol. Socs. Trans., v. 16, p. 159–174.

——— 1966b, Recent coastal sedimentation: central Louisiana coast: Louisiana State Univ. Coastal Studies Inst. Tech. Rept. 29, p. 1–73.

——— 1969, Brahmaputra River: channel processes and sedimentation: Sed. Geology, v. 3, no. 2–3, p. 129–239.

——— and S. M. Gagliano, 1964, Cyclic sedimentation in the Mississippi River deltaic plain: Louisiana State Univ. Coastal Studies Inst. Tech. Rept. 16, pt. G; also in Gulf Coast Assoc. Geol. Socs. Trans., v. 14.

——— and ——— 1965, Sedimentary structures: Mississippi River deltaic plain, in Primary sedimentary structures and their hydrodynamic interpretation: Soc. Econ. Paleontologists and Mineralogists Spec. Pub. 12, p. 133–148.

——— et al., 1964, Minor sedimentary structures in a prograding distributary: Marine Geology, v. 1, p. 240–258.

——— et al., 1970, Sedimentation in a Malaysian high tide tropical delta, in Deltaic sedimentation—modern and ancient: Soc. Econ. Paleontologists and Mineralogists Spec. Pub. 15, p. 185–197.

Collinson, J. D., 1968, Deltaic sedimentation units in the Upper Carboniferous of northern England: Sedimentology, v. 10, no. 4, p. 233–254.

——— 1968, The sedimentology of the Grindslow shales and the Kinderscant grit: a deltaic complex in the Namurian of northern England: Jour. Sed. Petrology, v. 39, no. 1, p. 194–221.

Conolly, J. R., 1965, Petrology and origin of the Hervey Group, Upper Devonian, central New South Wales: Geology Soc. Australia Jour., v. 12, pt. 1, p. 123–166.

——— 1968, Submarine canyons of the continental margin, East Bass Strait (Australia): Marine Geology, v. 6, p. 449–461.

——— and W. J. Cleary, 1971, Braided deep sea deltas and the origin of turbidite sands (abs.): Am. Geophys. Union Trans., v. 52, p. 244.

Cooper, W. S., 1958, Coastal sand dunes of Oregon and Washington: Geol. Soc. America Mem. 72, p. 1–169.

——— 1967, Coastal dunes of California: Geol. Soc. America Mem. 104, 131 p.

Corbeille, R. L., 1962, New Orleans barrier island, New Orleans area: Gulf Coast Assoc. Geol. Socs. Trans., v. 12, p. 223–229.

Credner, G. R., 1878, The deltas: Petermanns Geog. Mitt., v. 56, 74 p. (in German).

Cressey, C. B., 1928, The Indiana sand dunes and shorelines of the Lake Michigan basin: Geol. Soc. Chicago Bull. 8, p. 37.

Crickmay, C. H., 1955, Delta formation: Am. Assoc. Petroleum Geologists Bull., v. 39, p. 107–114.

——— 1960, Lateral activity in a river of northwestern Canada: Jour. Geology, v. 68, p. 377–391.

Croft, A., 1962, Some sedimentation phenomena along the Wasatch Mountain front: Jour. Geophys. Research, v. 67, p. 1511–1524.

Crosby, E. J., 1972, Classification of sedimentary environments, in J. K. Rigby and W. K. Hamblin, eds., Recognition of sedimentary environments: Soc. Econ. Paleontologists and Mineralogists Spec. Pub. 16, 340 p.

Crowell, J. C., 1952, Submarine canyons bordering central and southern California: Jour. Geology, v. 60, p. 58–83.

——— 1954, Geology of the Ridge Basin area: California Div. Mines Bull. 170, Map Sheet No. 7.

Culbertson, J. K., and D. R. Dawdy, 1964, A study of fluvial characteristics and hydraulic variables, middle Rio Grande, New Mexico: U.S. Geol. Survey Water-Supply Paper 1498-F, 74 p.

Curray, J. R., 1960, Sediments and history of Holocene transgression, continental shelf, northwest Gulf of Mexico, in F. P. Shepard et al., eds., Recent sediments, northwest Gulf of Mexico: Am. Assoc. Petroleum Geologists, p. 221–266.

——— 1964, Transgressions and regressions, in Papers in marine geology: New York, The Macmillan Co., p. 175–203.

——— 1969a, Estuaries, lagoons, tidal flats and deltas, in New concepts of continental margin sedimentation: Am. Geol. Inst. Short Course Notes, Nov. 7–9, p. III–1 to III–30.

——— 1969b, Shore zone sand bodies: barriers, cheniers and beach ridges, in New concepts of continental margin sedimentation: Am. Geol. Inst. Short Course Notes, Nov. 7–9, p. II–1 to II–18.

——— and D. G. Moore, 1963, Facies delineation by acoustic-reflection: northern Gulf of Mexico: Sedimentology, v. 2, p. 130–148.

——— and ——— 1964, Holocene regressive littoral sand, Costa de Nayarit, Mexico, in Deltaic and shallow marine deposits: New York, Elsevier Publishing Co., p. 176–182.

——— and ——— 1971, Growth of the Bengal deepsea fan and denudation in the Himalayas: Geol. Soc. America Bull., v. 82, p. 563–572.

——— et al., 1969, Holocene history of a strand plain, lagoonal coast, Nayarit, Mexico, in Coastal lagoons, a symposium: Internat. Symposium on Coastal Lagoons Mem., Mexico, D. F., p. 63–100.

——— et al., 1970, Late Quaternary sea-level studies in Micronesia: Carmarse expedition: Geol. Soc. America Bull., v. 81, p. 1865–1880.

Curtis, D. M., 1970, Miocene deltaic sedimentation,

Louisiana Gulf Coast, *in* Deltaic sedimentation—modern and ancient: Soc. Econ. Paleontologists and Mineralogists Spec. Pub. 15, p. 293–308.

Daboll, J. M., 1969, Holocene sediments of the Parker estuary, Massachusetts: Massachusetts Univ. Dept. Geology Contr. No. 3-CRG, 138 p.

Damuth, J. E., and R. W. Fairbridge, 1970, Equatorial Atlantic deep-sea arkosic sands and ice-age aridity in tropical South America: Geol. Soc. America Bull., v. 81, p. 189–206.

Davies, D. K., 1966, Sedimentary structures and subfacies of a Mississippi River point bar: Jour. Geology, v. 74, no. 2, p. 234–239.

—— *et al.*, 1971, Recognition of barrier environments: Am. Assoc. Petroleum Geologists Bull., v. 55, p. 550–565.

Davis, J. R., 1971, Sedimentation of Pliocene sandstones in Santa Barbara Channel, California (abs.): Am. Assoc. Petroleum Geologists Bull., v. 55, p. 335.

Davis, W. M., 1938, Sheetfloods and streamfloods: Geol. Soc. America Bull., v. 49, p. 1337–1416.

Dennison, J. M., 1971, Petroleum related to Middle and Upper Devonian deltaic facies in central Appalachia: Am. Assoc. Petroleum Geologists Bull., v. 55, p. 1179–1193.

Denny, C. S., 1965, Alluvial fans in the Death Valley region, California, Nevada: U.S. Geol. Survey Prof. Paper 466, 62 p.

—— 1967, Fans and pediments: Am. Jour. Sci., v. 265, p. 81–105.

Dickas, A. B., and J. L. Payne, 1967, Upper Paleocene buried channel in Sacramento Valley, California: Am. Assoc. Petroleum Geologists Bull., v. 51, no. 6, p. 873–882.

Dickinson, K. A., *et al.*, 1972, Criteria for recognizing ancient barrier coastlines, *in* J. K. Rigby and W. K. Hamblin, eds., Recognition of sedimentary environments: Soc. Econ. Paleontologists and Mineralogists Spec. Pub. 16, 340 p.

Dietz, R. J., and H. J. Knebel, 1971, Thou Sans Fond submarine canyon, Ivory Coast, Africa: Deep-Sea Research, v. 18, p. 441–447.

Dill, R. F., *et al.*, 1954, Deep-sea channels and deltas of the Monterey submarine canyon: Geol. Soc. America Bull., v. 65, p. 191–194.

Dillon, W. P., 1970, Submergence effects on a Rhode Island barrier and lagoon and inferences on migration of barriers: Jour. Geology, v. 78, no. 1, p. 94–106.

Dineley, D. L., and B. P. J. Williams, 1968, Sedimentation and paleoecology of the Devonian Escuminac Formation and related strata, Escuminac Bay, Quebec: Geol. Soc. America Spec. Paper 106, p. 241–264.

Dobby, E. H. G., 1936, The Ebro delta: Geog. Jour., v. 87, p. 455–469.

Dodge, C. F., 1965, Genesis of an Upper Cretaceous offshore bar near Arlington, Texas: Jour. Sed. Petrology, v. 35, no. 1, p. 22–35.

Doeglas, D. J., 1962, The structure of sedimentary deposits of braided rivers: Sedimentology, v. 1, p. 167–190.

Donaldson, A. C., 1966, Deltaic sands and sandstones, *in* Symposium on recently developed geologic principles and sedimentation of the Permo-Pennsylvanian of the Rocky Mountains: Wyoming Geol. Assoc. 20th Ann. Field Conf. Guidebook, p. 31–62.

—— 1969, Ancient deltaic sedimentation (Pennsylvanian) and its control on the distribution, thickness and quality of coals, *in* A. C. Donaldson, ed., Some Appalachian coals and carbonates: models of ancient shallow-water deposition: West Virginia Geol. and Econ. Survey, p. 93–123.

—— *et al.*, 1970, Holocene Guadalupe delta of Texas Gulf Coast, *in* Deltaic sedimentation—modern and ancient: Soc. Econ. Paleontologists and Mineralogists Spec. Pub. 15, p. 107–137.

Dondanville, R. R., 1963, The Fall River Formation, northwestern Black Hills, lithology and geologic history: Wyoming Geol. Assoc.–Billings Geol. Soc. Joint Field Conf. Guidebook, p. 87–99.

Dott, R. H., Jr., 1964, Ancient deltaic sedimentation in eugeosynclinal belts, *in* Deltaic and shallow marine deposits: Amsterdam, Elsevier Pub. Co., p. 105–113.

—— 1966, Eocene deltaic sedimentation at Coos Bay, Oregon: Jour. Geology, v. 74, p. 373–420.

Duncan, J. R., and L. D. Kulm, 1970, Mineralogy, provenance and dispersal history of late Quaternary deep-sea sands in Cascadia basin and Blanco fracture zone of Oregon: Jour. Sed. Petrology, v. 40, p. 874–887.

Eckis, R., 1928, Alluvial fans of the Cucamonga district, southern California: Jour. Geology, v. 36, p. 224–247.

Emery, K. O., 1960a, Basin plains and aprons off southern California: Jour. Geology, v. 68, p. 464–479.

—— 1960b, The sea off southern California: New York, John Wiley and Sons, Inc.

—— *et al.*, 1952, Submarine geology off San Diego, California: Jour. Geology, v. 60, no. 6, p. 511–548.

Ericson, D. B., M. Ewing, and B. C. Heezen, 1951, Deep-sea sands and submarine canyons: Geol. Soc. America Bull., v. 62, p. 961–965.

—— and —— 1952, Turbidity currents and sediments in the North Atlantic: Am. Assoc. Petroleum Geologists Bull., v. 36, p. 489–511.

—— *et al.*, 1955, Sediment deposition in deep Atlantic, p. 205–220, *in* Arie Poldervaart, ed., Crust of the earth: Geol. Soc. America Spec. Paper 62, 762 p.

—— *et al.*, 1961, Atlantic deep-sea cores: Geol. Soc. America Bull., v. 72, p. 193–285.

Evans, G., 1965, Intertidal flat sediments and their environments of deposition in the Wash: Geol. Soc. London Quart. Jour., v. 121, p. 209–245.

Ewing, M., *et al.*, 1958, Sediments and topography of the Gulf of Mexico, *in* L. G. Weeks, ed., Habitat of oil: Am. Assoc. Petroleum Geologists, p. 995–1053.

Exum, F. A., and J. C. Harms, 1968, Comparison of marine-bar with valley-fill stratigraphic traps, western Nebraska: Am. Assoc. Petroleum Geologists Bull., v. 52, no. 10, p. 1851–1868.

Eymann, J. L., 1953, A study of sand dunes in the Colorado and Mojave Deserts: Los Angeles, Univ. Southern California, MS thesis, 91 p.

Fahnestock, R. K., 1963, Morphology and hydrology of a glacial stream—White River, Mount Rainier, Washington: U.S. Geol. Survey Prof. Paper 422-A, 70 p.

Felix, D. W., and D. S. Gorsline, 1971, Newport submarine canyon, California: an example of the effects of shifting loci of sand supply upon canyon position: Marine Geology, v. 10, p. 177–198.

Ferguson, J., 1863, Delta of the Ganges: Geol. Soc. London Quart. Jour., v. 19, p. 321–354.

Ferm, J. C., 1970, Allegheny deltaic deposits, *in* Deltaic sedimentation—modern and ancient: Soc. Econ. Paleontologists and Mineralogists Spec. Pub. 15, p. 246–255.

—— and V. V. Cavaroc, Jr., 1969, A field guide to Allegheny deltaic deposits in the upper Ohio Valley:

Ohio Geol. Soc. and Pittsburgh Geol. Soc.

Field, M. E., and O. H. Pilkey, 1971, Deposition of deep-sea sands: comparison of two areas of the Carolina continental rise: Jour. Sed. Petrology, v. 41, p. 526–536.

Finch, W. I., 1959, Geology of uranium deposits in Triassic rocks of the Colorado Plateau region: U.S. Geol. Survey Bull. 1074-D, 164 p.

Finkel, H. J., 1959, The barchans of southern Peru: Jour. Geology, v. 67, p. 614–647.

Fischer, P. J., 1971, An ancient (upper Paleocene) submarine canyon and fan: the Meganos Channel, Sacramento Valley, California (abs.): 67th Ann. Geol. Soc. America Cordilleran Section Mtg. Prog., v. 3, no. 2, p. 120.

Fisher, W. L., 1969, Gulf Coast Basin Tertiary delta systems, in Delta systems in exploration for oil and gas: Texas Univ. Bur. Econ. Geology, p. 30–39.

———— and L. F. Brown, Jr., 1969, Delta systems and oil and gas occurrence, in Delta systems in exploration for oil and gas: Texas Univ. Bur. Econ. Geology, p. 54–66.

———— and J. H. McGowen, 1969, Depositional systems in Wilcox Group (Eocene) of Texas and their relationship to occurrence of oil and gas: Am. Assoc. Petroleum Geologists Bull., v. 53, no. 1, p. 30–54.

———— et al., 1969, Delta systems in the exploration for oil and gas: Texas Univ. Bur. Econ. Geology, 78 p.

———— et al., 1970, Depositional systems in the Jackson Group of Texas—their relationship to oil, gas and uranium: Texas Univ. Bur. Econ. Geology Geol. Circ. 70-4.

Fisk, H. N., 1940, Geology of Avoyelles and Rapides Parishes, Louisiana: Louisiana Geol. Survey Geol. Bull. No. 18.

———— 1944, Geological investigation of the alluvial valley of the lower Mississippi River: Mississippi River Commission, Vicksburg, Mississippi.

———— 1947, Fine-grained alluvial deposits and their effects on Mississippi River activity: Mississippi River Commission, Vicksburg, Mississippi.

———— 1952, Geological investigation of the Atchafalaya basin and problems of Mississippi River diversion: Mississippi River Commission, Vicksburg, Mississippi, p. 1–145.

———— 1955, Sand facies of Recent Mississippi delta deposits: 4th World Petroleum Cong. Proc., Sec. I/C, p. 1–21.

———— 1958, Recent Mississippi River sedimentation and peat accumulation, in Ernest Van Aelst, ed., 4th Congres pour l'avancement des etudes de stratigraphie et de geologie du Carbonifere, Heerlen, 1958: Compte Rendu, v. 1, p. 187–199.

———— 1959, Padre Island and the Laguna Madre flats, coastal south Texas, in 2d Coastal Geog. Conf., April 6-9: Natl. Acad. Sci.–Natl. Research Council Pub., p. 103–151.

———— 1961, Bar-finger sands of the Mississippi delta, in J. A. Peterson and J. C. Osmond, eds., Geometry of sandstone bodies: Am. Assoc. Petroleum Geologists, p. 29–52.

———— and E. McFarlan, Jr., 1955, Late Quaternary deltaic deposits of the Mississippi River, in A. Poldervaart, ed., The crust of the earth: Geol. Soc. America Spec. Paper 62, p. 279–302.

———— et al., 1954, Sedimentary framework of the modern Mississippi delta: Jour. Sed. Petrology, v. 24, no. 2, p. 76–99.

Flemal, R. C., 1967, Sedimentology of the Sespe Formation, southwestern California: Princeton, New Jersey, Princeton Univ., PhD thesis, 258 p.

Folk, R. L., 1971, Longitudinal dunes of the northwestern edge of the Simpson Desert, Northern Territory, Australia, Part 1. Geomorphology and grain size relationships: Sedimentology, v. 16, p. 5–54.

———— and W. C. Ward, 1957, Brazos River bar—a study of the significance of grain size parameters: Jour. Sed. Petrology, v. 27, p. 3–26.

Frazier, D. E., 1967, Recent deltaic deposits of the Mississippi River: their development and chronology: Gulf Coast Assoc. Geol. Socs. Trans., v. 17, p. 287–315.

———— and A. Osanik, 1961, Point-bar deposits, Old River locksite, Louisiana: Gulf Coast Assoc. Geol. Socs. Trans., v. 11, p. 121–137.

———— and ————, 1969, Recent peat deposits, Louisiana coastal plain: Geol. Soc. America Spec. Paper 114, p. 63–85.

Freeman, J. C., 1949, Strand-line accumulation of petroleum, Jim Hogg County, Texas: Am. Assoc. Petroleum Geologists Bull., v. 33, no. 7, p. 1260–1270.

Friedkin, J. F., 1945, A laboratory study of the meandering of alluvial rivers: U.S. Waterways Experiment Sta., Vicksburg, Mississippi, May.

Friedman, G. M., and K. G. Johnson, 1966, The Devonian Catskill deltaic complex of New York, type example of a "teutonic delta complex," in Deltas in their geologic framework: Houston Geol. Soc.

Galehouse, J. J., 1967, Provenance and paleocurrents of the Paso Robles Formation, California: Geol. Soc. America Bull., v. 78, p. 951–978.

Galloway, W. E., 1968, Depositional systems of lower Wilcox Group, north-central Gulf Coast basin: Gulf Coast Assoc. Geol. Socs. Trans., v. 18, p. 275–289.

Geijskes, D. C., 1952, On the structure and origin of the sandy ridges in the coastal zone of Surinam: Koninkl. Nederlandsch Aardrijksk. Genoot. Tijdschr., v. 69, no. 2, p. 225–237.

Gellert, J. F., 1968, The Yangtze River mouth and delta: Warsaw, Przegląd Geog., v. 40, p. 413–418 (in German).

Giddings, J. L., 1947, Mackenzie River delta chronology: Tree Ring Bull., v. 13, p. 26–29.

Gilbert, G. K., 1914, The transportation of debris by running water: U.S. Geol. Survey Prof. Paper 86.

Giles, R. T., and O. H. Pilkey, 1965, Atlantic beach and dune sediments of the southern United States: Jour. Sed. Petrology, v. 35, no. 4, p. 900–910.

Gilluly, J., 1927, Physiography of the Ajo region, Arizona: Geol. Soc. America Bull., v. 48, p. 323–348.

Glenn, J. L., and A. R. Dahl, 1959, Characteristics and distribution of some Missouri River deposits: Iowa Acad. Sci. Proc., v. 66, p. 302–311.

Glennie, K. W., 1970, Desert sedimentary environments —Developments in sedimentology, no. 14: Amsterdam, Elsevier Pub. Co., 222 p.

Gorsline, D. S., 1967, Sedimentologic studies of the Colorado delta: Southern California Univ. Geol. Rept. 67-1, 121 p.

———— and K. O. Emery, 1959, Turbidity-current deposits in San Pedro and Santa Monica basins off southern California: Geol. Soc. America Bull., v. 70, p. 279–290.

———— et al., 1967, Studies of submarine canyons and fans off southern California: Southern Calif. Univ. Geol. Rept. 67-3, 27 p.

Gould, H. R., 1970, The Mississippi delta complex, in Deltaic sedimentation—modern and ancient: Soc.

Econ. Paleontologists and Mineralogists Spec. Pub. 15, p. 3–30.
—— and E. McFarlan, Jr., 1959, Geologic history of the chenier plain, southwestern Louisiana: Gulf Coast Assoc. Geol. Socs. Trans., v. 9, p. 261–270.
Gourou, P., 1967, The Zambezi River delta: Acta Geog. Lovaniensia, v. 5, p. 31–36 (in French).
Greenman, N. N., and R. J. LeBlanc, 1956, Recent marine sediments and environments of northwest Gulf of Mexico: Am. Assoc. Petroleum Geologists Bull., v. 40, p. 813–847.
Greensmith, J. T., 1966, Carboniferous deltaic sedimentation in eastern Scotland, a review and reappraisal, in Deltas in their geologic framework: Houston Geol. Soc.
—— and E. V. Tucker, 1969, The origin of the Holocene shell deposits in the chenier plain facies of Essex (Great Britain): Marine Geology, v. 7, p. 403–425.
Gregory, J. L., 1966, A lower Oligocene delta in the subsurface of southeast Texas: Gulf Coast Assoc. Geol. Socs. Trans., v. 16, p. 227–241.
Gremillion, L. R., et al., 1964, Barrier and lagoonal sets on high terraces in the Florida panhandle: Southeastern Geology, v. 6, p. 31–36.
Griffith, E. G., 1966, Geology of Saber bar, Logan and Weld Counties, Colorado: Am. Assoc. Petroleum Geologists Bull., v. 50, p. 2112–2118.
Griggs, G. B., and L. D. Kulm, 1970, Sedimentation in Cascadia deep-sea channel: Geol. Soc. America Bull., v. 81, p. 1361–1384.
Guilcher, A., 1958, Coastal and submarine morphology: London, Methuen & Co., Ltd.: New York, John Wiley and Sons, Inc., 274 p.
—— 1959, Coastal sand ridges and marshes and their continental environment near Grand Popo and Ouidah, Dahomey: 2d Coastal Geog. Conf., U.S. Office Naval Research–Natl. Acad. Sci., Washington, D.C., p. 189–212.
Guy, H. P., et al., 1966, Summary of alluvial channel data from flume experiments, 1956–1961: U.S. Geol. Survey Prof. Paper 462-I, p. 1–96.
Gwinn, V. E., 1964, Deduction of flow regime from bedding character in conglomerates and sandstones: Jour. Sed. Petrology, v. 34, no. 3, p. 656–658.
—— and T. A. Mutch, 1965, Intertongued Upper Cretaceous volcanic and nonvolcanic rocks, central-western Montana: Geol. Soc. America Bull., v. 76, p. 1125–1144.
Hack, J. T., 1941, Dunes of the western Navajo country: Geog. Rev., v. 31, p. 240–263.
Hails, J. R., 1964, The coastal depositional features of southeastern Queensland: Australia Geog., v. 9, p. 208–217.
—— and J. H. Hoyt, 1969, The significance and limitations of statistical parameters for distinguishing ancient and modern sedimentary environments of the lower Georgia coastal plain: Jour. Sed. Petrology, v. 39, no. 2, p. 559–580.
Hale, L. A., 1961, Late Cretaceous (Montanan) stratigraphy, eastern Washakie basin, Carbon County, Wyoming: Wyoming Geol. Assoc. 16th Field Conf. Guidebook, p. 129–137.
Hamilton, E. L., 1967, Marine geology of the abyssal plains in the Gulf of Alaska: Jour. Geophys. Research, v. 72, p. 4189–4213.
Haner, B. E., 1971, Morphology and sediments of Redondo submarine fan, southern California: Geol. Soc. America Bull., v. 82, p. 2413–2432.
Hansen, H. J., 1969, Depositional environments of subsurface Potomac Group in southern Maryland: Am. Assoc. Petroleum Geologists Bull., v. 53, no. 9, p. 1923–1937.
Hantzschel, W., 1939, Tidal flat deposits (Wattenschlick), in P. D. Trask, ed., Recent marine sediments: Am. Assoc. Petroleum Geologists, p. 195–206.
Happ, S. C., et al., 1940, Some aspects of accelerated stream and valley sedimentation: U.S. Dept. Agriculture Tech. Bull., v. 695, 134 p.
Harms, J. C., 1966, Stratigraphic traps in a valley fill, western Nebraska: Am. Assoc. Petroleum Geologists Bull., v. 50, p. 2119–2149.
—— and R. K. Fahnestock, 1965, Stratification bed forms and flow phenomena (with an example from the Rio Grande), in Primary sedimentary structures and their hydrodynamic interpretation: Soc. Econ. Paleontologists and Mineralogists Spec. Pub. 12, p. 84–115.
—— et al., 1963, Stratification in modern sands of the Red River, Louisiana: Jour. Geology, v. 71, p. 566–580.
—— et al., 1965, Depositional environment of the Fox Hills sandstones near Rock Springs uplift, Wyoming: Wyoming Geol. Assoc. 19th Field Conf. Guidebook, p. 113–130.
Hattin, D. E., 1965, Stratigraphy of the Graneros shale (Upper Cretaceous) in central Kansas: Kansas Geol. Survey Bull. 178.
Heezen, B. C., and Marie Tharp, 1967, Indian Ocean floor: Natl. Geog. Mag., October.
—— and —— 1969, Atlantic Ocean floor: Natl. Geog. Mag., June.
—— et al., 1959, The floors of the oceans: Geol. Soc. America Spec. Paper 65, 122 p.
—— et al., 1964, Congo submarine canyon: Am. Assoc. Petroleum Geologists Bull., v. 48, p. 1126–1149.
Hewitt, C. H., et al., 1965, The Frye in situ combustion test—reservoir characteristics: Soc. Petroleum Engineers AIME Petroleum Trans., March, p. 337–342.
Holeman, J. N., 1970, The sediment yield of major rivers of the world: Water Resources Research, v. 4, no. 4, p. 737–747.
Hooke, R. L., 1967, Processes of arid-regional alluvial fans: Jour. Geology, v. 75, no. 4, p. 438–460.
Hoover, R. A., 1968, Physiography and surface sediment facies of a recent tidal delta, Harbor Island, central Texas coast: Austin, Texas, Texas Univ., PhD thesis, 223 p.
Hoppe, G., and S. R. Ekman, 1964, A note on the alluvial fans of Ladtjouagge, Swedish Lappland: Geog. Annaler, v. 46, p. 338–342.
Horn, D., 1965, Zur Geologischen entwicklung der sudlichen schleimundung in Holozan: Meyniana, v. 15, p. 41–58.
Horn, D. R., et al., 1971, Turbidites of the northeast Pacific: Sedimentology, v. 16, p. 55–69.
Horowitz, D. H., 1966, Evidence for deltaic origin of an Upper Ordovician sequence in the central Appalachians, in Deltas in their geologic framework: Houston Geol. Soc., p. 159–169.
Houbolt, J. J. H. C., and J. B. M. Jonker, 1968, Recent sediments in the eastern part of the Lake of Geneva (Lac Leman): Geologie en Mijnbouw, v. 47, p. 131–148.
Howard, J. D., 1966, Patterns of sediment dispersal in the Fountain Formation of Colorado: Mtn. Geologist, v. 3, p. 147–153.
Howe, H. V., and C. K. Moresi, 1931, Geology of

Iberia Parish: Louisiana Geol. Survey Geol. Bull. No. 1.

—— and ——— 1933, Geology of Lafayette and St. Martin Parishes: Louisiana Geol. Survey Geol. Bull. No. 3, 238 p.

—— et al., 1935, Reports on the geology of Cameron and Vermilion Parishes: Louisiana Geol. Survey Geol. Bull. No. 6, 242 p.

Hoyt, J. H., 1964, High angle beach stratification, Sapelo Island, Georgia: Jour. Sed. Petrology, v. 32, p. 309–311.

—— 1967, Barrier island formation: Geol. Soc. America Bull., v. 78, p. 1125–1136.

—— 1969, Chenier versus barrier, genetic and stratigraphic distinction: Am. Assoc. Petroleum Geologists Bull., v. 53, p. 299–306.

—— and V. J. Henry, Jr., 1965, Significance of inlet sedimentation in the recognition of ancient barrier islands, in Sedimentation of Late Cretaceous and Tertiary outcrops, Rock Springs uplift: Wyoming Geol. Assoc. 19th Field Conf. Guidebook, p. 190–194.

—— and ——— 1967, Influence of island migration on barrier island sedimentation: Geol. Soc. America Bull., v. 78, p. 77–86.

—— and ——— 1971, Origin of capes and shores along the southeastern coast of the United States: Geol. Soc. America Bull., v. 82, p. 59–66.

—— and R. J. Weimer, 1963, Comparison of modern and ancient beaches, central Georgia coast: Am. Assoc. Petroleum Geologists Bull., v. 47, p. 529–531.

—— et al., 1964, Late Pleistocene and Recent sedimentation on Georgia coast, in Deltaic and shallow marine deposits: New York, Elsevier Publishing Co., p. 170–176.

Hoyt, W. V., 1959, Erosional channel in the middle Wilcox near Yoakum, Lavaca County, Texas: Gulf Coast Assoc. Geol. Socs. Trans., v. 9, p. 41–50.

Hrabar, S. V., and P. E. Potter, 1969, Lower west Badeu (Mississippian) sandstone body of Owen and Green Counties, Indiana: Am. Assoc. Petroleum Geologists Bull., v. 53, no. 10, p. 2150–2160.

Huang, T., and H. G. Goodell, 1970, Sediments and sedimentary processes of eastern Mississippi cone, Gulf of Mexico: Am. Assoc. Petroleum Geologists Bull., v. 54, p. 2070–2100.

Hubbard, J. A. E. B., 1967, Facies patterns in the Carrowmoran Sandstone (Visean) of western County Sligo, Ireland: Geol. Assoc. Proc., v. 77, pt. 2, p. 233–254.

Hubert, C., et al., 1970, Deep-sea sediments in the lower Paleozoic Quebec supergroup: Geol. Assoc. Canada Spec. Paper 7, p. 103–125.

Hubert, J. F., 1960, Petrology of the Fountain and Lyons Formations, Front Range, Colorado: Colorado School Mines Quart., v. 55, p. 1–242.

—— 1964, Textural evidence for deposition of many western Atlantic deep-sea sands by ocean-bottom currents rather than turbidity currents: Jour. Geology, v. 72, p. 757–785.

Inman, D. L., and R. A. Bagnold, 1963, Littoral processes, in The sea, v. 2, chap. 2: New York, John Wiley and Sons, p. 529–553.

—— et al., 1966, Coastal sand dunes of Guerrero Negro, Baja California, Mexico: Geol. Soc. America Bull., v. 77, no. 8, p. 787–802.

Ives, R. L., 1936, Desert floods in Sonoyta Valley: Am. Jour. Sci., v. 32, p. 349–360.

Jacka, A. D., 1965, Depositional dynamics of the Almond Formation, Rock Springs uplift, Wyoming: Wyoming Geol. Assoc. 19th Field Conf. Guidebook, p. 81–100.

Jahns, R. H., 1947, Geological features of the Connecticut Valley, Massachusetts as related to recent floods: U.S. Geol. Survey Water-Supply Paper 996, 157 p.

Jelgersma, S., J. de Jong, and W. H. Zagwijn, 1970, The coastal dunes of the western Netherlands; geology, vegetational history and archeology: Geol. Stichting Med., new ser., no. 21, p. 93–154.

Johnson, A., 1965, A model for debris flow: University Park, Pennsylvania, Pennsylvania State Univ., PhD thesis, 233 p.

Johnson, R. B., 1967, The great sand dunes of southern Colorado: U.S. Geol. Survey Prof. Paper 575-C, p. C177–C183.

Johnson, W. A., 1921, Sedimentation of the Fraser River delta: Canada Geol. Survey Mem. 125, 46 p.

—— 1922, The character of the stratification of the sediments of the Recent delta of the Fraser River, British Columbia, Canada: Jour. Geology, v. 30, p. 115–129.

Kanes, W., 1969, Tidal inlets, tidal deltas, and barrier islands versus alluvial deposits: a discussion of sedimentary criteria, in A. C. Donaldson, ed., Some Appalachian coals and carbonates: models of ancient shallow water deposition: West Virginia Geol. and Econ. Survey, p. 259–263.

—— 1970, Facies and development of the Colorado River delta in Texas, in Deltaic sedimentation—modern and ancient: Soc. Econ. Paleontologists and Mineralogists Spec. Pub. 15, p. 78–106.

Kelling, G., 1968, Patterns of sedimentation in Rhondda Beds of South Wales: Am. Assoc. Petroleum Geologists Bull., v. 52, no. 12, p. 2369–2386.

—— and D. J. Stanley, 1970, Morphology and structure of Wilmington and Baltimore submarine canyons, eastern United States: Jour. Geology, v. 78, p. 637–660.

—— and M. A. Woollands, 1969, The stratigraphy and sedimentation of the Llandoverian rocks of the Rhoyader district, in A. Woods, ed., The Precambrian and lower Paleozoic rocks of Wales: Univ. Wales Press, p. 255–282.

Kessler, L. G., II, 1970, Bar development and aggradational sequences in the braided portions of the Canadian River, Hutchinson, Roberts and Hemphill Counties, Texas (abs.): 23d Ann. Rocky Mountain Sec. Geol. Soc. America Mtg. Prog., v. 2, no. 5, p. 338.

—— 1971, Characteristics of the braided stream depositional environment with examples from the South Canadian River, Texas: Earth Sci. Bull., v. 4, no. 1, p. 25–35.

King, C. A. M., 1959, Beaches and coasts: London, Edward Arnold Publications, Ltd., 403 p.

Klein, G. deV., 1962, Triassic sedimentation, Maritime Provinces, Canada: Geol. Soc. America Bull., v. 73, p. 1127–1146.

—— 1967, Comparison of Recent and ancient tidal flat and estuarine sediments, in Estuaries: Am. Assoc. Adv. Sci. Pub. 83, p. 207–218.

—— 1968, Sedimentology of Triassic rocks in the lower Connecticut Valley, in Guidebook for field trips in Connecticut, New England Internat. Collegiate Geol. Conf.: Connecticut Geol. and Nat. History Survey Guidebook 2, p. (c-1) 1–19.

—— 1970, Depositional and dispersal dynamics of intertidal sand bars: Jour. Sed. Petrology, v. 40, p. 1095–1127.

―――― 1971, A sedimentary model for determining paleotidal range: Geol. Soc. America Bull., v. 82, p. 2585–2592.

―――― and J. E. Sanders, 1964, Comparison of sediments from Bay of Fundy and Dutch Wadden Sea tidal flats: Jour. Sed. Petrology, v. 34, no. 1, p. 18–24.

Knight, W. V., 1969, Historical and economic geology of Silurian Clinton Sandstone of northwestern Ohio: Am. Assoc. Petroleum Geologists Bull., v. 53, no. 7, p. 1421–1452.

Kolb, C. R., 1962, Distribution of soils bordering the Mississippi River from Donaldsonville to Head of Passes: U.S. Army Corps Engineers, Waterways Expt. Sta. Tech. Rept. 3–601, p. 1–61.

―――― 1963, Sediments forming the bed and banks of the lower Mississippi River and their effect on river migration: Sedimentology, v. 2, p. 227–234.

―――― and R. I. Kaufman, 1967, Prodelta clays of southeast Louisiana: Internat. Marine Geotech. Research Conf. Proc., Monticello, Illinois, p. 3–21.

―――― and J. R. Van Lopik, 1966, Depositional environments of the Mississippi River deltaic plain, southeastern Louisiana, in Deltas in their geological framework: Houston Geol. Soc., p. 17–61.

―――― et al., 1968, Geological investigation of the Yazoo basin, lower Mississippi Valley: U.S. Army Corps Engineers, Waterways Expt. Sta. Tech. Rept. 3–480.

Komar, P. D., 1969, The channelized flow of turbidity currents with application to Monterey deep-sea channel: Jour. Geophys. Research, v. 74, p. 4544–4558.

―――― 1970, The competence of turbidity current flow: Geol. Soc. America Bull., v. 81, p. 1555–1562.

Krause, D. C., et al., 1970, Turbidity currents and cable breaks in the western New Britain trench: Geol. Soc. America Bull., v. 81, p. 2153–2160.

Krigstrom, A., 1962, Geomorphological studies of Sandur plains and their braided river in Iceland: Geog. Annals, v. 44, p. 328–346.

Kruit, C., 1955, Sediments of the Rhône delta; I, Grain size and microfauna: Koninkl. Nederlandse Geol. Mijnb. Genoot. Verh., Geol. Ser., v. 15, p. 357–514.

Kuenen, Ph. H., 1950, Marine geology: New York, John Wiley and Sons, Inc., 568 p.

Kukal, Z., 1964, Diagnostic features of Paleozoic deltaic sediments of central Bohemia: Sedimentology, v. 3, p. 109–113.

―――― 1971, Geology of Recent sediments: New York, Academic Press, 490 p.

Lagaaij, R., and F. P. H. W. Kopstein, 1964, Typical features on a fluvio-marine offlap sequence, in Developments in sedimentology, v. 1, Deltaic and shallow marine deposits: Amsterdam, Elsevier Publishing Co., p. 216–226.

Laming, D. J. C., 1966, Imbrication, paleocurrents, and other sedimentary features in the lower New Red Sandstone, Devonshire, England: Jour. Sed. Petrology, v. 36, no. 4, p. 940–959.

Land, L. S., 1964, Eolian cross bedding in the beach dune environment, Sapelo Island, Georgia: Jour. Sed. Petrology, v. 34, no. 2, p. 389–394.

Lane, E. W., 1957, A study of the shape of channels formed by natural streams flowing in erodible material: U.S. Army Corps Engineers, Missouri River Div., Omaha, Nebraska, Sediment Ser. 9, 106 p.

―――― 1963, Cross-stratification in San Bernard River, Texas, point bar deposits: Jour. Sed. Petrology, v. 33, p. 350–354.

―――― and E. W. Eden, 1940, Sand waves in the lower Mississippi River: Western Soc. Engineers Jour., v. 45, no. 6, p. 281–291.

Laporte, L. F., 1968, Ancient environments: Englewood Cliffs, New Jersey, Prentice-Hall, Inc., 116 p.

Laury, R. L., 1968, Sedimentology of the Pleasantview Sandstone, southern Iowa and western Illinois: Jour. Sed. Petrology, v. 38, no. 2, p. 568–599.

Lawson, A. C., 1913, The petrographic description of alluvial fan deposits: California Univ. Pubs. Geol. Sci., v. 7, p. 325–334.

LeBlanc, R. J., and H. A. Bernard, 1954, Resume of late Recent geological history of the Gulf Coast: Geologie en Mijnbouw, new series, v. 16c, p. 185–194.

―――― and W. D. Hodgson, 1959, Origin and development of the Texas shoreline: Louisiana State Univ. Coastal Studies Inst., 2d Coastal Geog. Conf., p. 57–101; EPR Pub. 219, Shell E&P Research, Houston.

LeBlanc, R. J., Jr., 1971, Environments of deposition of the Yegua Formation (Eocene), Brazos County, Texas: Corpus Christi Geol. Soc. Bull., v. 11, no. 6, 70 p.

Legget, R. F., R. J. E. Brown, and G. H. Johnston, 1966, Alluvial fan formation near Aklavik, Northwest Territories, Canada: Geol. Soc. America Bull., v. 77, p. 15–30.

Lené, G. W., and D. E. Owen, 1969, Grain orientation in a Berea Sandstone channel at south Amherst, Ohio: Jour. Sed. Petrology, v. 39, no. 2, p. 737–743.

Leopold, L. B., and M. G. Wolman, 1957, River channel patterns: braided, meandering and straight: U.S. Geol. Survey Prof. Paper 282-B, 85 p.

―――― and ―――― 1960, River meanders: Geol. Soc. America Bull., v. 71, p. 769–794.

―――― et al., 1966, Channel and hillslope processes in a semi-arid area, New Mexico: U.S. Geol. Survey Prof. Paper 352-G.

Lineback, J. A., 1968, Turbidites and other sandstone bodies in the Borden Siltstone (Mississippian) in Illinois: Illinois Geol. Survey Circ. 425, p. 1–29.

Lins, T. W., 1950, Origin and environment of the Tonganoxie Sandstone in northeastern Kansas: Kansas Geol. Survey Bull., v. 86, p. 107–140.

Liteanu, E., and A. Pricajan, 1963, The geological structure of the Danube delta: Bucharest, Hidrobiologia, v. 4, p. 57–82 (in Rumanian).

Logvinenko, N. V., and I. N. Remizov, 1964, Sedimentology of beaches on the north coast of the Sea of Azov, in Deltaic and shallow marine deposits, developments in sedimentology, v. 1: Amsterdam, Elsevier Publishing Co., p. 244–252.

Lucke, J. B., 1934, A study of Barnegat Inlet, New Jersey and related shoreline phenomena: Shore and Beach, v. 2, p. 45–93.

Lum, D., and H. T. Stearns, 1970, Pleistocene stratigraphy and eustatic history based on cores at Waimanalo, Oahu, Hawaii: Geol. Soc. America Bull., v. 81, p. 1–16.

Lumsden, D. N., and B. R. Pelletier, 1969, Petrology of the Grimsby Sandstone (Lower Silurian) of Ontario and New York: Jour. Sed. Petrology, v. 39, no. 2, p. 521–530.

Lustig, L. K., 1963, Competence of transport in alluvial fans: U.S. Geol. Survey Prof. Paper 475-C, p. 126–129.

―――― 1965, Clastic sedimentation in the Deep Springs Valley, California: U.S. Geol. Survey Prof. Paper 352-F.

MacIntyre, I. G., and J. D. Milliman, 1970, Physiographic features on the outer shelf and upper slope, Atlantic continental margin, southeastern United States: Geol. Soc. America Bull., v. 81, p. 2577–2589.

MacNeil, F. S., 1950, Pleistocene shore lines in Florida and Georgia: U.S. Geol. Survey Prof. Paper 221-F.

Mammerickx, J., 1970, Morphology of the Aleutian abyssal plain: Geol. Soc. America Bull., v. 81, p. 3457–3464.

Mann, C. J., and W. A. Thomas, 1968, Ancient Mississippi River: Gulf Coast Assoc. Geol. Socs. Trans., v. 18, p. 187–204.

Manos, C., 1967, Depositional environment of Sparland cyclothem (Pennsylvanian), Illinois and Forest City basins: Am. Assoc. Petroleum Geologists Bull., v. 51, no. 9, p. 1843–1861.

Martin, B. D., 1963, Rosedale channel—evidence for late Miocene submarine erosion in Great Valley of California: Am. Assoc. Petroleum Geologists Bull., v. 47, p. 441–456.

—— and K. O. Emery, 1967, Geology of the Monterey Canyon, California: Am. Assoc. Petroleum Geologists Bull., v. 51, p. 2281–2304.

Martini, I. P., 1971, Regional analysis of sedimentology of Medina Formation (Silurian), Ontario and New York: Am. Assoc. Petroleum Geologists Bull., v. 55, p. 1249–1261.

Mathews, W. H., and F. P. Shepard, 1962, Sedimentation of Fraser River delta, British Columbia: Am. Assoc. Petroleum Geologists Bull., v. 48, no. 8, p. 1416–1444.

Matthews, J. L., 1966, Sedimentation of the coastal dunes at Oceano, California: Univ. California at Los Angeles, PhD thesis, 138 p.

McBride, E. F., and M. O. Hayes, 1962, Dune cross bedding on Mustang Island, Texas: Am. Assoc. Petroleum Geologists Bull., v. 46, p. 546–552.

McCave, I. N., 1969, Correlation using a sedimentological model of part of the Hamilton group (Middle Devonian), New York State: Am. Jour. Sci., v. 267, no. 5, p. 567–591.

McCloy, J. L., 1969, Morphologic characteristics of the Blow River delta, Yukon Territory, Canada: Baton Rouge, Louisiana, Louisiana State Univ., PhD thesis, 161 p.

McEwen, M. C., 1969, Sedimentary facies of the modern Trinity delta, in Galveston Bay geology: Houston Geol. Soc., p. 53–77.

McGee, W. J., 1897, Sheetflood erosion: Geol. Soc. America Bull., v. 8, p. 87–112.

McGowen, J. H., and L. E. Garner, 1969, Comparison of Recent and ancient coarse-grained point bars (abs.): Am. Assoc. Petroleum Geologists Bull., v. 53, no. 3, p. 731–732.

—— and ——, 1970, Physiographic features and stratification types of coarse-grained point bars: modern and ancient examples: Sedimentology, v. 14, p. 77–111.

—— and C. G. Groat, 1971, Van Horn Sandstone, West Texas—an alluvial fan model for mineral exploration: Texas Univ. Bur. Econ. Geology, 57 p.

McKee, E. D., 1934, The Coconino Sandstone—its history and origin: Carnegie Inst. Washington Pub. 440, Contr. Paleontology, p. 77–115.

—— 1939, Some types of bedding in the Colorado River delta: Jour. Geology, v. 47, p. 64–81.

—— 1957, Primary structures of some Recent sediments: Am. Assoc. Petroleum Geologists Bull., v. 41, p. 1704–1747.

—— 1966, Structures of dunes at White Sands National Monument, New Mexico: Sedimentology, v. 7, no. 1, 69 p.

—— and G. C. Tibbitts, 1964, Primary structures of a seif dune and associated deposits of Libya: Jour. Sed. Petrology, v. 34, no. 1, p. 5–17.

McKenzie, P., 1958, The development of sand beach ridges: Australian Jour. Sci., v. 20, p. 213–214.

Melton, M. A., 1965, The geomorphic and paleoclimatic significance of alluvial deposits in southern Arizona: Jour. Geology, v. 73, p. 1–38.

Menard, H. W., 1955, Deep-sea channels, topography, and sedimentation: Am. Assoc. Petroleum Geologists Bull., v. 39, p. 236–255.

—— 1960, Possible pre-Pleistocene deep-sea fans off central California: Geol. Soc. America Bull., v. 71, p. 1271–1278.

—— 1964, Marine geology of the Pacific: New York, McGraw-Hill Book Co., Inc., 271 p.

—— et al., 1965, The Rhône deep-sea fan, in Submarine geology and geophysics: London, Butterworth, p. 271–286.

Miall, A. D., 1970, Continental marine transition in the Devonian of Prince of Wales Island, Northwest Territories: Canadian Jour. Earth Sci., v. 7, p. 125–144.

Miller, D. N., 1962, Patterns of barrier bar sedimentation and its similarity to Lower Cretaceous Fall River stratigraphy, in Symposium on Early Cretaceous rocks: Wyoming Geol. Assoc. 17th Field Conf. Guidebook, p. 232–247.

Milling, M. E., and E. W. Behrens, 1966, Sedimentary structures of beach and dune deposits: Mustang Island, Texas: Inst. Marine Sci. Pub., v. 11, p. 135–148.

Moody-Stuart, M., 1966, High and low sinuosity stream deposits, with examples from the Devonian of Spitsbergen: Jour. Sed. Petrology, v. 36, no. 4, p. 1102–1117.

Moore, D., 1966, Deltaic sedimentation: Earth Sci. Rev., v. 1, p. 87–104, Amsterdam, Elsevier Publishing Co.

Moore, D. G., 1970, Reflection profiling studies of the California borderland: structure and Quaternary turbidite basins: Geol. Soc. America Spec. Paper 107.

Moore, G. T., and D. O. Asquith, 1971, Delta: term and concept: Geol. Soc. America Bull., v. 82, p. 2563–2568.

Morgan, J. P., 1961, Mudlumps at the mouths of the Mississippi River, in Genesis and paleontology of the Mississippi River mudlumps: Louisiana Geol. Survey Geol. Bull. No. 35.

—— 1970, Depositional processes and products in the deltaic environment, in Deltaic sedimentation—modern and ancient: Soc. Econ. Paleontologists and Mineralogists Spec. Pub. 15, p. 31–47.

—— et al., 1953, Occurrence and development of mud flats along the western Louisiana coast: Louisiana State Univ. Coastal Studies Inst. Tech. Rept. No. 2.

—— et al., 1968, Mudlumps: diapiric structures in Mississippi delta sediments, in J. Braunstein and G. D. O'Brien, eds., Diapirism and diapirs: Am. Assoc. Petroleum Geologists Mem. 8, p. 145–161.

Muehlberger, F. B., 1955, Pismo beach–Point Sal dune field, California: Lawrence, Kansas, Univ. Kansas, Master's thesis, 106 p.

Muller, G., 1966, The new Rhine delta in Lake Constance, in M. L. Shirley, ed., Deltas in their geologic framework: Houston Geol. Soc., p. 107–124.

Mutch, T. A., 1968, Pennsylvanian non-marine sediments of the Narragansett Basin, Massachusetts and

Rhode Island: Geol. Soc. America Spec. Paper 106, p. 177–209.

Nagtegaal, P. J. C., 1966, Scour-and-fill structures from a fluvial piedmont environment: Geologie en Mijnbouw, v. 45, no. 10, p. 342–354.

Nanz, R. H., 1954, Genesis of Oligocene sandstone reservoir, Seeligson field, Jim Wells and Kleberg Counties, Texas: Am. Assoc. Petroleum Geologists Bull., v. 38, p. 96–117.

Nelson, B. W., 1970, Hydrography, sediment dispersal, and recent historical development of the Po River delta, Italy, in Deltaic sedimentation—modern and ancient: Soc. Econ. Paleontologists and Mineralogists Spec. Pub. 15, p. 152–184.

Nelson, C. H., 1968, Marine geology of Astoria deep-sea fan: Corvallis, Oregon, Oregon State Univ., PhD thesis, 287 p.

────── et al., 1970, Development of the Astoria canyon-fan physiography and comparison with similar systems: Marine Geology, v. 8, p. 259–291.

Nichols, M. M., 1964, Characteristics of sedimentary environments in Moriches Bay, in Papers in marine geology: New York, The Macmillan Co., p. 363–383.

Nilsen, T. H., 1969, Old Red sedimentation in the Buelandet-Vaerlandet Devonian district, western Norway: Sed. Geology, v. 3, no. 1.

Normark, W. R., 1970, Growth patterns of deep sea fans: Am. Assoc. Petroleum Geologists Bull., v. 54, p. 2170–2195.

────── and D. J. W. Piper, 1969, Deep-sea fan-valleys, past and present: Geol. Soc. America Bull., v. 80, p. 1859–1866.

Norris, R. M., 1966, Barchan dunes of Imperial Valley, California: Jour. Geology, v. 74, p. 292–306.

Nossin, J. J., 1965, The geomorphic history of the northern Padang delta: Jour. Tropical Geog., v. 20, p. 54–64.

Nunnally, N. R., 1967, Definition and identification of channel and overbank deposits and their respective roles in flood plain formation: Prof. Geographer, v. 19, no. 1, p. 1–4.

Odem, W. I., 1953, Subaerial growth of the delta of the diverted Brazos River, Texas: Compass, v. 30, p. 172–178.

Oomkens, E., 1967, Depositional sequences and sand distribution in a deltaic complex: Geologie en Mijnbouw, v. 46E, p. 265–278.

────── 1970, Depositional sequences and sand distribution in the post-glacial Rhone delta complex, in Deltaic sedimentation—modern and ancient: Soc. Econ. Paleontologists and Mineralogists Spec. Pub. 15, p. 198–212.

Ore, H. T., 1963, Some criteria for recognition of braided stream deposits: Contr. Geology, v. 3, no. 1, p. 1–14.

────── 1965, Characteristic deposits of rapidly aggrading streams, in Sedimentation of Late Cretaceous and Tertiary outcrops, Rock Springs uplift: Wyoming Geological Assoc. 19th Field Conf. Guidebook, p. 195–201.

Osterhoudt, W. J., 1946, The seismograph discovery of an ancient Mississippi River channel (abs.): Geophysics, v. 11, p. 417.

Otvos, E. G., Jr., 1970, Development and migration of barrier islands, northern Gulf of Mexico: Geol. Soc. America Bull., v. 81, p. 241–246.

Pack, F. J., 1923, Torrential potential of desert waters: Pan-Am. Geologist, v. 40, p. 349–356.

Paine, W. R., 1966, Stratigraphy and sedimentation of Hackberry shale and associated beds of southwestern Louisiana: Gulf Coast Assoc. Geol. Socs. Trans., v. 16.

Paull, R. A., 1962, Depositional history of the Muddy Sandstone, Bighorn basin, Wyoming: Wyoming Geol. Assoc. 17th Field Conf. Guidebook, p. 102–117.

Peterson, J. A., and J. C. Osmond, eds., 1961, Geometry of sandstone bodies: Am. Assoc. Petroleum Geologists, 240 p.

Petrescu, G., 1948, Le delta maritime du Danube; son evolution physico-geographique et les problems qui s'y posent: Rumania, Univ. Iaşi Ann. Sci., 2d sec. (Sc. Nat.), v. 31, p. 254-303.

Pettyjohn, W. A., 1967, New members of Upper Cretaceous Fox Hills Formation in South Dakota, representing delta deposits: Am. Assoc. Petroleum Geologists Bull., v. 51, no. 7, p. 1361–1367.

Pezzetta, J. R., 1968, The St. Clair River delta: Univ. Michigan, PhD thesis, 168 p.

Phleger, F. B, 1965, Sedimentology of Guerrero Negro Lagoon, Baja California, Mexico, in 17th Submarine geology and geophysics—Colston Research Society Symposium Proc.; Bristol, England, Paper 171, p. 205–237: London, Butterworths.

────── 1969, Some general features of coastal lagoons, in Coastal lagoons, a symposium: Internat. Symposium on Coastal Lagoons Mem., Mexico, D F., p. 2–25.

────── and G. C. Ewing, 1962, Sedimentology and oceanography of coastal lagoons in Baja California, Mexico: Geol. Soc. America Bull., v. 73, p. 145–182.

Pierce, J. W., 1970, Tidal inlets and washover fans: Jour. Geology, v. 78, no. 2, p. 230–234.

────── and D. J. Colquhoun, 1970, Holocene evolution of a portion of the North Carolina coast: Geol. Soc. America Bull., v. 81, no. 12, p. 3697–3714.

Piper, D. J. W., 1970, Transport and deposition of Holocene sediment of La Jolla deep sea fan, California: Marine Geology, v. 8, p. 211–227.

────── and W. R. Normark, 1971, Re-examination of a Miocene deep-sea fan and fan valley, southern California: Geol. Soc. America Bull., v. 82, p. 1823–1830.

Potter, P. E., 1967, Sand bodies and sedimentary environments, a review: Am. Assoc. Petroleum Geologists Bull., v. 51, p. 337–365.

Price, W. A., 1947, Equilibrium of form and forces in tidal basins of coast of Texas and Louisiana: Am. Assoc. Petroleum Geologists Bull., v. 31, no. 9, p. 1619–1663.

────── 1954, Environment and formation of the chenier plain: Texas A&M Research Found. Proj. 63, Dept. Oceanography, Texas A&M College.

────── 1963, Patterns of flow and channeling in tidal inlets: Jour. Sed. Petrology, v. 33, no. 2, p. 279–290.

Priddy, R. R., et al., 1955, Sediments of Mississippi Sound and inshore waters: Mississippi Geol. Survey Bull. 82.

Pryor, W. A., 1960, Cretaceous sedimentation in Upper Mississippi embayment: Am. Assoc. Petroleum Geologists Bull., v. 44, p. 1473–1504.

────── 1961, Sand trends and paleoslope in Illinois basin and Mississippi embayment, in J. A. Peterson and J. C. Osmond, Geometry of sandstone bodies: Am. Assoc. Petroleum Geologists, p. 119–133.

────── 1967, Biogenic directional features on several point bars: Sed. Geology, v. 1, p. 235–245.

Psuty, N. P., 1965, The geomorphology of beach ridges

in Tabasco, Mexico: Louisiana State Univ. Coastal Studies Inst. Tech. Rept. 30, p. 1–51.

Pugh, J. C., 1953, The Porto Novo-Badagri sand ridge complex: Nigeria, Univ. College of Ibadan Dept. Geography, Research Notes, no. 3, p. 1–14.

Rahn, P. H., 1967, Sheetflood, streamfloods, and the formation of pediments: Assoc. Am. Geographers Annals, v. 57, p. 593–604.

Rainwater, E. H., 1964, Transgressions and regressions in the Gulf Coast Tertiary: Gulf Coast Assoc. Geol. Socs. Trans., v. 14, p. 217–230.

Rapp, A., 1959, Avalanche boulder tongues in Lappland: Geog. Annaler, v. 41, p. 34–48.

Reineck, H. E., 1963, Sedimentgefuge im bereich der sudlichen Nordsee: Abhandl. Senckenberg: Naturf Gesell. 505, p. 1–136.

—— 1967, Layered sediments in tidal flats, beaches, and shelf bottoms of the North Sea, in Estuaries: Am. Assoc. Adv. Sci. Pub. 83, p. 191–206.

—— and I. B. Singh, 1967, Primary sedimentary structures in the Recent sediments of the Jade, North Sea: Marine Geology, v. 5, p. 227–235.

Reineck, H. R., 1972, Tidal flats, in J. K. Rigby and W. K. Hamblin, eds., Recognition of sedimentary environments: Soc. Econ. Paleontologists and Mineralogists Spec. Pub. 16, 340 p.

Revelle, R., and F. P. Shepard, 1939, Sediments off the California coast, in F. P. Shepard et al., eds., Recent marine sediments: Am. Assoc. Petroleum Geologists, p. 245–282.

Rex, R. W., 1964, Arctic beaches, Barrow, Alaska, in Papers in marine geology: New York, The Macmillan Co., p. 384–400.

Rickmers, W. R., 1913, The Duab of Turkestan: New York, Cambridge Univ. Press, 563 p.

Rigby, J. K., and W. K. Hamblin, eds., 1972, Recognition of sedimentary environments: Soc. Econ. Paleontologists and Mineralogists Spec. Pub. 16, 340 p.

Rona, P. A., 1970, Submarine canyon origin on upper continental slope off Cape Hatteras: Jour. Geology, v. 78, p. 141–152.

Royse, C. F., Jr., 1970, A sedimentologic analysis of the Tongue River–Sentinel Butte interval (Paleocene) of the Williston basin, western North Dakota: Sed. Geology, v. 4, p. 19–80.

Ruhe, R. U., 1964, Landscape morphology and alluvial deposits in southern New Mexico: Assoc. Am. Geographers Annals, v. 54, p. 147–159.

Rusnak, G. A., 1957, A fabric and petrographic study of the Pleasantview Sandstone: Jour. Sed. Petrology, v. 27, no. 1, p. 41–55.

—— 1960, Sediments of Laguna Madre, Texas, in Recent sediments, northwest Gulf of Mexico: Am. Assoc. Petroleum Geologists, p. 153–196.

Russell, R. J., 1936, Physiography of lower Mississippi River delta, in Lower Mississippi River delta: Louisiana Geol. Survey Bull. 8, p. 1–199.

—— 1942, Geomorphology of the Rhône delta: Assoc. Am. Geographers Annals, v. 32, p. 149–254.

—— 1967a, River and delta morphology: Louisiana State Univ. Coastal Studies Inst. Tech. Rept. 52, 49 p.

—— 1967b, Aspects of coastal morphology: Geog. Annaler, v. 49, ser. A, p. 299–309.

—— and H. V. Howe, 1935, Cheniers of southwestern Louisiana: Geog. Rev., v. 25, no. 3, p. 449–461.

—— and R. D. Russell, 1939, Mississippi River delta sedimentation, in P. D. Trask, ed., Recent marine sediments: Am. Assoc. Petroleum Geologists, p. 153–177.

Ryan, D. J., 1965, Cross-bedding formed by lateral accretion in the Catskill Formation near Jim Thorpe, Pennsylvania: Pennsylvania Acad. Sci. Proc., v. 38, no. 2, p. 154–156.

Ryder, J. M., 1971, The stratigraphy and morphology of para-glacial alluvial fans in south-central British Columbia: Canadian Jour. Earth Sci., v. 8, p. 279–298.

Sabate, R. W., 1968, Pleistocene oil and gas in coastal Louisiana: Gulf Coast Assoc. Geol. Socs. Trans., v. 18, p. 373–386.

Sabins, F. F., Jr., 1963, Anatomy of stratigraphic trap, Bisti field, New Mexico: Am. Assoc. Petroleum Geologists Bull., v. 47, p. 193–228.

Saucier, R. T., 1964, Geological investigation of the St. Francis basin: U.S. Army Corps Engineers, Waterways Expt. Sta. Tech. Rept. 3–659.

—— 1967, Geological investigation of the Boeuf-Tensas basin, lower Mississippi Valley: U.S. Army Corps Engineers, Waterways Expt. Sta. Tech. Rept. 3–757.

Schlee, J. S., and R. H. Moench, 1961, Properties and genesis of "Jackpile" Sandstone, Laguna, New Mexico, in J. A. Peterson and J. C. Osmond, eds., Geometry of sandstone bodies: Am. Assoc. Petroleum Geologists, p. 134–150.

Scholl, D. W., et al., 1970, The structure and origin of the large submarine canyons of the Bering Sea: Marine Geology, v. 8, p. 187–210.

Schumm, S. A., 1963, Sinuosity of alluvial rivers on the great plains: Geol. Soc. America Bull., v. 74, p. 1089–1100.

—— 1971, Fluvial paleochannels, in J. K. Rigby and W. K. Hamblin, eds., Recognition of sedimentary environments: Soc. Econ. Paleontologists and Mineralogists Spec. Pub. 16, 340 p.

Schwartz, M. L., 1971, The multiple causality of barrier islands: Jour. Geology, v. 79, no. 1, p. 91–94.

Scott, A. J., and W. L. Fisher, 1969, Delta systems and deltaic deposition, in Delta systems in exploration for oil and gas: Texas Univ. Bur. Econ. Geology, p. 10–39.

Scruton, P. C., 1960, Delta building and the deltaic sequence, in F. P. Shepard et al., eds., Recent sediments, northwest Gulf of Mexico: Am. Assoc. Petroleum Geologists, p. 82–102.

—— 1961, Rocky Mountain Cretaceous stratigraphy and regressive sandstones: Wyoming Geological Assoc. 16th Field Conf. Guidebook, p. 241–249.

Selley, R. C., 1970, Ancient sedimentary environments: Ithaca, New York, Cornell Univ. Press, 237 p.

Shannon, J. P., Jr., and A. R. Dahl, 1971, Deltaic stratigraphic traps in West Tuscola field, Taylor County, Texas: Am. Assoc. Petroleum Geologists Bull., v. 55, p. 1194–1205.

Sharp, R. P., 1942, Mudflow levees: Jour. Geomorphology, v. 5, p. 222–227.

—— and L. H. Nobles, 1953, Mudflow in 1941 at Wrightwood, southern California: Geol. Soc. America Bull., v. 64, p. 547–560.

Shawa, M. S., 1969, Sedimentary history of the Gilwood Sandstone (Devonian), Utikuma Lake area, Alberta, Canada: Bull. Canadian Petroleum Geology, v. 17, p. 392–409.

Shelton, J. W., 1965, Trend and genesis of lowermost sandstone unit of Eagle Sandstone at Billings, Montana: Am. Assoc. Petroleum Geologists Bull., v. 49, p. 1385–1397.

—— 1967, Stratigraphic models and general criteria for recognition of alluvial, barrier-bar, and turbidity-current sand deposits: Am. Assoc. Petroleum Geologists Bull., v. 51, no. 12, p. 2441–2461; also *in* Billings Geol. Soc. 17th Field Conf. Guidebook, p. 1–17.

────── 1972, Models of sand and sandstone deposits: Oklahoma Geol. Survey Bull. (in press).

Shepard, F. P., 1960, Gulf Coast barriers, *in* F. P. Shepard et al., eds., Recent sediments, northwest Gulf of Mexico: Am. Assoc. Petroleum Geologists, p. 197–220.

────── F. B Phleger, and Tj. H. van Andel, 1960, Recent sediments, northwest Gulf of Mexico: Am. Assoc. Petroleum Geologists, 394 p.

────── 1965, Importance of submarine valleys in funneling sediments to the deep sea, *in* Progress in oceanography, v. 3: New York, The Macmillan Co., p. 321–332.

────── and E. C. Buffington, 1968, La Jolla submarine fan-valley: Marine Geology, v. 6, p. 107–143.

────── and R. F. Dill, 1966, Submarine canyons and other sea valleys: Chicago, Rand McNally, 381 p.

────── and D. G. Moore, 1955, Sediment zones bordering the barrier islands of central Texas coast, *in* Finding ancient shorelines: Soc. Econ. Paleontologists and Mineralogists Spec. Pub. 3, p. 78–96.

────── and ────── 1960, Bays of central Texas coast, *in* F. P. Shepard et al., eds., Recent sediments, northwest Gulf of Mexico: Am. Assoc. Petroleum Geologists, p. 117–152.

────── et al., eds., 1960, Recent sediments, northwest Gulf of Mexico: Am. Assoc. Petroleum Geologists, 394 p.

────── et al., 1963, Submarine geology: New York, Harper and Row, 557 p.

────── et al., 1969, Physiography and sedimentary process of La Jolla submarine fan and fan-valley, California: Am. Assoc. Petroleum Geologists Bull., v. 53, p. 390–420.

Shepheard, W. W., and L. V. Hills, 1970, Depositional environments, Bearpaw-Horseshoe Canyon (Upper Cretaceous) transition zone, Drumheller "Badlands," Alberta: Bull. Canadian Petroleum Geology, v. 18, p. 166–215.

Shideler, G. L., 1969, Dispersal patterns of Pennsylvanian sandstones in the Michigan basin: Jour. Sed. Petrology, v. 39, no. 3, p. 1229–1237.

Shirley, M. L., ed., 1966, Deltas: Houston Geol. Soc., 251 p.

Shuiyak, B. A., and V. I. Boldyrev, 1966, The processes of beach ridge formation: Oceanology (Engl. ed.), v. 6, no. 1, p. 88–94.

Siebold, E., 1963, Geological investigation of nearshore sand transport—examples of methods and problems from Baltic and North Seas, *in* Progress in oceanography, v. 1: New York, The Macmillan Co., p. 3–70.

Simons, D. B., and E. V. Richardson, 1961, Forms of bed roughness in alluvial channels: Am. Soc. Civil Engineers Proc., Jour. Hydraulics Div., v. 87, no. HY 3, May, p. 87–105.

────── 1962, The effect of bed roughness on depth-discharge relations in alluvial channels: U.S. Geol. Survey Water-Supply Paper 1498-E.

────── 1966, Resistance to flow in alluvial channels: U.S. Geol. Survey Prof. Paper 422-J, 69 p.

────── et al., 1965, Sedimentary structures generated by flow in alluvial channels, *in* Primary sedimentary structures and their hydrodynamic interpretation: Soc. Econ. Paleontologists and Mineralogists Spec. Pub. 12, p. 34–52.

Sims, J. D., 1967, Geology and sedimentology of the Livingston Group, northern Crazy Mountains, Montana: Evanston, Illinois, Northwestern Univ., PhD thesis, 109 p.

Sioli, H., 1966, General features of the delta of the Amazon, *in* Scientific problems of the humid tropic zone deltas and their implications: UNESCO, Dacca Symposium Proc., p. 381–390.

Sloane, B. J., Jr., 1958, The subsurface Jurassic Bodcaw sand in Louisiana: Louisiana Geol. Survey Geol. Bull. 33.

Smith, A. E., Jr., 1971, Deltas of the world—modern and ancient, bibliography: Houston Geol. Soc. Delta Study Group, 42 p.

Smith, A. J., 1966, Modern deltas: comparison maps, *in* Deltas in their geologic framework: Houston Geol. Soc., p. 233–251.

Smith, H. T. U., 1954, Coastal dunes: Coastal Geog. Conf., February, U.S. Office Naval Research, p. 51–56.

Smith, N. D., 1970, The braided stream depositional environment: comparison of the Platte River with some Silurian clastic rocks, north central Appalachians: Geol. Soc. America Bull., v. 81, p. 2993–3014.

Snavely, P. D., Jr., et al., 1964, Rhythmic-bedded eugeosynclinal deposits of the Tyee Formation, Oregon Coast Range, *in* Symposium on cyclic sedimentation: Kansas Geol. Survey Bull. 169, p. 461–480.

Spearing, D. R., 1969, Stratigraphy, sedimentation, and tectonic history of the Paleocene-Eocene Hoback Formation of western Wyoming: Ann Arbor, Michigan, Univ. Michigan, PhD thesis, 179 p.

Stanley, D. J., 1967, Comparing patterns of sedimentation in some modern and ancient submarine canyons: Earth and Planetary Sci. Letters, v. 3, p. 371–380.

────── 1968, Graded bedding-sole marking-graywacke assemblage and related sedimentary structures in some Carboniferous flood deposits, eastern Massachusetts: Geol. Soc. America Spec. Paper 106, p. 211–239.

────── 1969, Submarine channel deposits and their fossil analogs ("fluxoturbidites"), *in* New concepts of continental margin sedimentation: Am. Geol. Inst. Short Course Notes, Nov. 7–9, p. DJS-9-1–DJS-9-17.

────── 1970, Flyschoid sedimentation on the outer Atlantic margin off northeast North America, *in* Flysch sedimentology in North America: Geol. Assoc. Canada Spec. Paper No. 7, p. 179–210.

────── 1972, Submarine channel deposits, fluxoturbidites and other indicators of slope and base-of-slope environments in modern and ancient marine basins, *in* J. K. Rigby and W. K. Hamblin, eds., Recognition of sedimentary environments: Soc. Econ. Paleontologists and Mineralogists Spec. Pub. 16, 340 p.

────── and N. Silverberg, 1969, Recent slumping on the continental slope off Sable Island bank, southeast Canada: Earth and Planetary Sci. Letters, v. 6, p. 123–133.

────── et al., 1971, Lower continental rise east of the middle Atlantic states: predominant sediment dispersal perpendicular to isobaths: Geol. Soc. America Bull., v. 82, p. 1831–1840.

Stetson, H. C., 1953, The sediments of the northwestern Gulf of Mexico—Part 1. The continental terrace of the western Gulf of Mexico: its surface sediments, origin and development, *in* Papers in physical oceanography and meteorology: Massachusetts Inst. Tech. Abs. Theses, v. 12, no. 4, p. 5–45.

Stokes, W. L., 1961, Fluvial and eolian sandstone bodies in Colorado Plateau, *in* J. A. Peterson and J. C. Osmond, eds., Geometry of sandstone bodies: Am. Assoc. Petroleum Geologists, p. 151–178.

——— 1964, Eolian varving in the Colorado Plateau: Jour. Sed. Petrology, v. 34, no. 2, p. 429–433.

——— 1968, Multiple parallel-truncation bedding planes—a feature of wind-deposited sandstone formations: Jour. Sed. Petrology, v. 38, no. 2, p. 510–515.

Sullwold, H. H., Jr., 1960, Tarzana fan, deep submarine fan of late Miocene age, Los Angeles County, California: Am. Assoc. Petroleum Geologists Bull., v. 44, p. 433–457.

Sundborg, A., 1956, The river Klarälven. A study of fluvial processes: Geog. Annaler, v. 38, p. 127–316.

Swann, D. H., 1964, Late Mississippian rhythmic sediments of Mississippi Valley: Am. Assoc. Petroleum Geologists Bull., v. 48, no. 5, p. 637–658.

——— *et al.*, 1965, The Borden Siltstone (Mississippian) delta in southwestern Illinois: Illinois Geol. Survey Circ. 386, 20 p.

Swift, D. J. P., 1969, Inner shelf sedimentation: processes and products, *in* New concepts of continental margin sedimentation: Am. Geol. Inst. Short Course Notes, Nov. 7–9, p. DS-4-1–DS-4-46.

——— and R. M. McMullen, 1968, Preliminary studies of intertidal sand bodies in the Minas Basin, Bay of Fundy, Nova Scotia: Canadian Jour. Earth Sci., v. 5, no. 2, p. 175–183.

Sykes, G., 1937, The Colorado delta: Carnegie Inst. Washington Pub. 460, 193 p.

Taylor, J. H., 1963, Sedimentary features of an ancient deltaic complex: the Wealden rocks of southeastern England: Sedimentology, v. 2, p. 2–28.

Thachuk, N. M., 1968, Geological study of the Middle Devonian Gilwood arkoses in the Nipisi area, Alberta: Petroleum Soc. of Canadian Inst. Mining and Metallurgy, 19th Ann. Tech. Mtg., Preprint No. 6828, 17 p.

Thom, B. G., 1967, Mangrove ecology and deltaic geomorphology: Tabasco, Mexico: Jour. Ecology, v. 55, p. 301–343.

Thomas, W. A., and C. J. Mann, 1966, Late Jurassic depositional environments, Louisiana and Arkansas: Am. Assoc. Petroleum Geologists Bull., v. 50, p. 178–182.

Thompson, R. W., 1968, Tidal flat sedimentation on the Colorado River delta, northwestern Gulf of California: Geol. Soc. America Mem. 107, 133 p.

Thompson, W. O., 1937, Original structures of beaches, bars, and dunes: Geol. Soc. America Bull., v. 48, p. 723–752.

Todd, T. W., 1968, Dynamic diversion: influence of longshore current-tidal flow interaction on chenier and barrier island plains: Jour. Sed. Petrology, v. 38, p. 734–746.

——— and W. A. Monroe, 1968, Petrology of Domengine Formation (Eocene) at Potrero Hills and Rio Vista, California: Jour. Sed. Petrology, v. 38, p. 1024–1039.

Trask, P. D., *et al.*, 1955, Recent marine sediments—a symposium: Soc. Econ. Paleontologists and Mineralogists Spec. Pub. 4, 736 p.

Treadwell, R. C., 1955, Sedimentology and ecology of southeast coastal Louisiana: Baton Rouge, Louisiana, Louisiana State Univ., unpub. thesis, p. 1–176.

Trimonis, E. S., and K. M. Shimkus, 1970, Sedimentation at the head of a submarine canyon: Acad. Sci. USSR, Oceanology (Engl. ed.), v. 10, p. 74–85.

Trowbridge, A. C., 1911, The terrestrial deposits of Owens Valley, California: Jour. Geology, v. 19, p. 709–747.

——— 1930, Building of Mississippi delta: Am. Assoc. Petroleum Geologists Bull., v. 14, p. 867–901.

van Andel, Tj. H., 1960, Sources and dispersion of Holocene sediments, northern Gulf of Mexico, *in* F. P. Shepard *et al.*, Recent sediments, northwest Gulf of Mexico: Am. Assoc. Petroleum Geologists, p. 34–55.

——— 1964, Recent marine sediments of Gulf of California, *in* Tj. H. van Andel and G. C. Shor, Jr., eds., Marine geology of the Gulf of California: Am. Assoc. Petroleum Geologists Mem. 3, p. 216–310.

——— 1967, The Orinoco delta: Jour. Sed. Petrology, v. 37, no. 2, p. 297–310.

——— and J. R. Curray, 1960, Regional aspects of modern sedimentation in northern Gulf of Mexico and similar basins, and paleogeographic significance, *in* F. P. Shepard *et al.*, eds., Recent sediments of northwest Gulf of Mexico: Am. Assoc. Petroleum Geologists, p. 345–364.

——— and G. C. Shor, Jr., eds., 1964, Marine geology of the Gulf of California: Am. Assoc. Petroleum Geologists Mem. 3, 408 p.

Van Lopik, J. R., 1955, Recent geology and geomorphic history of central coastal Louisiana: Louisiana State Univ. Coastal Studies Inst. Tech. Rept. 7, p. 1–89.

Vann, J. H., 1959, The geomorphology of the Guiana Coast: Louisiana State Univ. Coastal Studies Inst., 2d Coastal Geog. Conf. Proc., p. 153–187.

van Straaten, L. M. J. U., 1951, Texture and genesis of Dutch Wadden Sea sediments: 3d Internat. Cong. Sedimentology Proc., Netherlands, p. 225–244.

——— 1954a, Composition and structure of Recent marine sediments in the Netherlands: Leidsche Geol. Meded., pt. 19, p. 1–110.

——— 1954b, Sedimentology of Recent tidal flat deposits and the Psammites du Condroz (Devonian): Geologie en Mijnbouw, v. 16, p. 25–47.

——— 1959, Littoral and submarine morphology of the Rhône delta: Louisiana State Univ. Coastal Studies Inst., 2d Coastal Geog. Conf. Proc., p. 233–264.

——— 1961, Sedimentation in tidal flat areas: Alberta Soc. Petroleum Geologists Jour., v. 9, p. 203–226.

——— 1965, Coastal barrier deposits in south- and north-Holland . . .: Netherlands, Geol. Sticht. Meded., new series, no. 17, p. 41–75.

Venkatarathnam, K., 1970, Formation of the barrier spit and other sand ridges near Chilka Lake on the east coast of India: Marine Geology, v. 9, no. 2, p. 101–116.

Visher, G. S., 1965a, Use of vertical profile in environmental reconstruction: Am. Assoc. Petroleum Geologists Bull., v. 49, no. 1, p. 41–61.

——— 1965b, Fluvial processes as interpreted from ancient and Recent fluvial deposits, *in* Primary sedimentary structures and their hydrodynamic interpretation: Soc. Econ. Paleontologists and Mineralogists Spec. Pub. 12, p. 116–131.

——— 1968, Depositional framework of the Bluejacket-Bartlesville sandstone, *in* Guidebook to geology of the Bluejacket-Bartlesville sandstone, Oklahoma: Oklahoma City Geol. Soc., p. 32–51.

——— 1971, Physical characteristics of fluvial deposits, *in* J. K. Rigby and W. K. Hamblin, eds., Recognition of sedimentary environments: Soc. Econ. Paleontologists and Mineralogists Spec. Pub. 16, 340 p.

────── et al., 1971, Pennsylvanian delta patterns and petroleum occurrences in eastern Oklahoma: Am. Assoc. Petroleum Geologists Bull., v. 55, p. 1206–1230.

Volker, A., 1966, The deltaic area of the Irrawaddy River in Burma, in Scientific problems of the humid tropic zone delta and their implications: UNESCO, Dacca Symposium Proc., p. 373–379.

Waechter, N. D., 1970, Braided stream deposits of the Red River, Texas panhandle (abs.): Geol. Soc. America Abs. with Programs, v. 2, no. 7, p. 713.

Walker, J. R., and J. V. Massingill, 1970, Slump features on the Mississippi fan, northeastern Gulf of Mexico: Geol. Soc. America Bull., v. 81, p. 3101–3108.

Walker, K. R., 1962, Lithofacies maps of Lower Mississippian clastics of eastern and east-central United States: Am. Assoc. Petroleum Geologists Bull., v. 46, p. 105–111.

Walker, R. G., 1966, Deep channels in turbidite-bearing formations: Am. Assoc. Petroleum Geologists Bull., v. 50, p. 1899–1917.

Walton, W. C., 1970, Groundwater resource evaluation: New York, McGraw-Hill Co., 664 p.

Wanless, H. R., et al., 1970, Late Paleozoic deltas in the central and eastern United States, in Deltaic sedimentation—modern and ancient: Soc. Econ. Paleontologists and Mineralogists Spec. Pub. 15, p. 215–245.

Warme, J. E., 1966, Geology of Mugu Lagoon: Univ. California at Los Angeles, PhD dissert.

────── 1971, Paleoecological aspects of a modern coastal lagoon: California Univ. Pubs. Geol. Soc., v. 87, 134 p.

Way, J. H., Jr., 1968, Bed thickness analysis of some Carboniferous fluvial sedimentary rocks near Joggins, Nova Scotia: Jour. Sed. Petrology, v. 38, no. 2, p. 424–433.

Weidie, A. E., 1968, Bar and barrier-island sands: Gulf Coast Assoc. Geol. Socs. Trans., v. 18, p. 405–415.

Weimer, R. J., 1961a, Spatial dimensions of Upper Cretaceous sandstones, Rocky Mountain area, in J. A. Peterson and J. C. Osmond, eds., Geometry of sandstone bodies: Am. Assoc. Petroleum Geologists.

────── 1961b, Upper Cretaceous delta on tectonic foreland, northern Colorado and southern Wyoming (abs.): Am. Assoc. Petroleum Geologists Bull., v. 45, p. 417.

────── 1965, Late Cretaceous deltas, Rocky Mountain region (abs.): Am. Assoc. Petroleum Geologists Bull., v. 49, p. 363.

Welder, F. A., 1959, Processes of deltaic sedimentation in the lower Mississippi River: Louisiana State Univ. Coastal Studies Inst. Tech. Rept. No. 12, p. 1–90.

Wermund, E. G., and W. A. Jenkins, Jr., 1970, Recognition of deltas by fitting trend surfaces to Upper Pennsylvanian sandstones in north-central Texas, in Deltaic sedimentation—modern and ancient: Soc. Econ. Paleontologists and Mineralogists Spec. Pub. 15, p. 156–269.

Wessel, J. M., 1969, Sedimentary history of Upper Triassic alluvial fan complexes in north-central Massachusetts: Massachusetts Univ. Dept. Geology Contr. No. 2, 157 p.

Wier, C. E., and W. A. Girdley, 1963, Distribution of the Inglefield and Dicksburg Hills Sandstone members in Posey and Vanderburgh Counties, Indiana: Indiana Acad. Sci. Proc., v. 72, p. 212–217.

Wilde, P., 1965, Recent sediments of the Monterey deep-sea fan: Cambridge, Massachusetts, Harvard Univ., PhD dissert., 153 p.

Williams, G. E., 1966, Paleogeography of the Torridonian Applecross Group: Nature, v. 209, no. 5030, p. 1303–1306.

────── 1969, Characteristics and origin of a Pre-Cambrian pediment: Jour. Geology, v. 77, p. 183–207.

Williams, P. F., and B. R. Rust, 1969, The sedimentology of a braided river: Jour. Sed. Petrology, v. 39, no. 2, p. 649–679.

Wilson, M. D., 1967, The stratigraphy of the Beaverhead Group in the Lima area, southwestern Montana: Evanston, Illinois, Northwestern Univ., PhD dissert., 183 p.

────── 1970, Upper Cretaceous-Paleocene synorogenic conglomerates of southwestern Montana: Am. Assoc. Petroleum Geologists Bull., v. 54, p. 1843–1867.

Winder, C. G., 1965, Alluvial cone construction by alpine mudflow in a humid temperate region: Canadian Jour. Earth Sci., v. 2, p. 270–277.

Winterer, E. L., et al., 1968, Geological history of the pioneer fracture zone with the Delgada deep-sea fan northeast Pacific: Deep-Sea Research, v. 15, no. 5, p. 509–520.

Wolff, M. P., 1967, Deltaic sedimentation of the Middle Devonian Marcellus Formation in southeastern New York: Ithaca, New York, Cornell Univ., PhD dissert., 231 p.

Wolman, M. G., and L. M. Brush, 1961, Factors controlling the size and shape of stream channels in coarse noncohesive sands: U.S. Geol. Survey Prof. Paper 282-G, p. 183–210.

────── and L. B. Leopold, 1957, River flood plains: some observations on their formation: U.S. Geol. Survey Prof. Paper 282-C, p. 87–107.

Wurster, P., 1964, Delta sedimentation in the German Keuper Basin, in Deltaic and shallow marine deposition: Amsterdam, Elsevier Publishing Co., p. 436–446.

Zenkovich, V. P., 1964, Formation and burial of accumulative forms in littoral and nearshore marine environments: Marine Geology, v. 1, p. 175–180.

────── 1967, Processes of coastal development: New York, Interscience Publishers—Division of John Wiley and Sons, Inc.

Discussion

EDWARD N. WILSON, Kentucky Geological Survey, Lexington, Kentucky

You remarked that some of the deltaic bodies were rather thin and not very extensive. In the central United States, the Pennsylvania System contains several of these deltaic sequences and some of them are fairly thick. I should like to ask if there is anything inherently disadvantageous to these sandstone bodies for emplacement of limited volumes of waste?

R. J. LEBLANC

I do not think so, for the following reasons. Deltas of various sizes can prograde seaward into a basin over long periods of time. Thus, they can produce relatively thick deltaic sand bodies over extensive areas which consist of several individual genetic units stacked over each other. It is true that some of the Pennsylvanian sandstones are thick and occur over extensive areas. There is nothing wrong with these sandstones from the standpoint of the emplacement of limited amounts of waste into them. It is important to mention that deltaic sands grade seaward into prodelta silts and clays.

PAUL WITHERSPOON, University of California, Berkeley, California

First, I want to compliment you on a very excellent review of depositional conditions. I wanted to ask if you have looked at conditions such as the Mount Simon Sandstone of central United States; 1,000–2,000 ft thick, it can be traced all the way across Indiana, Illinois, Ohio, and up to New York, where it is called the "Potsdam," and to Minnesota and southern Illinois. Would the mechanisms you have described relate to accumulation of that thick sand body over those hundreds of miles?

R. J. LEBLANC

I cannot answer that specific question because I am not familiar with the Mount Simon Sandstone. However, I can comment on other sandstones which occur over very extensive areas. For example, the Castlegate Sandstone of northwestern Utah extends for many miles from west to east; but the Castlegate is not a uniform sandstone deposited in one environment. Actually, it consists of alluvial-fan, braided-stream, and deltaic sandstones. I believe that many other sandstones are similar to the Castlegate in that they are extensive but of multiple origin; therefore, the models I described can explain their origin.

JIM HALLORAN, Montana Water Resources Board, Helena, Montana

Can you give us some idea what this barrier-island model will look like after marine transgression or regression?

R. J. LEBLANC

One of the largest oil fields discovered in the United States during the past several years is the Bell Creek field of Montana. Two professors from Texas A&M University correctly interpreted this reservoir as a barrier-island sandstone body. I refer you to Dr. R. R. Berg's[1] excellent paper on this barrier-bar sandstone, because time does not permit a detailed answer to your question.

[1] Berg, R. R., and D. K. Davies, 1968, Origin of Lower Cretaceous Muddy Sandstone at Bell Creek field, Montana: Am. Assoc. Petroleum Geologists Bull., v. 52, no. 10, p. 1888–1898.